BARCODE

KU-188-081

# THE DUBLIN FIRE BRIGADE

# THE DUBLIN FIRE BRIGADE

*A history of the brigade,
the fires and the emergencies*

**Tom Geraghty and Trevor Whitehead**

Dublin City Council
Comhairle Cathrach Bhaile Átha Cliath

First published 2004 by
Dublin City Council
Dublin City Library & Archive
138–144 Pearse Street
Dublin 2

A catalogue record is available for this book
from the British Library
ISBN: 0-94684-170-5 [Casebound]
ISBN: 0-94684-171-3 [Paperback]

Design and origination by David Cooke
Printed by Colour Books Ltd.

Co-funded by LANPAG
Local Authority Workplace Partnership Group

# Contents

*Page*

Introduction ............................................................................................... xi
Réamhfhocail / Foreword ........................................................................... xiii
Acknowledgements ..................................................................................... xv

1  Earliest times ......................................................................................... 1
2  First organisation ................................................................................... 14
3  The municipal fire service ...................................................................... 25
4  Continuing progress .............................................................................. 41
5  A growing emergency service ................................................................. 55
6  Growth, deaths and reports ................................................................... 71
7  The end of the century .......................................................................... 86
8  The years of planning ............................................................................ 108
9  Motorisation and a full emergency service ............................................. 120
10 Civil unrest and revolution .................................................................... 135
11 The terrible years .................................................................................. 159
12 A new administration and amalgamation ............................................... 182
13 The great tragedy .................................................................................. 195
14 Re-organisation and the "Emergency" ................................................... 215
15 A modern fleet ...................................................................................... 240
16 Fires, bombs, death and progress .......................................................... 252
17 The Stardust tragedy and the new Fire Services Act ............................... 280
18 The modern fire brigade ........................................................................ 294

*Appendices*

1  Dublin Corporation Fire Brigade Act, 1862 ........................................... 298
2  Pembroke Fire Brigade .......................................................................... 302
3  Rathmines and Rathgar Fire Service ...................................................... 304
4  Chief Officers of Dublin Fire Brigade, 1862–2000 ................................. 307
5  Principal Mobile Firefighting Appliances in use by Dublin Fire Brigade,
   1863–2000 ............................................................................................ 308

*Bibliography* ............................................................................................ 313

*Index* ...................................................................................................... 317

# List of illustrations

Fire engine, from printed notice issued by Lord Mayor of Dublin, 1711.
   Postcard, © Dublin City Public Libraries. . . . . . . . . . . . . . . . . . . . . . . . . . . . .3
Early eighteenth-century manual fire engine, St Werburgh's Parish.
   Photograph, © Trevor Whitehead . . . . . . . . . . . . . . . . . . . . . . . . . . . . . . .5
Manual fire engine in use.
   Print, from Hibernian Insurance policy, issued 1790. © Trevor Whitehead . . . . . . . . . . . . . .7
Fire mark of Hibernian Fire Insurance Company, c. 1800.
   Ink drawing, © Trevor Whitehead . . . . . . . . . . . . . . . . . . . . . . . . . . . . . . .9
"The burning of Holmes' Emporium, College Green, Dublin" by W. Sadler.
   © National Gallery of Ireland . . . . . . . . . . . . . . . . . . . . . . . . . . . . . . . . 13
Domville Estate 1860s, Dublin Fire Brigade training estate workers in use of manual
   fire engine. Photograph, © National Library of Ireland . . . . . . . . . . . . . . . . . . . . . 40
Invoice from Dublin Fire Brigade to Messrs Gunn, Grafton Street, 31 Oct 1872: for the
   attendance of one Fire Brigade man in Gaiety Theatre, 27 Nov 1872 to 31 Oct 1872 . . . . . 52
Explosion at Hammond Lane foundry.
   Photograph, © National Library of Ireland . . . . . . . . . . . . . . . . . . . . . . . . . . 62
Fatal fire in Westmoreland Street, Dublin, May 1891.
   Print, © Dermot Dowling . . . . . . . . . . . . . . . . . . . . . . . . . . . . . . . . . . 78
Funeral of Charles Stewart Parnell, October 1891. Dublin Fire Brigade forming guard of
   honour at City Hall. Photograph, © National Library of Ireland . . . . . . . . . . . . . . . . . 85
Illuminated address presented to T.M. Healy from Dublin City firemen in gratitude
   to him for steering their pension scheme through the House of Commons.
   Illustration from Fire and Water, January 1892. . . . . . . . . . . . . . . . . . . . . . . . . 88
South City Markets fire, August 1892.
   Photograph, © Edward Chandler. . . . . . . . . . . . . . . . . . . . . . . . . . . . . . . 91
Large Shand Mason steamer commissioned by Dublin Fire Brigade, 1893.
   Print, © Trevor Whitehead . . . . . . . . . . . . . . . . . . . . . . . . . . . . . . . . . 94
Destruction of Arnott's following fire, 1902.
   Photograph, © National Library of Ireland . . . . . . . . . . . . . . . . . . . . . . . . . . 95
Dublin Fire Brigade at Stephen's Green West, 1897.
   Film still, © Lumiere Brothers . . . . . . . . . . . . . . . . . . . . . . . . . . . . . . . 101
Purcell's aerial ladder, from patent, 1899.
   Drawing, © Trevor Whitehead . . . . . . . . . . . . . . . . . . . . . . . . . . . . . . . 106
Purcell horse-drawn ladder.
   Photograph, © Dublin City Council . . . . . . . . . . . . . . . . . . . . . . . . . . . . . 106
Mercer's Hospital, showing the Purcell Escape, first-ever 100-foot extension ladder in Europe.
   Photograph, The Fire Call, July 1902. . . . . . . . . . . . . . . . . . . . . . . . . . . . . 107
Display of hook-ladder drill and jump-sheet, Tara Street Station, c. 1910.
   Photograph, © National Library of Ireland . . . . . . . . . . . . . . . . . . . . . . . . . 123
First motor fire engine, towing the steamer.
   Photograph, © National Library of Ireland . . . . . . . . . . . . . . . . . . . . . . . . . 136
Thomas Street fire station, 1913.
   Photograph, © Dublin City Council . . . . . . . . . . . . . . . . . . . . . . . . . . . . . 137
Dublin's first motor-driven ambulance.
   Photograph, © Dublin Fire Brigade Historical Society . . . . . . . . . . . . . . . . . . . . 138
Fireman and two horses, Tara Street, 1911.
   Photograph, © National Library of Ireland . . . . . . . . . . . . . . . . . . . . . . . . . 144
Shooting incident near Custom House, 1920, showing Cinema Ambulance.
   Photograph, © Edward Chandler. . . . . . . . . . . . . . . . . . . . . . . . . . . . . . 147
Dublin Fire Brigade fire fighting in O'Connell Street, 1916.
   Photograph, © National Library of Ireland . . . . . . . . . . . . . . . . . . . . . . . . . 149
Dublin Fire Brigade at ruins of General Post Office, 1916.
   Photograph, collection of Edward Chandler. © National Library of Ireland. . . . . . . . . . . . 152
Commendation, British Fire Prevention Committee to Chief Officer Purcell, 1916.
   © Dublin City Council . . . . . . . . . . . . . . . . . . . . . . . . . . . . . . . . . . . 154
Cork and Dublin fire brigades working together during the burning of Cork, 1920.
   Photograph, © National Museum of Ireland. . . . . . . . . . . . . . . . . . . . . . . . . 166

Burning of Custom House, 1921, with firemen tackling fire.
  Photograph, © Department of the Environment . . . . . . . . . . . . . . . . . . . . . . . . . . . . . . . . . . 168
Burning of Custom House, 1921: dome has vanished.
  Photograph, © Old Dublin Society . . . . . . . . . . . . . . . . . . . . . . . . . . . . . . . . . . . . . . . . . . . . 169
Dublin Fire Brigade at Four Courts, c. 1922, fireman Bernard Matthews (centre) with
  Chief Officer Myers (right). Photograph, © Paul Matthews . . . . . . . . . . . . . . . . . . . . . . . . . 172
Purcell Escape used for demolition work, O'Connell Street during Civil War.
  Photograph, © Edward Chandler. . . . . . . . . . . . . . . . . . . . . . . . . . . . . . . . . . . . . . . . . . . . . . 173
Funeral of Michael Collins passing Bank of Ireland, 1922.
  Photograph, © Michael Counihan. . . . . . . . . . . . . . . . . . . . . . . . . . . . . . . . . . . . . . . . . . . . . 174
Funeral of Michael Collins approaching O'Connell Bridge, 1922.
  Photograph, © Michael Counihan. . . . . . . . . . . . . . . . . . . . . . . . . . . . . . . . . . . . . . . . . . . . . 175
Dublin Fire Department, Leyland appliance supplied in 1912.
  Photograph, © British Commercial Museum Trust Archives . . . . . . . . . . . . . . . . . . . . . . . . 194
Dublin Fire Department, Leyland appliance supplied in 1939.
  Photograph, © Trevor Whitehead . . . . . . . . . . . . . . . . . . . . . . . . . . . . . . . . . . . . . . . . . . . . . 194
Dublin firemen, army officers and Garda detectives at scene of Terenure plane crash, 1950s.
  Photograph, © The Irish Times. Collection Old Dublin Society . . . . . . . . . . . . . . . . . . . . . . 203
Pearse Street Fire, 1936.
  Photograph, © the Carroll Family. . . . . . . . . . . . . . . . . . . . . . . . . . . . . . . . . . . . . . . . . . . . . . 208
Removal of coffin of Robert Malone from lying-in-state at City Hall. Pearse Street Fire, 1936.
  Photograph, © the Dublin Fire Brigade Historical Society . . . . . . . . . . . . . . . . . . . . . . . . . . 211
Auxiliary Fire Service under instruction from Dublin Fire Brigade c. 1940.
  Photograph, © Dublin Fire Brigade Historical Society . . . . . . . . . . . . . . . . . . . . . . . . . . . . . 222
Richmond Cottages, nos. 24, 25, 26, and 27 following North Strand bombing.
  Photograph, © Dublin City Council . . . . . . . . . . . . . . . . . . . . . . . . . . . . . . . . . . . . . . . . . . . . 233
Clarence Street North, nos. 33 and 34 (tenements) following North Strand bombing.
  Photograph, © Dublin City Council . . . . . . . . . . . . . . . . . . . . . . . . . . . . . . . . . . . . . . . . . . . . 234
Fire at the Stephen Street Bakery, early 1920s.
  Photograph, © National Library of Ireland . . . . . . . . . . . . . . . . . . . . . . . . . . . . . . . . . . . . . . 235
Dublin Fire Brigade's Merryweather turntable ladder used in attacking the flames as the
  Abbey Theatre is destroyed by fire in 1951. Photograph, © Trevor Whitehead . . . . . . . . . 241
Fire at premises of The Irish Times in Fleet Street in 1951.
  Photograph, © Trevor Whitehead . . . . . . . . . . . . . . . . . . . . . . . . . . . . . . . . . . . . . . . . . . . . . 242
Dennis F8 fire engine,1957: first limousine motor purchased by Dublin Fire Brigade.
  Film still, © Gael Linn. . . . . . . . . . . . . . . . . . . . . . . . . . . . . . . . . . . . . . . . . . . . . . . . . . . . . . 246
Cuffe Street Fire, 1955.
  Photograph, © The Irish Press . . . . . . . . . . . . . . . . . . . . . . . . . . . . . . . . . . . . . . . . . . . . . . . 247
Central control room, Tara Street fire station, 1957.
  Film still © Gael Linn . . . . . . . . . . . . . . . . . . . . . . . . . . . . . . . . . . . . . . . . . . . . . . . . . . . . . . . 249
Dublin Fire Brigade officers with retained firemen from Irish county brigades on training
  course, 1958 . . . . . . . . . . . . . . . . . . . . . . . . . . . . . . . . . . . . . . . . . . . . . . . . . . . . . . . . . . . . . 251
Fire in CIE depot.
  Photograph, © Independent Newspapers Ltd, Dublin.
  Collection Dublin Fire Brigade Historical Society . . . . . . . . . . . . . . . . . . . . . . . . . . . . . . . . 259
Fire at Noyek's, Parnell Street, 1972.
  Photograph, © Independent Newspapers Ltd, Dublin. Collection Trevor Whitehead . . . . . 267
The aftermath of car-bomb explosions at Sackville Place, 1972
  Photograph, © The Irish Times . . . . . . . . . . . . . . . . . . . . . . . . . . . . . . . . . . . . . . . . . . . . . . . 269
Aerial photographs of the Stardust complex, following the fire. © The Irish Press . . . . . . . . . . 293
Loreto Convent School, St Stephen's Green, destroyed by fire 1986.
  Photograph, © Collection Dublin Fire Brigade Historical Society . . . . . . . . . . . . . . . . . . . . 295
Fire engines belonging to Pembroke Fire Brigade.
  Photograph, © Las Fallon . . . . . . . . . . . . . . . . . . . . . . . . . . . . . . . . . . . . . . . . . . . . . . . . . . . 303
Rathmines Fire Brigade, with chief officer in civilian clothes.
  Photograph, © Collection Dublin Fire Brigade Historical Society . . . . . . . . . . . . . . . . . . . . 306

## Colour illustrations

1. Fire officer with the Royal Exchange Insurance Company, 1832. Print, Johnson Collection, © National Library of Ireland
2. "The Fireman's Polka" composed by Charles S. Macdona. Dedicated to Captain Boyle, Dublin Fire Brigade. Cover shows horse-drawn fire engine approaching O'Connell Bridge. Coloured lithograph, Dublin Civic Museum, collection Old Dublin Society
3. Shand Mason steamer and crew, with Chief John Boyle on the right. Photograph, Collection Trevor Whitehead, © National Library of Ireland
4. Captain Purcell's letter-book, mid 19th century. Collection Dublin Fire Brigade Historical Society
5. The first motor fire engine, made by Leyland to design of CFO Purcell, 1909. Coloured print, Collection Trevor Whitehead
6. Theatre Royal fire. Coloured lithograph, " National Library of Ireland
7. Detail of above, Captain Robert Ingram (right) talking to civic officials. © National Library of Ireland
8. Portrait of Captain Boyle. Watercolour, Dublin Civic Museum, collection Old Dublin Society
9. "Station Officer O'Brien" by James Conway. Painting, collection Dublin Fire Brigade Training Centre, Photograph Gerry Allwell
10. "Burning of the Custom House, 1921", artist unknown. Painting, Collection Dublin Fire Brigade Training Centre, Photograph Gerry Allwell
11. Tara Street Fire Station, late 1920s. Back row: N. Bohan, J. Byrne, J. Gibney, J. Leetch, T. Smart, P. Cobbe, M. Murphy, L. Kiernan, T. Coyne, T. Walsh. Second row: P. Kelly, N. Seaver, P. O'Grady. N. Doyle, P. Bruton, T. Kavanagh, J. O'Hara, J. Dariton, R. Matthews. Front row: W. Carroll, W. French, J. Kinsella, C. McDonagh (Foreman), Liuetenant J. Connolly, Station Officer J. Howard, T. Curran, J. Markey, A. Kennedy. Photograph, collection of the Carroll Family (William Carroll, district officer, Dublin Fire Brigade, deceased)
12. The funeral of the victims of the Pearse Street fire, 1936 with the flag-draped coffin of Robert Malone in the lead as they pass Tara Street fire station, with full staff of Dublin Fire Brigade, led by Chief Fire Officer Connolly, forming a guard of honour. Photograph, from Robert Malone
13. Unveiling of memorial stone to firefighters who died in Pearse Street Fire, 1936. Photograph, © Dublin Fire Brigade
14. A Dublin Firefighter. Portrait by James Conway. Collection Dublin Fire Brigade Museum
15. The Mansion House with Dublin Fire Brigade engines. Transparency, © Dublin City Council
16. Fire at premises of The Irish Times in Fleet Street in 1999. Photograph, collection Trevor Whitehead.
17. Water tender built by the Irish firm of Hughes on a German Magirus chassis in 1985. Photograph, collection Trevor Whitehead
18. Fire drill in Dublin Fire Brigade Training Centre. Photograph, © Dublin Fire Brigade
19. Hose drill with extension ladders in Dublin Fire Brigade Training Centre. Photograph, © Dublin Fire Brigade
20. Dennis turntable ladder in Dublin Fire Brigade Training Centre. Photograph, © Dublin Fire Brigade
21. The O'Brien Institute, Dublin Fire Brigade's Training Centre. Photograph, © David Barriscale
22. The Dublin Fire Brigade Museum at the O'Brien Institute. Photograph, © David Barriscale
23. Watch sitting on turntable ladder outside Tara Street Fire Station in December 1996, a few weeks before the building was vacated Photograph, © Alan Finn
24. One of Dublin Fire Brigade's fleet of Dennis/Magirus turntable ladders. Photograph, © Trevor Whitehead
25. Provision of ambulance service is an important function of Dublin Fire Brigade. Photograph, © Dublin Fire Brigade
26. Women firefighters have been an integral part of Dublin Fire Brigade since 1994. Photograph, © Dublin Fire Brigade
27. Author Tom Geraghty (right) with Cork firefighter Tom O'Brien at Madison Square Garden for memorial held on 12 October 2002 to honour New York Fire Department colleagues who died in 9/11 Twin Towers attack. Photograph courtesy the Author

And on the misfortunate poor fire-brigade men
Whose task it will be to shovel up our ashes and shovel
What is left of us into black plastic refuse sacks
Fire-brigade men are the salt of the earth.

(*Six Nuns Die in Convent Inferno* by Paul Durcan)

# Introduction

I am delighted to have the opportunity to welcome you to this history of Dublin Fire Brigade. Dublin City Council is proud of its emergency services and is proud to be the publisher of this important account of their history. The Fire Brigade and Ambulance are on call twenty-four hours a day, seven days a week, to provide rapid response to distress of all kinds. The professionalism of the men and women who provide these services is outstanding. A typical day's work calls for courage, determination and a willingness to risk life and limb to rescue others. The courtesy and kindness of our emergency staff has often been remarked upon, when they comfort and console people who have been devastated by disaster and bereaved by death.

This history reminds us that Dublin City's emergency services have been providing this dedicated response for generations, the Fire Brigade since 1862 and the Ambulance since the beginning of the 20th century. They have borne witness to many formative events in the history of our city and country, including the 1916 Rising and the War of Independence. During World War II, Dublin Fire Brigade responded to requests for help from Belfast during the Blitz and fought devastating fires in that city side-by-side with Belfast Fire Brigade. And at home, in the same year, our Fire Brigade went to the aid of Dubliners caught up in the North Strand Bombing. These feats are well known and are recounted here, but this book also contains many more accounts of heroism, and provides a detailed record of the accomplishments of both services. We owe a debt of gratitude to the two authors who have researched and written this fine book. Tom Geraghty was a firefighter with Dublin Fire Brigade for many years, and brings a special insight to this history, while Trevor Whitehead has published extensively about the history of firefighting in both Britain and Ireland. I congratulate both authors for their dedication in bringing the story of Dublin Fire Brigade to a wider audience.

The publication of this book is one of the first achievements of the partnership process within Dublin City Council. I am pleased to acknowledge the generous assistance of LAMPAG in providing a grant in aid of publication, with matching funding being provided by Dublin City Council. We are also grateful for a grant from the Dublin Fire Brigade Social Club towards the publication of photographs in this volume, while Dublin City Library and Archive has overseen the production of this book.

As we face into the 21st century, this history of past achievements renews our confidence in the future and in the capacity of our emergency services to respond as they have always done to the people of Dublin.

*John Fitzgerald*
*Dublin City Manager*
*Spring 2004*

# Réamhfhocal / Foreword

Chun na mílte míle a thaisteal ní mór ar dtús an chéad coiscéim a thógaint. Ní ró fhada an lá é ó bhí an Bhriogáid ag brath ar capaill agus clogíní lena ceann scribhe a bhaint amach. San lá atá inniú an úsáidtear innil fior-chomhachtúil agus séoltar iad ag an láthair le cabhair ón ríomhaire agus an raideo. Bíonn an lár-ionad I dteagmháil leanúnach leis an inneal dóiteáin agus é ar an mbealach chun an eachtra. Tá corás ríomhaireachta nua-aimseartha gluaiseachta ar fáil don Serbhís. Sar i bhfad beidh an lár-ionad i mBaile Átha Cliath in ann déileáil le glaonna ó Cúige Laighean ar fad chomh maith le Conndae Cabháin is Conndae Muinneacháin.

Buníu an Bhriogáid tímpeall céad is a dachad blian ó shoin. De réir a chéile tá sí ag dul i láidireacht. Tá gaiscíocht is suáilceas na Briogáide ar eolas ag gach aon duine. Ní nach íonadh a bhfuil muinntear Átha Cliatha ana-shásta is morálach as an dea-scéal.

From humble origins Dublin Fire Brigade has progressed to modern sophistication. The tolling of the parish bell to summon help has given way to computerised mobilisation. The horse has made way for the diesel engine. The firefighter is still a finite variable. Today Dublin's firefighters are better trained and equipped than their antecedents. They are also unique in these islands in so far as they operate an emergency ambulance service. This service is for the greater Dublin area and all firefighters are qualified Emergency Medical Technicians.

Fire, the enemy, also has new dimensions. With the advent of hazardous substances alone, fire has become more insidious. Exacting command and control measures have to be exercised by the Fire Officer. Risk assessment is constantly changing as the incident progresses. Training skills and experience of the firefighters are stretched even beyond their elastic limit. On the national scene there are no Standards of Fire Cover. Britain had a war. Ireland had an emergency. The former gave standardisation in terms of fire cover, brigade establishment and equipment to Britain. Development in the Irish Fire Service remained almost static. The watershed was the *Report of the Tribunal of Inquiry on the Fire at the Stardust, Artane, Dublin on the 14th February 1981* issued in June 1982. A great deal of investment was made in the Service by way of new stations, new equipment and centralised training courses. The year 2002 saw the issue of the Farrell, Grant, Sparks report *Review of Fire Safety and Fire Services in Ireland*. The report is a beacon of hope and expectancy for the Irish Fire Service. An interesting aside is that all the stakeholders in the service gave it unqualified support.

Computerised mobilisation has been introduced into Ireland for the Fire Service, using three counties, Limerick (serving all of Munster), Castlebar (serving all of Connaught and Co. Donegal), and Dublin (serving all of Leinster and the Counties of Cavan and Monaghan). The Dublin system is the last one to go live and is unique in so far as it is shared with the Ambulance Service of what was formerly the Eastern Health Board.

Most of the hard work has been done. It is hoped that the Farrell, Grant, Sparks report will address the rest. The timing could not be better as the report has the support of all the players.

This magnum opus by Tom Geraghty and Trevor Whitehead is a major achievement, involving an enormous amount of energy and industry. It is a tribute to the authors as it is to the Dublin Fire Brigade. Such works are painstaking and rare. Generally they have a specialised readership. Taking on the task confronted by the negatives is in essence a labour of love. Dublin will be the richer for it. We are in their debt.

Táim fíor bhuíoch as an deis dom-sa cúpla focal a scríobh ar son na Seiribhíse Doiteáin agus go mór-mhór na Briogáide Dóiteáin Baile Átha Cliath. Sceál fada atá sa leabhair seo, scéal ó chroí, scéal na suáilcí is na duáilcí. Tá siúl agam to mbainfidh na léightheoirí sult is taithneamh as.

Beír bua.

*Micheál Breathnach*  *Michael Walsh*
*Priomh-Oifigeach Dóiteáin Gníomhach*  *Acting Chief Fire Officer*
*Briogáid Dóiteáin Bhaile Átha Cliath*  *Dublin Fire Brigade*
*Baile Átha Cliath 2004*  *Dublin 2004*

# Acknowledgements

In researching this history of fire fighting in Dublin we have been helped by so many people. Many retired and now deceased veterans of the brigade have provided us with vivid details on the years of the "Emergency" and the historic journey to Belfast. Since the idea of compiling this history originated some thirty years ago we have both been lucky in the people we have kept in contact with or worked with over those years. Because very few of the firefighters involved in the above years have written about their experiences much checking on matters referred to in conversations had to take place. In this respect Fergus Flanagan's retentive mind has been invaluable, backed up by Paddy Walshe and Tom Coleman. Since the recent death of Jackie Conroy these men are the last survivors of 1941. Over the years Larry Carroll, Eddy Finlay, John A. Kelly, Peter Hanley, Mickey Conroy, Billy Carroll and Dennis O'Leary were great fonts of information on those dramatic days; now all are gone and with them so much history of the Dublin Fire Brigade.

We owe an enormous debt of gratitude to those who conceived and worked in putting together the Dublin Fire Brigade Museum. Its collection of materials has been invaluable in supporting the research for this book; so too has been the co-operation of the staff in the O'Brien Institute, in particular John L'Estrange, Aidan Carroll and Frank Kenny. Las Fallon, Liam Clarke and Gerry Alwell of the present operational staff must be mentioned because of their continued help throughout the writing. Robert Malone, Paul Matthews and Eddy King have given us personal papers and valuable details about their fathers.

Outside of the brigade, Dennis Hill, Mike Smith, Willie Doyle and many other members of the Fire Brigade Society have continued to be helpful while Liam O'Lunaigh, Pat Poland and Ian Scott along with the late Tom Larkin have been assisting in research for over thirty years. Michael Corcoran, Catherine Byrne, Mairead Mullaney and E.H. Gledhill have been supportive in directing research on a wider basis.

In SIPTU Maria Worth, who typed a first draft of the early chapters, was responsible for the first major advance in getting the story together, and Joe Davis and the late Ken O'Callaghan in copying and binding the manuscript for distribution to possible publishers ensured publication at some stage. Francis Devine and Jack McGinley of the Irish Labour History Society were always helpful in providing advice and encouragement.

We wish to acknowledge the courteous and efficient service readily received from the following:
Berkely Library, Trinity College Dublin
British Commercial Vehicle Museum Archives
British Library: Newspaper Library and Patents Information
Dublin City Archives
Dublin Civic Museum
Dublin Fire Brigade Museum and Research Centre
Dublin & Irish Collections (Gilbert Library)
Fire Brigade Society Library
Irish Film Archive
National Library of Ireland
National Photographic Archive
Old Dublin Society

Science Museum Library

Transport Society of Ireland

This book would not have been possible were it not for the support and skill of Bernie Mc Loughlin who pored over pages of hand written manuscript turning them into legible print and The Dublin Fire Brigade Sports & Social Club who early in the venture agreed to provide £2,000 towards the research and publication of the manuscript. The Acting Chief Fire Officer of the the Dublin Fire Brigade, Michael J. Walsh, has given his encouragement at all times and has kindly written the Foreword.

We are indebted to Dublin City Council for agreeing to publish this book and to Matt Twomey, Assistant City Manager, Engineering Department, Dublin City Council, and to Matt Merrigan, LANPAG, for providing funding for the publication. Our thanks go to Deirdre Ellis King, Dublin City Librarian; Mary Clark, Dublin City Archivist; Alastair Smeaton, Divisional Librarian; and to Frank Kelly, Personnel Manager, Dublin City Council, for their determination to have this book published. Finally, we are most grateful to John Fitzgerald, Dublin City Manager, for supporting this venture and for kindly supplying an introduction to the book.

*Tom Geraghty*
*Trevor Whitehead*
*Spring, 2004*

# 1 Earliest times

Uncontrolled fire in mediaeval times was an ever-present hazard. The flimsy wooden houses and lack of an adequate water supply meant that fires, once started, spread rapidly. Large areas of Dublin city were destroyed by fire in 1190 and 1283. St Mary's Abbey was burnt down in 1304 and the valuable chancery rolls stored there were destroyed in the flames. The following year an ordinance issued by the common council left little doubt as to the seriousness of its intent.

> *For fire taking place in any house from which flames issue not, the householder, after the fire has been extinguished, is liable to a fine of twenty shillings. If the flames be visible externally the fine is forty shillings. Any person answerable for the burning of a street shall be arrested, cast into the middle of the fire, or pay a fine of one hundred shillings.*

Despite this threat of drastic punishment to the householder the number of outbreaks of fire does not appear to have decreased and, more surprisingly, the severe punishment clause was not revoked until 1715.

Two and a half centuries after the 1305 ordinance the council is calling for implements to be made to assist in fire fighting. It agreed in 1546 that "twelve graps of iron shall be made for pulling down houses that shall chance to be afire, forty buckets of leather for carrying of water". In 1592 the mayor was made responsible for ensuring that water was brought to the scene of a fire, and every parish was to provide six buckets and two ladders. These would be kept in the parish churches. Although these measures were well intentioned the common council was complaining in 1600 about the misuse of funds. Apparently the money collected to purchase buckets and ladders had not been used for that purpose.

The first mention of a fire engine in the Assembly Rolls occurs in 1637. The common council requested that "an instrument called a 'Water Spout' be obtained from England, very necessary for quenching of any great fire suddenly". It is not known exactly what sort of "instrument" this was, or whether it ever arrived in Dublin. Machines for fire fighting imported from Germany were certainly in use in London in the early years of the seventeenth century, and the first patent to be issued for a fire engine in London had been granted to Roger Jones in 1625.

The Dublin municipal records at the time show that the need was recognised for some control to be exercised in combating the scourge of fire, and that engines were needed. However, little progress seems to have been made throughout the next sixty years. The following resolution was passed in 1653.

> *Desiring that a course be laid down for certain engines to be used upon occasion of fire, it is therefore ordered that if any Corporation will disburse so much as the said engines shall cost, not exceeding eighty pounds, the said moneys shall be repaid by the treasurer upon Mr. Mayor's warrant.*

It failed to produce the desired result.

Prior to the seventeenth century there had been many large fires in London and in most of the cities and towns throughout the British Isles. But the worst one was yet to come. On 2 September 1666 a fire broke out which would eventually prove to be the catalyst for a new era of organised fire fighting. Known ever since as the Great Fire of London, it destroyed 13,000 houses, St Paul's Cathedral and eighty-seven parish churches, the Royal Exchange and many other civic buildings as well as leaving 200,000 people homeless. Only six deaths were attributable to this fire but, in terms of financial loss and total disruption of the commercial life of the capital, its effect was catastrophic. It had burned for four days and nights and destroyed almost half of the wealth of the entire kingdom.

Following the Great Fire of London changes in the law were enacted resulting in new building regulations whereby the old timber-framed houses with fronts of lath and plaster gave way to brick buildings. As the rebuilding of London progressed and urgent consideration was being given to town planning, water supplies and fire prevention matters, another aspect of dealing with financial losses from fires was the subject of much discussion. The businessmen of the city, the losers of great wealth in the 1666 fire, were only too aware that bankruptcy from fire destruction could still occur even with the new brick buildings being erected. Marine insurance had existed for some years, but insuring a ship and cargo for a specific voyage entailed a relatively short-term risk, whereas insuring a house was a matter of indefinite duration. The possibility of insuring against fire losses had been considered in recent years, but it was not until 1680 that a speculative builder, Nicholas Barbon, set up the first office in London for insuring houses and other buildings. This was known as The Fire Office, and to help reduce his losses as well as to attract new customers he retained the services of Thames watermen to act as firemen. How these developments affected the situation in Dublin will be considered in a later chapter.

In 1670 the Lord Lieutenant of Ireland, undoubtedly aware of the action already taken by English cities and towns following the Great Fire of London, warned Dublin's lord mayor to ensure that the city was properly provided with fire engines. But no positive action seems to have been taken until 1705.

> *Whereas certain of the Commons petitioned the Assembly setting forth that the city is very much enlarged and in great danger from fire for want of engines, etc, and fit persons to make use of same, a committee was appointed to consider proper ways and means for extinguishing fires and proper persons necessary for such work, after considering the matters recommended to be forthwith provided two water engines of the best sort, and that the treasurer do immediately send for one of them to London, and the other to be made here, and that a house be built to keep the engines.*

Six years later, as a result of these momentous decisions, we find the following entry in the Dublin Assembly Roll for 4 July 1711.

> *Ordered, by the authority of the said Assembly on the petition of John Oates, water engine maker, that he, the said John Oates, be allowed six pounds per annum during the city's pleasure to keep in order the water engine belonging to the city and that the same*

*be so kept, that he cause six men once a quarter, at his own expense, to play on the said engine, and on any extraordinary accident of fire to have twelve men or more in readiness, to be paid for their service as the committee think fit, his salary to commence from Easter last and paid by half-yearly payments on my Lord Mayor's warrant.*

At the same time the treasurer was ordered to pay Mr Oates fifty pounds for one large water engine. Just over one year later the annual payment to Oates was raised to twenty pounds for keeping no fewer than three engines in order. Incidentally, it is a rather curious paradox that in those days the machine which was used to quench fire was termed a "water engine" while the equivalent machine today is called a fire engine!

Shortly after the meeting of the Assembly in July 1711 a broadsheet was issued to inform the public that decisive action had indeed been taken in regard to fire fighting. This piece of printed publicity is of particular value to historians in a number of ways. Most importantly, it includes the earliest known illustration of a Dublin fire engine. It seems likely that this is the one "already made for the Honourable City of Dublin" by John Oates. The drawing shows a machine quite similar in general design to the ones made by John Keeling and other English makers towards the end of the seventeenth century. It presents an unintentionally humorous impression by including what looks like a midget (leprechaun?) perched on top of the engine.

In fact many fire engines of that date were eqipped with a fixed metal branch-pipe and nozzle, and a man had to stand on the engine to direct the jet of water. This particular drawing, however, clearly shows flexible hose being used, with two men operating the pump handles, and the three flaming castles of the city arms painted on the side of the engine. It is possible that, as instructed in 1705, one engine was obtained from a London maker and John Oates then constructed his own engine copying the English design. He is quick to point out that he makes his engines "to as great perfection as in London".

Fire engine, from printed notice issued by Lord Mayor of Dublin, 1711.
Postcard, © Dublin City Public Libraries

A significant point of interest is that the city now had fire engines, someone paid to look after them, and "twelve men or more in readiness" to fight fires. Here was the beginning of a rudimentary form of what today is called a retained fire brigade, ie men who are normally engaged in a job or trade who are prepared to down tools at any time if called upon to put out a fire.

The Dublin fire fighting service at this time was probably no different from that in London. In 1718 a Dutchman, writing to a Londoner, observed the following.

*From the knowledge I have of the great confusion and impertinent crowd that is always at a fire in London, nobody governs, nobody will obey, very few will work and the many, the very many, only look on and encumber the ground to the hindrance of all that would work. I am amazed that the English should so often suffer by the calamity without even entering into proper method to cure it.*

When fires broke out, large numbers of people began to gather, some to seek selection as pumpers on the manual engines, some simply to enjoy themselves and others ready to pilfer anything they could get their hands on. It required from eight to twenty men to operate the pump handles on either side of the machine, and it was very hard work.

The volunteers would soon grow tired and usually expected to receive liquid refreshment, so the cry of "Beer Oh!" became a regular chant as the pump handles went up and down. The crew of each engine, on arrival at the scene of the fire, tried to manoeuvre their machine into the most favourable position to obtain water – without any consultation with other crews. This resulted frequently in the water pressure being reduced to such a degree for all concerned that the jets of water were well-nigh useless. The Dublin Council was conscious of these problems and directed the lord mayor to attend all fires in the city so that, by his office, crowds might be controlled and some of the damaging competition on the fireground might be eliminated.

In the early eighteenth century, laws were enacted relating to fire prevention measures, but they applied only to London. Similar laws were however passed some years later by the parliament in Dublin. The following extract from an Act of 1715 (2 George I) gives a vivid account of the events at a city fire.

*Upon the breaking out of any fire within the City of Dublin the Chief Magistrate, Constables and Beadles shall immediately repair to the place where the fire shall happen, with the badges of their authority, and be aiding as well in extinguishing the fire and causing people to work at the engines for throwing up the water, as preventing goods being stolen; and shall seize and apprehend all ill-disposed persons that they shall find stealing and pilfering from the inhabitants. Also the said Magistrates, Constables and Beadles shall give their utmost assistance to help the inhabitants to remove their goods.*

In 1719 another Local Act (6 George I) was passed which ordered the churchwardens of all the city parishes to provide one large and one small fire engine (for each parish) and to appoint an engine keeper. The parishes were also authorised to make payments to the first, second and third engine keepers who arrived at the scene of a fire. The wording of these examples of Irish legislation relating to fires is identical to that of an Act relevant to London which was passed in 1707 by the English parliament.

Two interesting references in the *Calendar of Ancient Records of Dublin* mention the building of an engine house and the abolition of the office of city engine keeper or fire master as he was called. In 1738 "the first engine was exposed to weather, but following the purchase of the late John Molineaux's yard, it was resolved that the engine house be erected to the south western end of the market house, which is of no use to the City and which place the committee apprehend will be convenient for the engine house". The same source outlines a petition in 1740 by Robert Betagh for admission to the office of fire master in place of Wm. Taplin, deceased. The committee responsible for disposing of city employments found that "such an employment is entirely useless to the City and therefore recommend that the office be discontinued for the future".

During the eighteenth century Dublin became a great centre of wealth and fashion. Trade was increasing, docks were built, and two great inland waterways – the Grand Canal and the Royal Canal – were constructed to link Dublin with the river Shannon. The design and construction of manual fire engines was improved, notably by Richard Newsham of London, but progress remained slow. The early machines were very crude – merely a pair of plunger pumps on wheels or sledge with a cistern below that had to be filled by buckets. Directly connected to the outlet was a swivelling goosenecked branch-pipe from which the water was directed in a series of pulsations as each pump plunger made its delivery stroke. The wheels were often made of solid wood, and even when spoked were so small as to make speedy transport difficult.

By the second half of the eighteenth century various improvements had been made, the chief being the introduction of an air vessel enabling a jet of water to be operated continuously instead of in spurts. The other major advance was the more general use of leather hose, which meant that a jet could be brought much closer to the fire than was possible with the fixed branch-pipe on the engine.

It is fortunate that one of the old Dublin parish fire engines is still extant in St Werburgh's church. It is a large implement with solid wooden wheels and was the parish engine in the eighteenth and early nineteenth centuries. This historic engine is in reasonable condition considering its age, and is worthy of complete restoration.

In 1753 the churchwardens of St Michael's parish purchased a fire engine for sixteen pounds which was probably small in size and without wheels. In 1767 the engine keeper was paid four guineas quarterly "to play the two engines four times in the year at his own expense and to fix an in-scription board over his door setting forth that he is engine keeper of St. Michael's parish". In Dublin two hundred and fifty years ago such fire engines as were available were located either in the watch houses or in the parish churches. An ordinance by the common council in 1729 granted land

**Early eighteenth-century manual fire engine, St Werburgh's Parish.
Photograph, © Trevor Whitehead**

to the parishioners of St Mary's to enlarge the water house at the north end of Essex Bridge (the main north/south thoroughfare at that time) to accommodate the watchmen and fire engines.

Manual fire engines were still imported from London but there were now one or two Irish makers at work in Dublin, following in the footsteps of John Oates. Engines were being built by John Bolton of Pill Lane on the north side of the city, and later by John Smith of Moore Street.

The parish of St Paul's bought one of Bolton's engines in 1748, and erected a shed to house it. In 1749 the parishioners of St Brigid's were granted leave to build an engine house "on the south side of the present Watch House in St Stephen's Street".

The following year George Robinson petitioned "to be employed as Water Engineer in order to assist Mr. Bolton, due to the distance the latter had to travel to put out fires, by which time the said fires had greatly spread". The petition was granted by the city and both men vowed "constantly to aid and assist each other in extinguishing all accidental fires that may happen and that they be paid a salary of £20 a year". The standing committee of the city assembly in 1751 agreed to reward "such persons as shall be serviceable from time to time in extinguishing such accidental fires as may happen in this city and the Treasurer do pay such sums for the said services". Although the office of "Fire Master" appears to have been abolished in 1740 Thomas Barrington was appointed in 1751 "Engine Keeper" (in place of Mr Bolton, deceased) and "Assistant Fire Extinguisher" under Leo Robinson "during the city's pleasure at an annual salary of ten pounds".

The impression gained from all the corporation records so far quoted is that for many centuries fire fighting was a very haphazard affair. Rules were drawn up but frequently ignored. Responsibilities were allocated but not always carried out. For instance in 1764 the city assembly

> ordered that there be immediately provided, at the expense of the City, fire engines, with ladders, hooks, buckets and other utensils for the extinguishing of accidental fires; and that the several officers and servants of the City, with the labourers employed at the Ballast Office, and at the Pipe-Water Office, do attend the Magistrates on the first alarm of the Fire Bell; to be liberally rewarded by the City for the same.

Even though fire regulations had been specified in the Act of 1715, and it was now fifty years since the city fathers had purchased the Oates fire engines, there is obvious concern that further action is needed. A letter to the *Freeman's Journal* in January 1764, giving an eyewitness account of a fire, highlights just one of the problems.

> The alarm bell immediately led me to enquire for the fire! Being informed (as I had some friends in that neighbourhood) I resorted there. How dreadful the appearance! How shocking the reflection! The Chief Magistrate, High Sheriffs, a proper Guard, and several engines were there with numbers of people to work them; but a want of water to abate the conflagration was the general cry. The Liffey contiguous, and from the late heavy rain the aqueducts of the town full of that element. I was amazed at the

*cry, and enquired for the cause; that the Overseers of the Pipe Water Works were not to be found for two hours, without whom the water plugs (if any) on the main pipes could not be ascertained, so that the little water procured was by mere accident found on picking up the pavement. A poor shift indeed and a very ineffectual supply.*

This account illustrates the crucial role of the overseers of the Pipe Water Works and the disastrous consequences of their absence.

On 27 February 1792 the Irish parliament was in session in the House of Commons at College Green. John Townsend was in the chair when Henry Grattan introduced a discussion on the regulation of breweries and distilleries. A Mr Fether on behalf of the brewers was being examined by Henry Grattan when the Clerk of the House suddenly cried out "the House is completely on fire, in less than five minutes the whole dome will fall in on you!".

Manual fire engine in use. Print, from Hibernian Insurance policy, issued 1790. © Trevor Whitehead

According to the *Freeman's Journal* "… the House was immediately adjourned and every member escaped as well as he could, the spectators in the gallery retreated with the most violent perception. In ten minutes the whole place was in a conflagration". In its editorial the newspaper had complained that

*several incendiary hand bills of a dangerous tendency have been put in circulation by porter house politicians, holding out the conduct of America and France as models for imitation. The disseminators of such factious principles merit the most severe punishment, which is probable some of them may very shortly meet with, as their persons and motives are equally well known.*

7

By the time all the members had evacuated the building the whole of the western part of the roof was ablaze, with the flames spreading rapidly to the other sections. Within an hour or so the dome collapsed. All available fire engines in the city were called out, a regiment of soldiers and a detachment of cavalry arrived from barracks "to keep off the multitude". The fire took over eight hours to extinguish, but was confined to the commons area by the massive stone walls of the chamber. But the newspapers were not satisfied that the fire was accidental. The *Freeman's Journal* speculated: "Suspicions are strongly entertained that the conflagration was not the effect of an accident, but of design". It went on in the manner of present-day tabloids, employing hyperbole such as must have helped sell papers in those days as it does today.

> *It is hard, however, to suppose any human being so hellishly depraved as to meditate the destruction of a thousand fellow creatures in cold blood, for the House and gallery held about that number. Yet the manner in which things occurred, by the tumbling of the dome, must be admitted to have been as efficacious as that projected by Guy Fawkes, of atrocious memory, for annihilation of the English House of Commons members and mansion in the reign of James 1st.*

As for the Irish brewers, whose case was being pleaded when the fire broke out, the paper considered their product defective and inferior.

> *In order to circumvent each other in their customers, they undersell each other; hence their liquor is scandalously impoverished – and while the customer can purchase a pot of sound invigorating British Porter for three pence halfpenny, it is absurd to suppose that he will give a selfish preference to the trash they too frequently feed to the market under the name of Irish Porter.*

The Guinness Brewery, founded in 1759, may have been involved in having this Bill enacted into law, because it would undoubtedly lead to the closure of smaller breweries on health grounds. No one was prosecuted or hanged for the parliamentary fire. The final word may be given to the *Freeman's Journal*.

> *During the course of yesterday, great number of labourers have been employed at the Parliament House in removing rubbish and conveying water, in order to prevent any further danger from the vast mass of ignited timber and smouldering ruins occasioned by the dreadful fire. The Lords yesterday sat as usual in the Upper Chamber which fortunately received no injury.*

At the end of the seventeenth century as we have seen, in the aftermath of the Great Fire of London, the first fire insurance companies were formed in London. These companies maintained their own fire brigades, providing for the first time some properly organised firefighters available to the public. During the 1700s the number of fire insurance companies grew rapidly throughout Great Britain. The Royal Exchange Assurance was the first company to transact business in Ireland. Established in London in 1720, an office was opened in Dublin in 1722. (See colour section, illustration number 1.)

The business of the English companies in Ireland was gradually developed and

8

increased by their agents, but the first Irish insurance company, the Hibernian Fire Insurance Company, was not established in Dublin until 1771. The company included many of the most important members of the business community in the city, which had a population at that time of 150,000. It was intent on furthering its business, and was very aware both of the shortcomings of the parish system of fire prevention and the poor state of the parish fire engines. In 1776 the Hibernian offered to take over the responsibility for fire fighting in return for the amount being spent by the parishes. This particular offer was not accepted, but it was a remarkably early instance of an idea which was accepted in London many years later.

The first half of the nineteenth century saw some improvements and changes in the way fires were fought in Dublin. The insurance companies became in fact the principal fire fighting force in the city. It is known that at some period between 1800 and 1860 at least seventeen insurance companies each operated a fire engine, and one company, the Royal Exchange, for a number of years had two engines. These fire engines were larger and more powerful than the older ones owned by the parishes. More significantly, however, they were maintained in good repair and manned by more competent crews. Each company clothed its firemen in brightly coloured uniforms with a large badge on the left arm, and advertisements invariably called attention to its own brigade which was always described as "in constant readiness". The Royal Exchange proclaimed

*In case of fire, application to be made at the Engine House, Crown Alley, immediately at rear of the Commercial Buildings where the fire engines and firemen are in constant readiness, or to the Engineer, No. 5 College Green.*

The National Assurance Company of Ireland had its "powerful engine always in readiness" at the company's engine house in Anglesea Street, and the West of England Company engine was stationed in Crown Alley. In 1801 the Globe Insurance Company sent a "sixth size engine" from London to Dublin: "it was mounted on springs, with branch and suction complete £120, with extra gear amounting to a total of £200".

In 1806 the Royal Exchange ordered an "eighth size engine" from London, and in 1833 a newspaper advertisement by the Phoenix office stated: "fire engine and active firemen are in constant requisition to afford assistance in event of fire".

Each company fixed a sign called a Fire Mark at the front of its premises to identify itself to the firemen if there was a fire. The earliest marks were made of lead and bore the policy number (see illustration); later marks were made of copper, iron or tin. Some of these marks are in the Civic Museum in South William Street, Dublin, and a few may still be seen on city

**Fire mark of Hibernian Fire Insurance Company, c. 1800. Ink drawing, © Trevor Whitehead**

buildings, for example in Leeson Street, Pembroke Road and Gardiner Street. Although the insurance brigades now owned the majority of the available fire pumps, some parish engines still turned up at fires to claim the attendance payment. The decision as to who was first on the scene was sometimes a source of quarrelling and a certificate had to be signed before payment was authorised. A certificate issued in 1838 reads as follows.

> *I Stephen Brown, Engine Keeper of the Parish of St. Nicholas Within, do solemnly and sincerely declare that I was third in attendance at the house No. 13 Werburgh Street, with the above engine on the 16th inst. the chimney of the said house being on fire. This declaration is made for the satisfaction of the churchwardens of St. Werburgh's Parish and for the purpose of claiming 4/7½d. the sum allowed by Act of Parliament.*

The churchwardens' accounts of St Mary's parish, Donnybrook, contain an item typical of many parishes: "Premium for Enginemen for attending fire, £1".

Newspaper reports of those days often quoted the order in which the various engines arrived at a fire, as in this report: "The police engine arrived at 1.55 followed by the Corporation engine five minutes later. The National Assurance engine arrived at 2.15, Royal Exchange engine at 2.20 and the West of England engine at 2.40". At that particular fire the insurance company engines appear to have been slower than usual, for whatever reason.

### Custom House Docks warehouse fire of 1833

The 1830s saw some tragic and disastrous fires. Perhaps the most spectacular was at the great Custom House Docks warehouse. At about half-past one on the morning of 10 August 1833 flames were seen coming through the roof. The building, which stood between George's Dock and the Custom House, stored tobacco, sugar and spices, and it also contained a huge hoard of whiskey in barrels in the vaults below. Within a short period of time "twelve distinct columns of flames were seen to shoot up into the sky, illuminating the whole port area right down to the mouth of the bay". The fire engines were unable to prevent rapid spread of fire and many tons of timber stacked on the dockside had to be quickly moved to prevent an even greater conflagration. The 59th and 74th regiments were called from their barracks to assist in the battle with the fiercely growing inferno. One report of the night's events maintained that the efforts of the parish engines were "to no more avail than pop-guns on a volcano". By three o'clock the fire was raging out of control "with unabated fury, mocking every human effort to control its devastating influence". Six engines were in operation at the time, worked mainly by the soldiers with a plentiful supply of water, when the roof of the warehouse collapsed bringing with it part of the walls. Up to one hundred puncheons of whiskey had been pushed out of the warehouse area and tumbled into the docks. Many of them had broken open and the whiskey flowed on the river, catching fire, causing other casks to explode, and illuminating the river view of Gandon's magnificent building.

Light dragoons and mounted police were called out to control the crowds and to prevent looting. Ships had to be hosed down to prevent them catching fire, while others were moved down stream. It was five o'clock in the morning before an on-the-spot reporter could record "no fear is longer apprehended for the safety of the Custom House, the shipping in the river or the houses adjoining". The warehouse was by then only a burning shell.

This fire was probably the most destructive seen in the city up to that time. It could have been a great deal worse. The extensive vaults containing hundreds of casks of spirits actually survived the flames, though it was three days before some merchants finally opened the vault doors to allow air to enter and to enable them to inspect what had become a dangerous area.

The Phoenix Fire Office was quick to remind the public of the advantages of insurance. The Dublin newspapers on 12 August carried a well-illustrated advertisement.

> *This Office has attained the highest eminence, by the magnitude of its transactions, liberality and promptitude with which it has fulfilled its engagements, it has paid out £3m sterling … fire engines and active men in constant requisition to afford assistance in the event of an accident.*

More than 4,000 hogsheads of sugar and 300 casks of whiskey had been destroyed along with a considerable amount of tallow, hemp and tan. The estimated loss to the city merchants was upwards of £80,000 in addition to structural damage to one ship.

The warehouse, being crown property, was not insured. An investigation took place into the cause of the fire, and the local parish pleaded with the government to cover the losses if the blaze was found to be malicious. The merchants decided to petition parliament for recompense, the case to be argued by Daniel O'Connell, who acted as legal representative for many of those who sustained loss. O'Connell claimed that the fire should be treated as a national calamity and the losses by the city merchants should be made good out of the public purse.

The captain of a vessel moored in the river immediately opposite the warehouse claimed that he saw lights moving about in the premises just before the place erupted into flames. Others maintained that when the fire was first noticed it was breaking out in several locations at the same time. The investigation's findings were inconclusive, however, and the general opinion came down in favour of spontaneous combustion because the sugar was stored in an area which also housed palm oil and cotton waste. The government refused to rescue the merchants from their huge losses, but some years later, for political reasons, the treasury granted £60,000 as a contribution towards some of their losses. The merchants learned an expensive lesson. In future they insured their warehoused goods.

The warehouse was rebuilt and continued in use for over 150 more years of Dublin trade and commerce. Its location was to be the scene of the great millennium fireworks display on the last night of 1999 but no reporting of that colourful event came near to emulating the eloquence of the *Dublin Evening Freeman* of 10 August 1833 when describing the Custom House warehouse fire.

*We have witnessed a grand spectacle and its sublimity was heightened by the terrific impressions which it conveyed. The spiral flames shooting to the sky, their deep blood red colour occasioned by the peculiar qualities of the fuel … a scene which, if the lookers-on could maintain an indifference to the ravages of the destructive element, might be pronounced as the grandest in nature.*

### The Royal Arcade fire of 1837

In 1819 an attractive new shopping arcade was built at a cost of £16,000 to connect College Green with Suffolk Street. It consisted of a number of small, attractive shops selling clothing, millinery, jewellery, books and other materials, and there was also a wax works located there. The arcade had cast iron gates at both street ends that were closed at 10 pm while above the shops was a gentlemen's billiards hall. Fire broke out in this "Royal Arcade" in the early hours of 24 April 1837. Clouds of smoke billowed forth from both ends as the fire spread rapidly from shop to shop, the flames fanned as a result of the tunnel effect of the construction.

Bells in the city were soon tolling to alert the fire brigades, the first of which arrived in time to see the roof catch fire. There was much delay before water was eventually played onto the blazing edifice as the fire engines were manned by civilians and soldiers assisting the insurance crews. Soon most of the insurance interests with their fire engines were on the scene, the National, Globe, London Union, Royal Exchange, Imperial, Scottish Union, Hibernian and Dublin Steam Packet Company.

The crew members from Trinity College were particularly praised because they "worked with a degree of discipline and regularly with power and efficacy". Such was the inferno at the height of the fire that the military discussed using cannon to prevent it spreading.

The roof of the arcade crashed down, venting the fire as it fiercely spread to the Royal Hotel and other adjacent premises. Furniture and personal effects were removed from threatened buildings as huge crowds gathered along the perimeter of chains set by the police and military to keep the inhabitants back from the fire. A wall collapsed on one of the manual fire pumps causing the only casualty as one of the civilian helpers was buried in the debris and sustained serious leg injuries. Soon four premises in Suffolk Street together with the hotel and three buildings on College Green were burning shells. Rain began to fall and this assisted the struggling fire crews. The artillery regiment in attendance began pulling down with the use of ropes walls left standing unsupported and liable to crash down any minute. By 11 am, some nine hours after the fire was first noticed, a smouldering ruin was all that remained of the arcade and adjoining premises.

The *Freeman's Journal* of 27 April reported on three men who had been brought before the courts charged with the theft of articles saved from the fire. The editorial on the same day contained the following statement.

*The enormous loss of property incurred by the late calamity in College Green and the destitution of many industrious families may in a great measure be attributed to the*

"The burning of Holmes' Emporium, College Green, Dublin" by W. Sadler.
© National Gallery of Ireland

*shameful neglect of the Corporation of Dublin … the entire Arcade was totally destroyed before the engines could be supplied with a drop of water. In no city in the three kingdoms are the inhabitants taxed at so exorbitant a rate for pipe water as Dublin and in no other town are there so just complaints of a scanty supply. The Corporation are bound by law to lay down fire plugs at convenient distances and still refuse to do so and the consequences are awfully exemplified in the smoking ruins of the most beautiful and valuable part of the City.*

A local Act of 1719, already mentioned, had empowered churchwardens to fix fire plugs on the water mains within their own parishes if they wished to do so. But in 1788 another Act ordered the lord mayor and corporation to fix fire plugs throughout the city. Obviously they had failed to do this.

When the lord mayor tried to introduce a tax for fitting the plugs in 1837, the churchwardens of one parish were so incensed by the corporation's attitude to its responsibilities that they prepared a petition

*on the liability of the Corporation of Dublin to affix fire plugs on all the public water mains laid down by them within the boundary of the Circular Road, with the opinion and advice of the Rt. Hon. John Richards, H.M. Attorney General, as given to the parishioners of St. Mark and respectfully addressed by them to the Lord Mayor, Sheriffs, Commons and Citizens of Dublin, urgently requesting that they may affix the fire plugs on the mains, and so afford that protection to the persons and properties of their fellow citizens, which they are legally bound to do.*

As we shall see, some progress in this matter was not too far off.

# 2 First organisation

Lessons learned in London from the Great Fire of 1666 led to private investment in the fight against fire. The London experience also led to important improvements in city water supply systems. Dublin, however, was slow in changing its attitude to the need for effective fire fighting services and proper water supplies to assist in this essential work. In 1555 the mayor had taken responsibility for the existing watercourse and in return "was to be entitled to have corn from all the Mills in or about the city who are supplied with water". He was also charged "to care for the water that came into the city throughout its whole course from the Dodder to Dolphin's Barn, and to repair and mend that course from time to time as might be requisite". Between 1594 and 1600 the city authorities appear not to have carried out their part of the arrangement, and the stonemasons and others involved in repair of the watercourse were owed large sums of money and were often unpaid for work carried out. As a result the watercourse became greatly decayed. The Down Survey map (1656) shows that the water supply for the people of Dublin ran from the river Dodder near Balrothery opposite Firhouse, taking a north-easterly direction in a "tongue" down by the walls of Templeogue Church to Mount Down Flour Mills where it was joined by the river Poddle. It flowed on to Kimmage under the present Kimmage Cross Road and the waters divided with another tongue at Larkfield Mills, the Poddle continuing through Harold's Cross towards the city. The branch flowed in a north-westerly direction bounding Mount Argus and through Rutland Mills to reach Dolphin's Barn, then in "the back of the pipes" to the City Basin, near James's Gate. This Basin was enlarged in 1670 and a new watercourse, known as the Glib river, was built to run through Thomas Street to a sunken cistern at the south end of New Row.

From there a new main was laid to convey water over the old Liffey bridge at Church Street to supply the north city area. However, all was not as it should be. Mill owners on the watercourse and owners of land through which the water flowed claimed absolute rights to the water. At times they took the law into their own hands by stopping the supply of water until they were compensated.

In the year 1735 the watercourse was found to be "much choked with weeds" and there was seepage through the banks. In some places above Dolphin's Barn breaches had been made in the bank which were only stopped by sods, the intention being that they would serve as sluices in summer to let water run in small channels through the adjacent fields for private use! It is quite clear that the system of water supply for Dublin could never have been adequate to deal with any major fire, even if the city had contrived to organise the parish and insurance brigades to work together.

With the construction of the canals in the 1770s new reservoirs were built at Blessington Street and Portobello. These were supplied from the canals, while the Poddle supply to James Street was augmented from the same source. But piped water from these sources was soon seen to be wholly inadequate. The canal companies were requested to increase their supply to the new city area which had grown up following the developments of the Dublin Wide Streets Commission (set up in 1757) and was spreading eastwards, with the major construction on the north side of the city. This was the period of the Irish Parliament when Dublin was considered the second city in the Empire.

In the early part of the eighteenth century there were still wooden water mains in much of the city. A section of one of the mains is preserved in the Dublin Fire Brigade Museum. In 1726 elm timber pipes were purchased by the corporation at £4 per ton in twelve foot lengths and eight inches diameter. But because the corporation failed to acquire sufficient supply of the elm it was agreed that "round red fir, the bark on" would do for the same purpose, so 200 tons of fir were ordered in lengths of sixteen and twenty feet, with no piece less than twelve inches in diameter. It must not be thought that this small diameter piping provided a satisfactory supply of water for fighting fires. When a fire broke out the procedure for obtaining water was troublesome and time-consuming. First the street had to be dug up to locate the main, then the pipe was fractured with an axe and eventually the water filled the pit which had been dug. The suction hose from a fire engine could then be laid in the water, but the rate of flow and quantity of water must often have been insufficient for even those early engines. It is recorded that in 1798 the corporation was enjoined as follows: "from the very great delay that occurs in opening the ground in case of fire … ten sets of proper tools for opening mains be provided. One set to be lodged in the pipe water stores … the engineer to have marks put in each street denoting the situation of fire plugs". How effectually these instructions were carried out is not certain.

By 1802 it was found that the wooden pipes were decaying, causing flooding in certain areas and contamination of the water supply. In that year metal mains were laid in several areas, so that from then onwards there was a combination of wood and metal pipes delivering the water from the Basin throughout the city. A lack of funds obviously prevented the conversion of the entire system to metal.

Towards the middle of the nineteenth century a significant improvement was made in the method of controlling the outlet of water from the mains. Charles Gavin, a Dublin Corporation employee, designed a fire plug for metal pipes which was adopted in Dublin, though initially it was manufactured by the Hammond Lane Foundry for sale to Liverpool. Gavin's invention was patented in 1844 and received much praise from the borough engineer and Mr Crofton, and also from the National Insurance Company. Before the new fire plug was available "the tail pipe [ie the suction inlet] of the engines sucked up sand and gravel, impeding the working, to repair which was attended with considerable expense". The water supplies at this time were under the control of the Paving Board whose responsibilities were taken over by the Waterworks Committee after the 1849 Dublin Improvement Act. The use of Gavin's fire plug by the corporation resulted in an appreciable saving of both money and water, but when he sought some compensation for the outlay of his own money on the initial project he was refused any payment.

Dublin lifestyle at this time was changing rapidly. The population had stabilised after a period of decrease, and the coming of the railway in so few years after the building of the canals dramatically altered public transport. The Dublin and Kingstown Railway was formally opened in December 1834 and the horse omnibus service was established in the city in 1840. Port expansion and increasing trade had led to a growing population and in 1836 total goods handled through the Port of Dublin amounted to 400,000 tons.

Colliers arriving in port daily reflected the increasing use of coal for both domestic and industrial purposes and this, together with traffic congestion, seemed likely to lead to a rise in the number of accidental fires. Beresford Ellis in his history of the working class claimed that Dublin housed up to 4,500 landlords who were absentees from rural Ireland. This privileged class lived in and owned extensive property in the city. This period of rapid growth in the city coincided with the time when the majority of parish fire engines were worn out and becoming obsolete. It seemed inevitable that major fires would occur more frequently than before and that there would be increasing loss of life.

The use of water increased greatly with the changes in lifestyle and the proliferation of hotels, stores and large business premises. The city now had numerous breweries and distilleries which needed voluminous amounts of water in the processing of their products. It is certain that the city's growth had long outstripped its piped water system, the supply being inadequate and the piping needing upgrading or replacement. Also, there were still large areas of the city, mainly back streets, lanes or courts, where there was no water connection to individual premises. Eventually the passing of the 1849 Dublin Improvement Act vested sufficient powers in the city council to tackle the problem of providing an adequate water supply to cope with the size and demands of the expanding city. The complete implementation of this Act was to take some time, much acrimony, opposition and obstruction, but the wheels were now in motion, the day was coming when the Dublin water supply would match that of the other important cities of the United Kingdom. The 1849 legislation was the first decisive move towards providing real authority to Dublin Corporation to raise rates and to take power from the parishes. The establishment of a waterworks committee with powers to take over control of the city water supplies from the Paving Board was a massive step forward towards securing a high pressure system. Real progress for the city, including its future fire fighting capability, was vitally dependent on this essential improvement in its water supply system.

The borough engineer was instructed to carry out an examination of the entire system. His report in 1851 recommended a complete upgrading of the pipe system. This entailed replacement of most of the obsolete water mains. The basins at Blessington Street, Basin Lane and Portobello contained the city water supply, but the pipes from these three sources were in most cases too small, too old or simply in a state of decay. The pipes failed to supply water to as many as 4,000 of the houses along the main thoroughfares, and none was supplied to cottages or slums in courts and small lanes. The engineer did, however, request the corporation not to go ahead with any major change until a decision had been made on the source of a new, satisfactory supply of water. It was the responsibility of the waterworks committee to examine the options. The engineer had pointed out that the seven and a half inch mains then in place for draining the system from the basins should be replaced by eighteen inch mains, and a twelve inch main should run from the James Street basin to Dublin Castle. He estimated that to implement his proposals would cost almost £17,000.

The waterworks committee set about its task of employing new staff and deciding their pay and conditions. Working hours for labourers were fixed at 6 am to 6 pm, with one and a half hours allowed for breakfast and dinner, from March to October, and from daylight to dusk in the other four months. Any employees taking presents or gratuities for any duty would be dismissed.

One particular group among the labourers – the turncocks – was directly concerned with fire brigade matters. The turncocks were to be "sober, clean, careful and civil; to have charge of all cocks, hydrants and fire plugs in their districts". They were held accountable for every case of waste of water, and had to

*attend promptly by day or by night at every fire, and to render the fire brigade all the assistance in their power by opening and shutting cocks and valves, placing standpipes etc, and to take steps to ensure that the greatest possible pressure of water be brought on the mains in the street and neighbourhood in which a fire may be. The Inspector of Turncocks is to attend all fires and direct the turncocks in procuring the best supply of water.*

In 1854 the corporation purchased two fire engines from J.& R. Mallet of Ryder's Row, a firm which described itself as "Plumber, Hydraulic Engine Maker and Iron Foundry". The firm carried out a great diversity of engineering work including "small fire engines", but unfortunately no contemporary illustration of these engines has been found. One of the two engines was located in Whitehorse Yard, off Winetavern Street, under the control of a corporation waterworks inspector. The other engine, although owned by the corporation, was placed under the control of the Dublin Metropolitan Police and kept at their Kevin Street barracks which housed the mounted troop of the horse police. This fire engine together with 300 feet of leather hose was handed over to the police on the understanding that they would keep it in efficient working order free of any expense to the corporation.

Whitehorse Yard, where the corporation stored waterworks implements and equipment, was obviously a convenient place in which to locate fire engines, but it did not include accommodation for those who would have to operate the fire fighting equipment. In December 1854 the corporation took leases on No. 5 and No. 6 Cook Street for an annual rent of £22-10-0 with the intention, once the houses had been renovated, of using them to house labourers from the Pipe Water Works who would act as firemen. At the city council meeting on 15 December 1854 it was agreed to purchase clothes for the firemen at a cost of £2-15-0, hats at 7 shillings and boots at 9 shillings and sixpence, to be issued to those who dwell in the Cook Street houses. At the same meeting payment to the canal companies of £2,254-6-6 was sanctioned for water then being supplied to the city.

It was not until March 1856 that a final decision was made on the selection of men to occupy the houses as caretakers in preference to charging them a weekly rent: "The men in your employment to be near the stores in Winetavern Street, and ready to turn-out in cases of fire with the engine kept there".

Some time prior to this the overseer of the waterworks, Mervyn Crofton, had been given the title of "Assistant Inspector of Parish Engines" and was instructed to check every one of the fire engines held by the city's twenty parishes. He was also put in charge of the staff attached to the fire brigade depot at Whitehorse Yard. His work in the parishes was considered as interference by many of the churchwardens who resented any civic involvement in their little kingdoms. Mr Crofton later found himself in confrontation with both the police and the army when he tried to assume control at fires where they were engaged in assisting as volunteers in the operation of the fire brigade manual engines.

The lord mayor, reporting to the city council in September 1856, stated that following the inspection of the parish fire engines it was found (rather surprisingly) that "most were properly cared for", but that the St Werburgh's engine was 150 years old! He also told the council that

> the Corporation engine under management of the officers and men of the Pipe Water Department was also examined and promises to be a most useful auxiliary to the Police brigade ... although only recently formed, the engine and brigade have rendered efficient service at several fires.

It seems from this report that the lord mayor and the council accepted that their fire brigade role was merely that of assistance to the police operated engine. Following some debate on the corporation's right or duty to maintain a fire engine, the law agent was requested to examine the legal situation and report to the next council meeting. The agent's report is of particular historical interest because it quotes from the Act of 1719 concerning the purchase of fire engines by the parishes. In addition to one large engine and one small engine

> each parish shall provide, keep and maintain one leather pipe and jacket of the same size as the plug or fire cock, and in default thereof it shall be lawful for the respective Grand Juries of the respective counties where such parish lies, at the Quarter Session to be held for such county, to present and charge upon the several inhabitants of such parish such sums of money as may enable the churchwardens to buy and provide within the said parish in like manner as other public moneys levied within the said county.

The Act went on to instruct each parish to employ "one or more persons to take care of said engines" with a fine of £10 for failure to do so.

The law agent advised the council that it had no obligation to provide engines for any parish not having them, and "the penalty enacted under the 9th Section of the said (1719) Act against engine keepers for neglect of repairs could not reasonably be enforced unless they have engines in their charge". But having received representations from some of the parishes Dublin Corporation did agree to pay five shillings for each horse used "for removal of fire engines upon occasions of fire". This was for the return of engines to their church depots.

It was at this time that the members of the corporation became aware of a "Justices and Police Force (Dublin) Bill" which was about to be processed through parliament. In July 1857 the corporation petitioned parliament regarding this proposed legislation. The Bill proposed an extension of the Dublin Metropolitan Police District to cover those portions of the County of Dublin south of the Liffey within eight miles of Dublin Castle … "thereby enabling the Government to dispense with the Constabulary Force at present stationed therein and paid out of the Consolidated Fund". The corporation stated that

> *in the opinion of your petitioners it is unjust to charge the citizens of Dublin with any portion of the cost of maintaining the Dublin Metropolitan Police, which is as much a Government Force as the Royal Irish Constabulary which is supported out of the Consolidated Fund, while cities like Cork, Limerick and Waterford and all counties in Ireland do not contribute towards the police establishment.*

For many centuries all the efforts of those engaged in fighting fire had been directed towards saving property, with little constructive thought being given to saving human life.

By the middle of the nineteenth century it was becoming clear that the citizens, increasingly alarmed by the number of fires involving loss of life, were seriously concerned at the absence of suitable ladders which could be used for rescue purposes. This matter was discussed at a meeting of the city council in March 1857 and it was agreed that two escape ladders were needed – one for the north side of the city and one for the south – but there were doubts as to whether corporation funds were applicable for such purpose. The problem was then examined by the waterworks committee and its report was adopted in April by the full council on a vote of seventeen to fourteen. The split vote came after the council had heard the law agent's opinion on the purchase of two escapes. His opinion was that although the Dublin Improvement Act did not in its terms include the purchase or maintenance of apparatus for preservation of life in cases of fire, the section on "provisions for preservation against drowning" might be cited. (The corporation had been authorised to purchase rescue equipment for placing at strategic points on the lower reaches of the Liffey to assist in river rescues.) Thus the cost of providing and maintaining one or more fire escapes could be regarded as a proper municipal expenditure to be paid out of the Improvement Fund. The price of a fire escape was £56 plus delivery cost, and it was stated that it would require a crew of five men to propel it to a fire and then operate it. It was then mentioned that members of the waterworks committee had met the police commissioners "with regard to the care and management of the fire escapes and have been informed that there is no objection to the mounted police (at Kevin Street) taking charge of one of the escapes to attend at fires in the city, but they are of the opinion that one could not be kept on the north side as there are seldom more than two men at each police station". It was the proposal to purchase only one escape that produced the close, split vote. A significant aspect of the waterworks committee report was the decision to discuss the proposed purchase of fire escapes with the police, and the agreement to hand control of these appliances to the police indicated a reluctance on the part of the corporation to actually organise and operate a fire brigade. With the ongoing concerns over high taxation of the citizens it may be that the conservative councillors did not want the cost of a fire brigade to be forced on them as well.

Three months after the consideration of fire escapes the councillors found themselves confronted by a sub-committee report on a fire which had occurred at Lincoln Place in June 1857. The report described in detail the circumstances of the outbreak and the attitude of the police and the insurance companies. The fire was discovered at 11.05 pm by a police constable "but before he could proceed to the police station another constable arrived … they were occupied for up to 15 minutes in endeavouring to alarm the inmates of the house". The two constables then "hastened to Lad Lane and College Street police stations both arriving at about 11.30". Intelligence was dispatched to the different insurance companies, corporation officers, and Whitehorse Yard and Kevin Street. The time of arrival of various engines and personages was given as follows.

*Royal Exchange 11.54: Kevin Street (police) 11.55: National Insurance 11.55: Turncocks 12.00: Sun Insurance 12.03: Atlas Insurance 12.10: Whitehorse Yard (Corporation engine) 12.12: Mr. Crofton 12.15.*

The turncocks opened the water mains and six water carts turned up. Nevertheless the report claimed that there was total confusion at the scene:

*various fire engines endeavouring to procure a supply of water to the extent that the whole building was totally destroyed and serious damage caused to the adjoining property. Life and property cannot be properly protected from the consequences of accidental fire by the present services … the intermittent water supply and the water carts being of little use … in the absence of the Lord Mayor one person of experience and judgement should be clothed with sufficient authority to direct and control the operation of all fire engines.*

It was the general opinion of the sub-committee, and also the opinion of the police commissioner and the insurance companies, that some arrangement should be adopted for the more judicious and effective control and regulation of the working of all engines at fires and that "one person shall have power or ordering them to take the position from which he may consider they may be best adapted".

The insurance agents had already decided to recommend that their respective companies jointly appoint and pay a competent superintendent to act at all fires and to have the full control of all insurance engines. In fact, they sought co-operation for this move from both the police and the corporation. These proposals, if acted upon, would have brought about a system of fire fighting that existed in some other areas of the United Kingdom, and it would also have meant a continuation of the insurance companies' considerable financial contribution to the operation of the fire brigade. But the Council meeting adjourned without any decision being made on the proposals in the report.

In 1853 the waterworks committee had initiated a wide-ranging survey of the future changes needed in the city's water supply system, the work to be carried out by Mr Parke Neville the borough engineer, assisted by Mr Hawksley, a consultant engineer who had been involved in the recent upgrading of the water supplies to both London and Glasgow.

These men carried out detailed surveys of the Grand and Royal canals, and the rivers Dodder and Liffey. Various options were examined based on the needs of a population of 400,000, this being one half more than the existing population of Dublin and its neighbourhood. The following four options were considered:

- the current position, where three million gallons were taken from the Dodder, supplemented by supplies from both canals
- a supply from the Dodder and Grand Canal
- a system from the Liffey and the Dodder
- a move to have a major system from the Vartry river in County Wicklow.

Eventually the two engineers favoured the second option, which would provide six million gallons from the Grand Canal at the Twelfth Lock (Clondalkin) and in excess of five million gallons from the city water source on the Dodder to a reservoir at Templeogue. There was however at that time a major problem. Only one-third of the water coming from the Dodder was directly under the control of Dublin Corporation. The remaining two-thirds belonged to mill owners and other influential persons who would have to be bought out. The Earl of Meath, for instance, owned a large amount of property in the Liberties area and controlled his own water supply. He would hardly be likely to welcome the move by the corporation to take full control of the water supplies throughout the city.

Nevertheless it was urgent that the corporation make a decision and proceed with a new system because the current deal with the Grand Canal Company was due to expire in 1865. Accordingly, the waterworks committee took the bold decision to recommend that a Water Supply Bill be promoted in parliament. The committee stipulated that

> *permissive powers be taken in the Bill for purchasing the Rights of mill owners on the Dodder above the city weir, and below the weir on the city water course, and the Earl of Meath's water course also for supplying with water the Earl of Meath's Liberty.*

During the second half of the 1850s, while the extensive survey of the water supply system was being carried out, the public was again calling for the acquisition of escape ladders to be made available in the streets. In April 1859 the chairman of St Andrew's parish group wrote to Dublin Corporation "… it is indispensable that a number of fire escapes be procured for the city of Dublin which should be placed in prominent positions". The group had been in contact with the Royal Society for the Protection of Life from Fire regarding the situation in London (the RSPLF was formed in London in 1836). The parish group's statement continued: "In London only one trained conductor attends each escape, assistance at fires being obtained from policemen or strangers, three of whom receive payment for their services – the caller two shillings and sixpence and the others one shilling – if lives are saved five shillings, and if a late call so that the escape arrives after the engine the caller gets only one shilling and sixpence".

The Dublin borough engineer was instructed by the waterworks committee to check with London and his report confirmed what the parish group had said. He received a letter from Mr Sampson Low, Secretary of the RSPLF, which stated "… experience proves to be the only way of effectively working fire escapes and it will not answer to place them in charge of a Police Force, as in cases of such emergency there must be undivided responsibility to ensure rapidity and decision". This advice was soon to be heeded in Dublin.

The streetscape was gradually changing in Dublin. In 1859 the corporation drew up contracts with the Hibernian Gas Company for the lighting of public lamps on both the north and south sides of the city. The agreement was that one hundred and ninety new lamps would be provided. In December of the same year the Gas Company was the victim of a particularly hazardous outbreak of fire when tar from a burst tank flowed into the retort house and was ignited. The flames spread to two gas holders causing them to explode. Due to freezing conditions the firemen had difficulty obtaining a water supply and, according to one report, "that which was used speedily congealed, and the dresses of the firemen soon became covered with spray, which becoming frozen presented the appearance of glittering armour".

In the early hours of the morning of 11 November 1860 there occurred one of the most damaging fires of the century and one which showed how courage, fortitude and daring could, in the face of disaster, save lives. The fire was probably the final spur needed to force the corporation into forming a municipal fire service. It seems public bodies universally throughout history become actively focused only in the aftermath of disasters.

The fire in question cost three lives and at least one other person was seriously disabled for life. These human tragedies alone however may not have been sufficient to stir the corporation into positive action, because many other fires at the time had also led to loss of life. What was different in this case was that the fire was in the Kildare Street Club, haunt and home for many of the gentry and the judiciary; it could not easily be passed over as just another fatal conflagration. The ineptitude of the parish brigades, with their outmoded equipment and organisation, was an embarrassment to the gentry and left them with the conviction that something drastic would have to take place to prevent a similar occurrence in the future.

*One of the most fearful fires that have been witnessed in Dublin for many years took place in the premises of the Kildare Street Club, by which the entire structure of the Club-house building was reduced to a smouldering heap of ruins. The corpse of one sufferer had been extricated from the smouldering ruins which a few hours previous had been the centre of ease and luxury and rendezvous of a large section of the nobility and gentry of Ireland when in town. The alarm was raised by a Constable when he saw dense volumes of smoke accompanied by occasional flakes of flame issuing from the windows of the Club-house. Another Constable soon on the scene heard two loud reports as if a quantity of gunpowder had exploded.*

Two hundred soldiers of the 96th regiment were sent to the fireground. St Ann's engine was the first to arrive on the scene but later at the coroner's court was said to be "useless". In the presence of many onlookers, soldiers, firemen and police "A man and two female servants were seen calling for help from the top floor window. Beneath them the flames were raging with frightful violence, cutting off all escape … their screams and frantic cries were drowned in the roaring flames, the bursting crash of heated glass, and the falling of burning timber." When all hope for the three seemed lost, the man, a Mr Hughes

> *was seen to climb upon the upper section of the window. He climbed up onto the shoulders of one of the women, with a sheet in his mouth, by dint of his exertions and venturous climbing he achieved the roof. Standing on the parapet seventy feet above the street, he lowered the sheet to one of the two women below and in a few minutes she was seen waving to and fro in the air … and impelled by momentary strength resulting from terror and by the courage emanating from sheer despair, she succeeded with the aid of the man in getting up beside him on the parapet.*

With the aid of the first woman they succeeded in hauling the second woman from the window onto the parapet. Mr Hughes, although suffering from burns and injuries to his hands and feet caused when breaking the window glass, went and forced the skylight to release the ladder attached to the ceiling, which he had been unable to reach when inside the building. He led a total of nine people to safety across the roofs to the house next door. When the fire broke out there were three men and eleven women of the staff on the premises. Three of the women perished in the flames. After the fire had been extinguished major businesses in the city, led by Todd Burns, set up a fund for the seriously injured Mr Hughes.

The fire had been fought by the corporation engines assisted by the engines of the police, insurance companies and St Ann's and St Mark's parishes. However, much criticism was levied at all concerned and particularly at the police for their failure to turn out the escape from Kevin Street station, and at St Mark's parish engine, being the fifth to arrive in spite of the fact that it was housed much nearer than most of the others.

The police engine was under the control of Inspector Boyle who, more than twenty years later, was to become Dublin Fire Brigade's second chief. He claimed that the Kevin Street escape did not attend because it "was unwieldy as compared with modern London escapes and could not be worked". This was refuted by Dublin Corporation's Mr Crofton in evidence to the coroner's court: "I attend fires in the city when not ill, I attend all fires. I was not at the last fire as I was in bed unwell with rheumatic gout, but I have seen the fire escape in Kevin Street tried for trial sake, but never at a fire. It was never brought to a fire."

He went on to claim that the three inch water main in Kildare Street "could supply either the Corporation or Police engine but not both" and that the cock controlling the water supply was never turned off. The other engines in attendance could not work together because of "differing coupling sizes". The inquest report observed: "We regret to find that the usual deficiency of water in cases of fire was deplorably evidenced in this instance – the Police fire brigade not being able to obtain water for their engine until 22 minutes after their arrival!". Valuable silver plate was saved by two men on the morning of the fire, but an extensive library collection was destroyed and it was not until 16 November that the third body was recovered from the ruined building. The letter columns of the newspapers were filled with accusations and criticism levelled at the authorities, deploring the fact that Dublin, the second city of the Empire, had no fire escapes available for the rescue of citizens and no centrally controlled organisation for the fire services.

On the night of 11 November at 9 pm another large fire broke out at the haberdashery premises of Reilly's in Henry Street.

> *The house might be said to have been entirely gutted before a drop of water was poured upon it. Had the wind been high nothing could have saved the entire block of buildings between Henry Street and Prince's Street.*

It was claimed by the papers that if this fire had occurred later into the night there would have been deaths there as well as in Kildare Street: "Water from mains and water carts was in plentiful supply AFTER THE FATAL HALF HOUR by which time thousands of pounds of loss had been sustained, and the fire engines were only used to prevent the fire spreading to the adjoining buildings". The feelings of the public are summed up in the following newspaper comment.

> *Our fire engines are the most contemptible and our supply of water quite inadequate. Another calamitous fire, or series of fires, is to be added to the Kildare Street catastrophe, for no fewer than five occurred within the short space of twenty-four hours. Reilly's of Henry Street took fire in a few minutes and in less than an hour was a smouldering ruin. Water, water was the cry, but water there was none.*

The situation could no longer be ignored, but over twelve months were to elapse before the city fathers did finally take the only action which could help prevent recurrences of these events.

Following a commission of enquiry into the conditions relating to Dublin's water supply, and as a result of the waterworks committee decision (in spite of much opposition) to promote a Bill in Parliament, the Dublin Corporation Waterworks Act was passed in 1861. This enabled a public water rate to be levied and it required, among other essentials, the provision of "a supply of water for better security against fire".

Having secured this legal authority the waterworks committee fixed 1862 as the year to commence the project. The vast engineering and building work involved in providing a new water distribution system would be a huge long-term undertaking.

In fact it was not to be completed until 1868. After many debates at full council meetings the original proposal to supply the city's new water system from the Dodder and the Grand Canal was finally rejected, and the option to construct a major system from the Vartry river in County Wicklow was adopted.

The Waterworks Act empowered the corporation to borrow £300,000 for the execution of the necessary work and "for the purposes of the compensating of parties affected". A decision was made at this time to draw down at least £150,000 before December 1862.

# 3   The municipal fire service

The urgent need for a single fire brigade, efficiently trained and fully equipped, was now becoming painfully obvious. The Kildare Street Club fire had provoked much public discussion and speculation, and the large number of separate fire fighting units in the city led to confusion and annoyance. A correspondent to the *Dublin Builder* suggested a plan whereby the insurance companies could club together and purchase two or three steam fire engines of the type in use so effectively at the time in America. He wrote "… at present the fire engines of Dublin are absolutely useless, and do little or no good when a fire takes place".

On 9 January 1860 the military commander of the city had written to the lord mayor objecting to the manner in which his soldiers were being called out and utilised at fires: "General Lord Seaton desires to point out the urgent necessity there seems to be for the formation of an efficient fire brigade in the city of Dublin". The letter claimed that the soldiers were being called out "not to protect property but for the purpose it would seem of working the engines" and that this claim on the troops was "by a civilian, supposed to be a Mr. Crofton, an official of the Dublin Corporation". It went on to state "nor was any magistrate present during the time the troops were thus employed". The Mr Crofton referred to was of course the overseer of the water works and assistant inspector of parish engines! There would appear to have also been contact between the military and the police on this matter because Mr Atwell-Lake of the Dublin Metropolitan Police in Dublin Castle wrote to the Chief Secretary on 25 January 1860 outlining what was happening to the police.

> *When a fire breaks out in the city constables are despatched on cars from the nearest station to give assistance and to call out the different fire engines. This so far, although attended with considerable expense in car hire, must be considered the legitimate duty of the police and one which they at all times perform with alacrity.*

Referring to the fire engine in Kevin Street station he pointed out that "this engine, the duty of working which is wholly voluntary on the part of the police, has been at almost every fire that has taken place in the city, normally crewed by a sergeant and men in training". It is interesting, incidentally, to note the ancillary equipment associated with the engine: "2 keys for fire plugs; 10 wrenches for hose; 2 hand ropes for engine; connection hose; 1 goose neck; 1 copper elbow; 1 pair of lamps for engine; 10 fireman's hatchets.

Mr Atwell-Lake's letter continued as follows.

> *The Commissioners desire to state that they have long been of the opinion expressed by General Lord Seaton as to the necessity for the formation of an efficient fire brigade. The consequences of that want of such a force are too evident to need any remark. They may be seen at every fire that takes place, not only in the inconvenience arising*

*from each engine working independently, but in no small amount of jealousy which exists, more or less, among many of the (insurance) agents.*

The letter went on to state that the commissioners had advised the lord mayor that the fire engine should be taken out of Kevin Street station "as they anticipate that it will help considerably to facilitate the formation of the desired brigade, as the Corporation would then be compelled by the force of public opinion to move effectively in the matter."

Mr Atwell-Lake stated further that he was "not aware of any Act of Parliament making it obligatory on the police to act as firemen" and it was "a duty which cannot possibly be considered as appertaining to a constable." He also claimed that some constables had sustained injuries when acting as firemen and that the police were blamed by the press when something went wrong. He suggested that a Bill might be prepared repealing the powers of the parishes to levy money for fire engines, and giving powers to the corporation to raise through the collector general a rate of one penny in the pound on all houses, stores, etc. The amount so levied could be applied to establishing a fire brigade.

On 12 June Atwell-Lake sent another lengthy letter claiming that "the Corporation, as chosen by the citizens, appears to be the proper body in which should be vested, as in Liverpool, Manchester, etc the organisation and control of the fire brigade". He expressed the opinion that since fires were less frequent in Dublin than in the other cities cited, and bearing in mind the prospect of a new high pressure water supply, a chief fire officer and twenty-one men would be sufficient for the brigade. He also suggested that a steam fire engine should be purchased immediately. An intriguing request in a further letter of 19 June was that if the steam engine was purchased prior to the setting up of the city brigade it should be given to the police: they did not consider however that they should have to purchase it. The thinking behind the request seems to have been that a fire brigade should be organised by the corporation as soon as the new water system was completed but if such a brigade was not set up, then the police would run the fire service.

A letter from even higher authority, the Lord Lieutenant, was read at the waterworks committee meeting on 19 November: "I am directed by the Lord Lieutenant to transmit herewith copies of reports received from the Commissioners of the Metropolitan Police relative to the establishment of a Fire Service in Dublin". The enclosed reports were responses to enquiries concerning the organisation and administration of the fire brigades in New York and London. The letter finally stated "His Excellency trusts your Lordship (the lord mayor) will invite the attention of the Corporation to the subject, and on his part desires to express his anxious wish to further by any means in his power the success of the measure so necessary to the safety of the lives and properties of the citizens". The year ended, however, without the corporation making any major political or legal moves to establish a civic fire brigade. It was recorded that there were twenty-seven fires in 1860 and the lord mayor had attended twenty-four of them; three did not require his presence. Mr Crofton and the inspector of the Dublin Metropolitan Police attended all but one of the fires.

The fire fighting equipment available in Dublin at that time consisted of a number of engines: there were about twenty parish fire engines, the majority of which were very old, many undoubtedly in a poor state of repair; there were two engines owned by the corporation, one of which was operated by the police; the police themselves probably had another engine; there were about nine engines in the ownership of private establishments including the Bank of Ireland, the Guinness Brewery and the Custom House – almost all of these machines would have been small in size with limited capacity and, in some cases at least, probably inadequately maintained. In addition the insurance companies still had nine or ten engines available, although the number of insurance brigades had decreased since the 1820s. The competition created by the formation of many new fire insurance offices meant that some companies, such as the Shamrock and the St Patrick, were wound up after only three or four years in business. The Commercial Insurance Company of Dublin sold its fire engine by public auction in November 1826, and in the same year the Royal Exchange reduced its fire equipment in Dublin from two engines to one.

Following the Kildare Street Club fire and the introduction of the Waterworks Act there was a great deal of public speculation as to the specific requirements of a fire brigade. The following table shows comparative data for some of the other principal cities in the United Kingdom supplied by Police Commissioner Larcom.

| | Engines | Escapes | Firemen | Annual Fires | Population | Annual Cost of Brigade |
|---|---|---|---|---|---|---|
| London | 39 | 70 | 110 | 782 | 3 million | £18,000 |
| Liverpool | 15 | 11 | 123 | 161 | 500,000 | £ 1,854 |
| Manchester | 10 | 2 | 50 | 180 | 358,000 | £ 1,500 |
| Birmingham | 10 | 1 | 21 | 130 | 300,000 | £ 900 |
| Glasgow | 10 | 0 | 70 | 250 | 400,000 | £ 3,000 |

In November 1860 the waterworks committee received correspondence from the Irish Society for the Protection of Life from Fire. This organisation had been recently set up by worried citizens, many of them business people, who were demanding the purchase of fire escapes, and were unhappy at the inactivity of the corporation. The society requested that the Kevin Street escape should be given to them as soon as they employed a conductor who was trained to look after it. The committee agreed that the escape should be transferred from the police to the new society on a temporary basis. The corporation later agreed to give the society a fireman to operate the escape, and notices were placed on lamp posts denoting its location: "Fire Escape Station. St. Catherine's Church". The society purchased three further escapes in 1861.

It was the specific duty of the escape conductor to attempt to rescue anyone trapped in a fire, and this could sometimes mean having to wheel the heavy, cumbersome machine a considerable distance, hopefully with the help of a policeman or members of the public. These fire escapes, although not telescopic, were the forerunners of the escapes carried on Dublin fire engines up to the 1970s.

27

The main ladder, which was thirty-five feet in length, was mounted on a spring carriage with large wheels. Ten feet from the top of the main ladder was a twenty foot fly-ladder hinged on a bracket. This normally hung down along the main ladder and when required had to be swung up and over by means of ropes attached to iron levers. If still further height was needed then a separate sixteen foot ladder could be attached to the fly-ladder to give a total height of about sixty feet. The latter was known as the back fly-ladder because it was carried underneath the main ladder.

A canvas chute into which rescued persons could be dropped was attached permanently to the underside of the main ladder and strengthened with copper gauze. This life-saving appliance, invented by Abraham Wivell of the RSPLF weighed nearly half a ton. It is surprising to see from the statistics quoted that the city of Glasgow did not possess any escapes!

Some people were strongly opposed to the idea of the corporation having to shoulder sole responsibility for the fire protection of the city. There was a view that the fire brigade should be run by the police, as was to be the case in many towns and cities in England.

Some interesting ideas and suggestions were put forward in 1861 by Mr Wilson, the secretary of the Patriotic Assurance Company. He suggested that a brigade should consist of a captain and forty men, that there should be eight engines housed in a central fire station in College Green and that there should be at least six other stations. The captain must have overall command of all the brigades attending a fire whether the firemen are full-time or volunteers. He also emphasised the importance of properly maintained fire plugs which should be sited every seventy yards and be clearly indicated. Mr Wilson felt it was desirable to have the brigade composed of police constables and that it should be under the control of the corporation or the commissioner of police. Another of his ideas was the setting up of a "fire escape and salvage corps". This would be manned by certain members of the brigade, who should be sent to either Liverpool or London to be trained. (A salvage corps had been formed in Liverpool in 1842, but London did not have a salvage corps separate from the fire brigade until 1866.) The organisation envisaged by Mr Wilson, as we shall see, was considerably larger than that which actually materialised.

By the end of November 1861 there had been forty-five fires in Dublin, a large increase on the previous year. The lord mayor was supposed to attend all fires but in 1861 he failed to attend any, though his secretary did attend six. Mr Crofton had attended thirty-eight that year and the police inspector thirty-nine. It was said that these increased occurrences of fire, the imperfect supply of water and the inadequate means of extinguishing them "excited much discontent with the present arrangements" – surely an understatement. The corporation staff allocated to fire brigade duties comprised sixteen skilled and twelve unskilled men. Of the skilled, eight were regarded as firemen who had charge of the fire engine kept at Winetavern Street where they lodged. There were six turncocks and two officers who had responsibility for the water mains. The unskilled men were drivers of the ten water carts kept constantly ready for use, and there were two hosemen who assisted in filling the carts from the mains when called out on duty.

The waterworks committee now openly admitted that "the present fire brigade and the arrangements connected with it are the remnant of a bygone age, suited to the past". A proposal put forward by John Dwyer, the chairman of the committee, recommended

> that the present staff be enlarged and that they all be placed under a responsible chief who, having no other duties to perform, will devote himself to the complete training of all the members of his corps, instructing them in the opening of fire cocks, adjusting hose and hydrants, the directing of hand pipes, and imparting to them general information with regard to fires, the best method of applying water for extinguishing them, the use of fire escapes, and all the other duties appertaining to the office of fireman.

The members of the committee were at last becoming aware of the complexities of the job of fire fighting. They advised that "the Chief Fire Officer of the proposed brigade should be vested with very large powers when on active duty, and all fire engines or other implements used at fires, and the men in charge of them, should be under his absolute control". They still felt that the chief magistrate should continue to attend all fires because of his legal authority and standing.

The committee recommended that the two manual fire engines owned by the corporation should be maintained but that the city should also be provided with a steam fire engine. The latter recommendation is interesting because the use of steam-driven pumps in England was still in its infancy, although this type of appliance had become widely used in American cities over the previous decade. Information regarding steam engines had come from the police, as we have seen, who pointed out how efficient and cost-effective the new machines were compared with the old manual fire engines.

> The best hand engine will throw water to a height of 60 or 70 feet, but the practical effective work of such engines is very limited, usually no more than 40 or 50 feet during continuous duty, whereas the steam engine will throw a column of water 130 feet in height and 175 feet in horizontal distance.

Having discussed the type of fire engine which would be required the committee turned its attention to the need for swift communication in cases of fire. The members held a meeting with the manager of the Magnetic Telegraph Company, who assured them that a telegraph system could be installed enabling all stations to have instant contact with a control station for a relatively small cost; police stations could also be connected into the system. The members of the committee felt sure that both the police and the insurance companies would contribute to the cost of laying down the wires.

Progress towards establishing a properly organised civic fire brigade was undoubtedly gathering momentum. The first requirement in the process of setting up a municipal fire brigade was for the corporation to obtain the necessary statutory powers. It already had powers to improve the existing brigade and levy a rate for water, but it did not have powers to assume command of engines which were the property of private companies or public bodies when in attendance at fires. At the city council meeting on 7 December 1861 John Gray proposed

*That the waterworks committee be empowered to take steps necessary to enable the Municipal Council to apply for an Act of Parliament in the coming session to establish a more effective fire brigade, and to alter, explain and amend existing statutes relative to extinguishing fires and supply of water for such purposes, and to confer new powers on the Corporation.*

This was followed up at a meeting on 17 December at which Gray sought to have a specific Fire Bill printed and lodged in parliament. After lengthy discussion of the implications the committee agreed to publish the Bill. It was introduced to parliament early in the following year, but by then opposition to the corporation's policies had surfaced. On 10 January 1862 it was brought to the notice of John Gray and the corporation that a petition had been deposited in the Private Bill Office of the House of Commons to enable the commissioners of police within the Dublin Metropolitan District to establish a fire brigade. The corporation challenged the legality of the Bill because they had not been informed in advance of its publication and because it would interfere with the rights of the corporation by seeking powers under the Land Clauses Consolidation Act 1845 "to purchase houses and lands by compulsion" in order to establish a fire brigade. The Bill was intended to cover the Dublin Police District and to impose three farthings on the rate in that area in support of the fire brigade.

As a result of the corporation objections the Bill was rejected, but only by a vote of nineteen to nine, at the January council meeting. The Dublin Police District was a much large area than the City of Dublin and obviously there was support for the Bill from various interests both inside and outside the municipal boundary.

While the council was dealing with the Police Bill problem it was confronted by another pressure group in the form of the main fire insurance companies in Ireland. The insurers were demanding that the stamp duty on fire insurance should be repealed – having previously made an attempt to push a Bill through parliament on this issue, which failed. They were now pressing the corporation to support their case, claiming that nearly 60% of all property in the city was not insured against fire because of the high cost of insurance. But the council was not deflected from its preoccupation with fire brigade matters. On 5 May 1862 it dealt with a letter received from the Irish Society for the Protection of Life from Fire.

The society had been forced to suspend its activities because the parishes had not supported it financially, and the letter was a request to the corporation to take over responsibility for its fire escapes. In view of the fact that an Act was about to be passed authorising a municipal brigade, the society was anxious to take advantage of the situation and divest itself of the responsibility of providing a fire rescue service. It emphasised that it was in favour of the corporation's Bill.

*In consequence of statements which have appeared in the newspapers, and made before the committee of the House of Commons, that it be intimated to the Lord Mayor that*

*this Society disclaim opposition to the Bill for providing a fire brigade promised by the Corporation, and declare that no such opposition was ever authorised by this Society.*

Those who were opposed to the Fire Brigade Bill were using every conceivable tactic to have it defeated, but they were soon to fail in their efforts. The chairman of the ISPLF informed the corporation that the society would transfer to it all the fire escapes and other implements, which had cost £596. A sum of £244 was still owed by the society but the corporation agreed to cover this cost. The items to be handed over were as follows: eight escape ladders with covers and jumping sheets; maps; brushes; eight sentry boxes; helmets; hatchets; lamps; rattles; uniforms.

After many months of severe and costly opposition the Dublin Corporation Fire Brigade Act received the Royal Assent on 3 June 1862. It was a truly historic day for Dublin. The title of the Act read as follows.

*An Act to extend and define the powers of the Rt. Hon. the Lord Mayor, Aldermen and Burgesses of Dublin in respect to the extinguishing of fires, and the protection of life and property against fire, and for other purposes.*

The new Act, which made the Dublin Corporation legally responsible for the fire protection of the whole city, contained thirteen clauses and incorporated the Waterworks Act of 1861. Under its terms the corporation was authorised to erect and provide accommodation for a fire brigade and to provide fire engines and escapes. The costs and expenses of the brigade were to be paid out of the water rate provided that the amount did not exceed one and a half pence in the pound (for the first ten years) of the annual value of the property covered by that rate, and one penny for each succeeding year. This clause was to give rise to a serious problem of shortage of funding for the brigade in the 1870s. The corporation was also authorised to grant reasonable compensation to any fireman injured while on duty, and to establish telegraphic communication between fire stations.

It is of particular interest to note that a portion of the expenses incurred in extinguishing a fire within the city were to be borne by the owner or occupier of the affected premises, who was liable to pay a sum of £15 to the corporation "or whatever lesser sum is equal to one half of the said expenses". This was described as a contribution towards the expenses incurred. The fire brigade was also authorised to attend fires beyond the city boundary but when that occurred it was the responsibility of the owner of the affected premises to pay all of the expenses. Under Section 11 of the Act the corporation could (by mutual agreement) acquire any fire escapes or other appliances currently owned by insurance companies, societies or others. (The Dublin Corporation Fire Brigade Act 1862 is reproduced in full in Appendix I.)

The new Act was formally adopted at the corporation meeting of 27 June and it was immediately decided that the "Fire Extinction Depot" in Whitehorse Yard, that haphazard organisation that had been in place since 1854, should now form part of the new civic fire brigade. The waterworks committee agreed that because the organisation and operation of the brigade was so closely related to the water supply, the brigade's control and management should be formally referred to the waterworks committee as part of the future duties of that committee. Unfortunately those in the council who had opposed the corporation's drafting of the Fire Brigade Bill in the first place were still bent on some form of opposition, and they questioned the cost of seeing the Bill through parliament. The law agent gave a detailed breakdown of the costs, including travel, accommodation, printing, etc – they amounted to £833-0-3. This was surely a modest price to pay, even for those days, to ensure that Dublin would have a legally established municipal fire brigade!

The first vital step for the committee was to find a man capable of commanding the new brigade, but even here there was a problem. A move was made by some councillors to have Mr Crofton, the overseer of the waterworks, appointed as chief of the brigade. Fortunately reason prevailed, though the proposal was defeated by only six votes – twenty-six votes to twenty. The way was now clear to advertise in the national newspapers and the technical journals for what was described as the post of superintendent of the fire brigade. It was specified that the corporation required someone "with fire fighting experience plus knowledge". The committee, realising that the person appointed as superintendent would determine the character of the brigade, advised that a liberal salary should be allocated to the office "as may induce men of capacity". Its recommendation was £200 per annum plus accommodation. The waterworks committee was also careful to state that the salary should not be debited to the fire brigade account because the superintendent should also be responsible for the turncocks and for supervision of the water supply to prevent waste.

The response to the advertisement produced no less than twenty-two applicants. Only thirteen however presented themselves. A written examination was compiled jointly by Dr Dwyer, chairman of the waterworks committee, Mr Parke Neville, the borough engineer, and Captain Eyre Massey Shaw. (Shaw had filled the combined post of superintendent of the police and the fire brigade in Belfast from June 1860 until he arrived in London in September 1861 to take up his current position as superintendent of the London fire engine establishment.) The thirty-six questions on the exam paper included such items as "knowledge of working parts of a fire engine; the working of steam fire engines; use of high and low pressure hydrants; the operation of a 7 inch manual engine; the difference between suction and delivery hose".

Another searching question was was the following: "Dublin may be described as an oval or oblong circle, three miles one way and two another; how many stations, firemen, horses, engines would you need for an effective Brigade?" The city at this time was contained within the boundary of the North Circular Road, the Royal Canal, East Wall Road and the Grand Canal. Six of the candidates were deemed to have passed the written exam and were called for interview. Two were subsequently eliminated on the basis of their age, so at the council meeting in August following the interviews a vote was taken on the remaining four. The result was: Ingram 24; Acheson 19: Hart 14; Hodgens 4. In a somewhat strange decision, considering no second count took place, Mr Ingram was then proposed for the job and unanimously agreed. He was to be employed on a probationary basis from month to month until the members of the waterworks committee were satisfied that he should be given the post permanently.

James Robert Ingram was born in Dublin but emigrated to America in 1851 and stayed there until 1861. He was employed as an engraver with the American Banknote Company. He served as a volunteer fireman for eight years, having joined the Niagara No. 2 Company in New York in 1853: he also had experience as a volunteer firefighter in London. Ingram was now destined to lead the new Dublin Fire Brigade for the next twenty years. In the nineteenth century the most common title given to the person in charge of the fire brigade was captain, a title that did not of course automatically indicate an ex-military man. Other titles used in corporation minutes and official documents were superintendent, chief superintendent, and chief. The modern rank of chief fire officer did not come into official use in Dublin until about the 1950s.

At its meeting on 24 December 1862 the waterworks committee expressed its satisfaction with Ingram, reporting that he had attended several fires and invariably displayed technical skill, zeal, steadiness and devotion. "Your committee cannot abstain from expressing their unanimous opinion that the selection of Captain Ingram will prove most satisfactory to the Corporation". The full council meeting on 10 January 1863 duly confirmed James Ingram as permanently appointed to the position of Superintendent of the Dublin Fire Brigade.

After the Dublin Corporation Fire Brigade Act became law the corporation passed the following resolution.

> That this Council avail themselves of the earliest opportunity to express their strong sense of, and obligation for, the valuable services rendered to the Corporation and citizens of Dublin by Captain Shaw, the Superintendent of the London Fire Brigade, preceding and during investigation of the Fire Brigade Bill promoted by the Corporation; and also for the assistance and information afforded by him to the chairman of the waterworks committee in reference to the effective formation and working of the Fire Brigade for this city.

The fire fighting organisation in London at that time was an amalgamation of insurance company fire brigades. The Act which set up the Metropolitan Fire Brigade was not passed until 1865.

The Dublin waterworks committee, together with their superintendent, was now engaged in planning the organisation of the new brigade in detail – the recruiting of staff, the purchase of new fire engines and equipment, finding suitable premises, etc. Simultaneously it was dealing with the financial and physical aspects of constructing the new water supply system. The most pressing brigade issue was the provision of accommodation for the men and the appliances. It was decided to convert what was known as the City Assembly House, a building situated on the corner of South William Street and Coppinger Row which was formerly used by the lord mayor and the city council for their meetings. Between 1809 and 1852 it had in effect been the city hall, before the corporation moved to the old Royal Exchange building on Cork Hill. The building conversion work was almost complete when Captain Shaw arrived from London to inspect the Dublin Fire Brigade.

Being an Irishman – born in county Cork in 1828 – and having spent a year with the fire brigade in Belfast, Shaw was naturally following with interest the efforts being made to establish a municipal fire brigade in Dublin. As we have seen, he helped to draft the questions put to candidates for the position of fire brigade superintendent in the city. So his visit in August 1863 was no surprise, and by that time he had gained two years of valuable experience commanding a large fire fighting organisation in London. By all accounts he was satisfied with what he saw in Dublin. Having examined the three manual fire engines at Winetavern Street he then went to South William Street to inspect progress there. The building was almost ready for occupation, an engine room for two fire engines and other equipment having been constructed, together with stabling in the yard for two horses. The ground floor included a guard room accessible to the public and there was direct access from this room to the engine room and stables. A bell was connected to all other areas of the premises.

The upper storey was converted to accommodate up to twelve single men. There was a dormitory, a sitting room for men off duty and another for those on duty, kitchen, lavatory, wash room and storage space. The accommodation provided for the superintendent consisted of an office, living room and bedroom. Initially there was a problem in providing the superintendent with his own kitchen, but this was later solved when the complement of men was reduced from twelve to nine and the space needed to accommodate them was curtailed.

A ladder was fixed to the external wall in the yard together with a pulley for hose drying. A signboard inscribed "Fire Station" was erected above the entrance in Coppinger Row, lit by a gas lamp. On 3 November 1863 the men of Dublin Fire Brigade took over their first headquarters. South William Street became No.1 fire station and Winetavern Street became No. 2. Today nothing remains of No. 2 station – the building in Winetavern Street was eventually demolished and the site of Whitehorse Yard is now part of the Dublin Corporation civic offices. No. 1 station continued to be used by the brigade until 1887, after which it was again converted internally for other uses and in 1953 assumed its present use as the Dublin Civic Museum.

Earlier discussions with the Magnetic Telegraph Company resulted in the installation in 1863 of a communication system which was to prove very effective. The telegraph apparatus enabled messages to be passed to and from the two fire stations relatively speedily. All telegraph communications were transmitted through a central alarm control room in the city hall, and talks with the police commissioners were begun with a view to eventually linking all police stations into the system. The sending instrument consisted of a dial on which each letter of the alphabet had to be spelled out by moving a pointer; it appears crude compared to the telephone, but it was a significant advance at the time. The alphabetic system was already in use successfully in New York and Shaw introduced it to London when the Metropolitan Fire Brigade came into existence in 1866. It has been claimed that Dublin was the first European fire brigade to use the telegraph system to transmit fire messages, but in fact the telegraph was used by the brigades in Berlin as early as 1849 and in Glasgow in 1861.

Dublin Fire Brigade's first annual report was issued by Superintendent Ingram in January 1864. It provides a comprehensive account of what was accomplished in establishing the new brigade up to the end of 1863. The statistics in the report detail a total of 174 calls received which included 12 serious fires and 106 chimney fires. The brigade attended two fires outside the city boundary, at Santry and Drumcondra.

There were two slight injuries to firemen and Ingram was "happy to report magnetic telegraph lines and equipment in good order with instant communication, and all engines in excellent order". On 7 September 1862 advertisements for recruits had been published: they specified "preferably sailors" but it is not known if any of the fourteen who were appointed as firemen in October had been sailors. In addition seven escape conductors from the Irish Society for the Protection of Life from Fire joined the brigade. By the end of 1863 there were nine men stationed at South William Street and fifteen at Winetavern Street.

Seven of the wheeled fire escapes originally owned by the ISPLF were purchased by the brigade and these were stationed at the Bank of Ireland in Foster Place, St George's Church, St Catherine's Church, Kildare Street, Nelson's Pillar, Kevin Street and St Paul's Church. Five of the escapes were of London design and two telescopic escapes came from Dublin makers. The escape conductors were on duty from 8 pm until 6 am in the summer and from 8 pm until 7 am in the winter.

Agreement had been reached earlier that four new manual pumps and a steam fire engine would be purchased from Shand, Mason & Company in London. Of the four manual engines, one was located at William Street and the other three were at Winetavern Street. The steamer was due to arrive the following year. These appliances were of course horse-drawn and a Mr Ramsbottom put in a tender to the waterworks committee to supply a pair of horses at twelve shillings a week plus one shilling and sixpence per hour when actually engaged on fire duty. This tender was accepted because it saved the corporation the cost of buying these horses for fire fighting duties. The corporation had purchased accommodation in Cook Street in 1854 for the firemen attached to Whitehorse Yard. It now purchased a house in Winetavern Street, in front of the corporation depot, to cope with the expansion of the brigade. In addition an amount of up to £400 was made available in the hope of procuring premises for a temporary fire station on the north side of the city at Blessington Street. Also, an engineer, a Mr Laurence, was employed as a member of the permanent staff of the brigade to control and operate a workshop for the repair and maintenance of fire engines in the Winetavern Street station.

Byelaws for the control and management of the fire brigade were prepared in 1863 and covered among other matters the conditions of employment of members. The brigade was to consist of forty firemen and the chief fire officer, plus one hundred "supernumeraries" – these as we shall see included conductors and part-time firemen and all of them were to be trained to perform the full duties of fireman.

The byelaws contained twenty-eight regulations relating to firemen. The main provisions were as follows.

- Must at all times obey orders and work under the control of the chief superintendent or other nominated officer.
- Must devote all of their time to the brigade and must reside within the city or its suburbs.
- Shall not take any money while performing their duty without permission of the waterworks committee.
- Must give fourteen days notice when quitting the brigade.
- While on duty must at all times wear their uniform with badge.
- Must hand back all gear, uniform, badge on leaving the brigade, or face prosecution.
- Payment will be at fifteen shillings per week, but with one shilling to be deducted for a reserve fund controlled by the committee.
- When sickness renders a man unfit for duty the committee to have the right to deduct part of his wages.
- Each member of the brigade shall give a letter of guarantee not exceeding £10 as security for good conduct and honesty.
- Chief superintendent empowered to fine men for misconduct. A misconduct book to be maintained. Any man not on the book for one year to be given an increase of one shilling per week with a further sixpence per week in each subsequent year up to £1 per week.
- No fireman shall leave his residence between 9 pm and 7 am, except to go to his district station or out to a fire, without leave of his superior officer.
- The chief to attend all fires, take control, operate discipline, visit engine houses every day or night. Empowered to call on the police or military to keep order, protect property, and in cases of emergency may call for and receive the assistance of those forces in the working of fire engines. (This latter point was the issue which had led both the police commissioner and the military to raise objections and to petition the Lord Lieutenant in 1859 and 1860.)
- The chief also empowered to close roads or remove buildings to prevent the spread of fire.

The escape conductors were instructed to keep their sentry box doors open, dispatch a messenger to the nearest fire station in the event of a fire call, and get three persons (police preferably) to assist in wheeling the escape to the site of the incident. They must leave a man in charge of the lever while ascending the ladder to effect a search or rescue, no one else to be allowed to mount the ladder (the relevance of this was evident at a fatal fire in 1866). A conductor was responsible for cleaning the escape once a week and for removing the wheels and greasing them once a month. Before leaving an incident he was obliged to have a report form signed by two local housekeepers, the form to be handed in to the corporation before 9 am on the following day.

The byelaws envisaged a force of forty wholetime firemen, but this number was not reached until the year 1890. The total complement at the end of 1863 was twenty-five. A new official category entitled "supernumerary" was created to augment the wholetime members of the brigade. The supernumeraries were in fact part-time firemen whose principal role was to operate the manual pumps, thus saving the corporation from having to depend on the police, soldiers or members of the public to carry out this work. A study of the annual reports issued over the thirty years from the 1860s to the 1890s shows that the number of supernumeraries used at fires ranged from one to thirty-five; the average was fewer than ten. The use of supernumeraries decreased over the years as the wholetime staff slowly increased. Certainly the permitted number of one hundred was never reached. These men were paid one shilling per week for lodgings in lieu of accommodation being provided in a fire station. They received sixpence for each turn-out they attended and another sixpence for each hour they worked at the incident. The brigade's red tunic, badge and cap had to be worn while on duty. Vacancies for firemen were filled on a probationary basis by one of the supernumeraries or by turncocks.

It is worth describing briefly the five major fires fought in the Dublin brigade's first year. On 30 January 1863 a fire broke out at about 3 am in the carriage work-shop at the terminus of the Dublin & Drogheda Railway in Seville Place. The entire property was destroyed including wagons, carriages and machinery. First on the scene were an Inspector Armstrong and a constable, followed by Supterintendant Ingram. The fire was dealt with by engines from the corporation, St Mark's parish and the Royal Exchange Company, but the water supply was feeble and could maintain only one engine in operation. The fire was prevented from spreading by the fire brigade men removing a hay rick and railway sleepers which were adjacent to the burning premises.

The second major fire started in the shop of Messrs McDowell, hosiers and outfitters, in Lower Sackville Street at 1.50 am on 25 April. The alarm was raised by residents upstairs who found themselves trapped by the fire. They escaped through a skylight onto the roof of a next-door premises. It was a night of high winds, and flames rose rapidly throughout the house, consuming the entire property. Ingram was in charge with two brigade engines and those from St Mary's parish and the Royal Exchange. Again the firemen's work was handicapped by the shortage of water. When the corporation water carts from Winetavern Street arrived at 2.30 am the brigade was struggling with the growing flames and high winds. With the assistance of some civilian help the fire was prevented from spreading to adjoining premises. The building was finally gutted at a loss of £1,275 and the fire engines had to remain at the scene until nearly mid-day.

The next serious blaze was on 28 July at about 8.30 pm in a toy and fancy goods manufacturers at Lower Sackville Street. Because fireworks were stored on the premises part of the surrounding area had to be evacuated. Adjoining the premises on one side was a chemical store housing paints and oils, and on the other side was a grocery and wine store. The flames spread rapidly and threatened the rear portion of the *Irish Times* building. With the assistance of the police, firemen removed barrels of paint, pitch, turpentine and resin from adjoining premises. The Dublin Fire Brigade was assisted by engines from the National and Royal Exchange insurance companies and by those from St Mark's and St George's parishes. Mr Crofton also attended to supervise the water supply.

After two hours the fire was subdued but the building was virtually consumed. At about one o'clock in the morning the fire broke out again following a loud explosion which brought the entire shop front down in the street. Luckily one of the brigade engines had been left at the site and it succeeded in extinguishing the fresh outbreak which was caused by the ignition of a crate of fireworks. Damage was estimated at £2,000.

The next big fire was discovered at 9.30 pm on 19 September in a general hardware warehouse at 40 Mary Street. It originated in the lower floor of the premises and raged furiously for a time. However, owing to the speedy arrival of Ingram and his men the flames were brought under control relatively quickly. Engines from St Mary's parish and from the Royal Exchange and the National also arrived to give assistance. Water carts under the control of Mr Crofton were in attendance as well, but the warehouse was badly damaged. The lord mayor arrived on the scene while the fire was still in progress and stated that

> too much praise cannot be given to Chief Officer Ingram and the men acting under him, or to the keepers of the other engines, for the promptness they displayed. It is sufficient to remark that within twenty minutes after the alarm was raised the engines were at work, throwing water on the burning premises and thereby arresting the progress of the fire.

The last major conflagration of 1863 was in the early morning of 7 November. It involved a trunkmaker's premises on the corner of Wellington Quay and Eustace Street. The brigade engines were again supported by insurance company engines and one from St Mark's parish. All their exertions however failed to save the building which was entirely destroyed. The lord mayor with his usual promptitude was at the scene and the police assisted in crowd control. Despite the fact that on this occasion the water supply was satisfactory, the brigade still needed the help of the other engines to extinguish the fire.

James Ingram's first annual report concluded with the following statement.

> It is pleasing to bear testimony to the general efficiency and excellent conduct of the Brigade, three of whom I had the honour to recommend to your committee for badges of distinction; and I firmly trust that the steadiness, fearlessness and zeal with which they devote themselves to the duties of the department, will entitle them to the goodwill of your committee and the confidence of the citizens of which they have already received such special recognition.

This was a time when Dublin Corporation was at the forefront of many important and diverse issues, some of national character, which were occupying most of the council debates. Following documentation from Waterford Council seeking Dublin's support, the corporation set about drafting a Bill for parliament aimed addressing the taxation inequity between England and Ireland which had grown since the passing of the Act of Union in 1801. The minutes of the council meetings appear to show that whereas the equivalent taxing rate between England and Ireland just after the Union was at a ratio of 7 to 1, by 1860 that ratio had become 3 to 1. This indicated that the citizens of Ireland were paying more than twice as much tax as their counterparts in England. The corporation also complained in 1864 of the exceptionally high cost of the Dublin Metropolitan Police, a cost to Dublin ratepayers of £26,709. The council had agreed to support the petition by the insurance companies for the abolition, or reduction, of stamp duty on fire protection policies. Jointly with these companies it drafted a Bill. It was also involved in drawing up an agreement to compensate former parish officers whose posts had been suppressed following the passing of the 1861 Dublin Improvement Act.

It was estimated that the new fire brigade was going to cost about £3,000 a year, but the following figures show the actual cost to be somewhat less.

| | |
|---|---|
| September 1861 to 1862 | £353 |
| (before the formation of new brigade) | |
| 1 September 1862 to 1863 | £2,697 |
| 1 September 1863 to 31 March 1864 | £2,655 |
| TOTAL | £5,705 |

The municipal authority was entitled to levy a number of rates in addition to the domestic rate. It could levy a public water rate at three pence in the pound on the annual value of all rateable property within the borough. This was estimated to produce a sum in excess of £6,500 which with a domestic water rate of one shilling in the pound would produce a total collectable in 1863-1864 of £32,135. It is interesting to note that while Dublin Corporation, representing as it did a population exceeding 300,000, had great trouble securing a £250,000 grant loan from the Exchequer to build a waterworks for its citizens, the Midland Great Western Railway was able to secure £500,000 to enable the directors of the railway to build a line from Athlone to Galway.

The waterworks committee was full of praise for the initial progress made by James Ingram in organising the new fire brigade, which was described as the most popular organisation connected with the corporation. But there were "some cases of irregularity". Drunkeness was reported to the committee on occasion and some men had to be dismissed for the offence. The committee insisted that a more rigid enforcement of the penalty of expulsion should be adhered to in all cases of inebriation of men on duty. The committee also revealed that, following detailed investigations, none of the hydrants in the city reached the standards set down of cheapness, security from grit and frost, and facility for being brought into immediate use by firemen.

Work on the new high pressure water supply system had begun and in July 1863 the waterworks committee reported to the council that it had been enabled to complete negotiations with the loan fund commissioners and the Treasury for a loan of £250,000. The scheme entailed the building of a primary reservoir at Roundwood and a secondary reservoir at Stillorgan, with the diversion of three-quarters of the water in the Vartry River into the supply for the city. The Roundwood reservoir would be capable of holding 2.4 billion gallons, the estimated daily requirement for the city being 12 million. Construction involved a conduit and tunnel thirty-three inches in diameter to carry the water seventeen miles to Stillorgan, and two twenty-seven inch pipes from there into the city, with Stillorgan having a capacity of ten days' supply. An agreement was entered with the Dublin Wicklow and Wexford Railway to facilitate the laying of water pipes along the railway line, thereby delivering a substantial saving for the corporation in progressing the project.

The Pembroke and Blackrock townships then attempted to promote a Bill to amend the 1861 Waterworks Act in order to give them powers "to enter into controls with other bodies than the Corporation of Dublin for supplying water to the districts within the boundaries of the respective districts". The Bill was opposed by the corporation and finally withdrawn after an agreement was entered into by the city representatives to supply both of the townships at an agreed rate. It was decided by the waterworks committee at its July 1863 meeting that a sum of £6,000 would be paid to the Earl of Meath as compensation for the right of the corporation to levy £1,000 a year water rates from the Liberties.

In June 1864 John Gray, chairman of the waterworks committee, reported as follows.

*Valves and hydrants will have to be of the most improved make, and all existing valves must be removed. The new valves will be equally light on either side so as to allow for the pressure being reversed at will, and the hydrants will be of not less than three inches in diameter. The couplings for the fire service will be 2½ inches diameter, with the fire hose three inches and hand pipes having nozzles from ½ to 1½ inches. Arrangements throughout the city are to be such as to allow two or more hydrants being easily brought to bear on any fire.*

Domville Estate 1860s, Dublin Fire Brigade training estate workers in use of manual fire engine.
Photograph, © National Library of Ireland

# 4    Continuing progress

Dublin Fire Brigade received its first steam fire engine early in 1864 from the London maker Shand Mason. On 3 February a large crowd assembled in the grounds of Trinity College to see the new machine being demonstrated before the Lord Lieutenant and civic dignitaries. Twenty-eight firemen and forty super-numeraries attended with three manual engines as well as the steamer. There was a strong wind at the time and, despite the spectators being drenched by the spray, the display was judged to be very successful. The previous week, however, the Shand Mason company had suffered the embarrassment of a temporary break-down. The steamer, en route to Dublin, was being demonstrated to engineers and members of Manchester City Council when the injector and feed pump failed to work and the trial had to be discontinued.

This new type of fire engine received a considerable boost from public trials held at the International Exhibition in London in 1862, and more particularly following the steam fire engine competition at the Crystal Palace in 1863. This was the beginning of a fifty-year period when the steamer was the most power-ful and efficient fire pump available. Its main operating advantages over those of manual pumps were a) the reduction in manpower required – from twenty-two men to one or two, b) its much greater power available continuously, and c) its cheaper running costs – the quantity of coal required for the boiler fire was considerably cheaper than paying for a team of pumpers. Dublin's new steamer, which cost £350, could pump water at the rate of 300 gallons per minute, and it was allocated to No.1 Station. The other equipment at this station consisted of a ladder truck and a hose jumper. The latter appliance was a large hose reel on a two-wheel carriage seating a crew of four and drawn by one horse. There were two horses at No. 1 Station and in 1864 a further two were procured for No. 2 Station.

Although a municipal fire brigade was now in being not all the insurance companies had disbanded their brigades. The Dublin Fire Brigade's annual report for 1865 acknowledges the help received from the men and engines of the Royal Exchange and the National. This assistance at fires had been requested by Supterintendant Ingram. The same report mentions that two old parish engines had been donated by the churchwardens of St Audoen's and St Andrew's. Not surprisingly Ingram comments that "they are totally unsuited for present day fire brigade work". At the end of that year the appliances owned by the brigade, in addition to the seven fire escapes located at different points in the city, were one steamer, one manual engine, one hose jumper and one ladder truck at William Street, and three manual engines and one hose jumper at Winetavern Street. Some parishes, as well as a few insurance companies, continued to maintain fire engines for a few years after the 1862 Act made the corporation responsible for providing an adequate fire service.

The problem of low water pressure persisted. The height of a jet obtainable for fire fighting was often no more than ten or twelve feet above street level and, according to the 1860 commission report, sometimes it was as low as three feet. This situation did not change in any meaningful way until the completion of the Vartry system in 1868. The parish churchwardens, those responsible under early legislation for fire fighting, also carried out the tedious role of assessment of each house in the parish for rates.

From their records one discovers what extraordinary congestion existed in much of the city at this time. Illness and disease were prevalent in the over-crowded dwellings, proper sanitary arrangements were as scarce as the water supply, and when fires broke out in these areas of the city loss of life frequently occurred.

When fires were confined to the back streets or alleyways, and the loss of life took place in these dank areas, regrettably very little public notice was taken. The duty of the parish and the insurance firefighters was to protect property. Little or no responsibility for death was taken by either of these two groups who fought fires. Today's firefighter is of an entirely different hue. The first duty is to save life, to check inside buildings even when burning or collapsing, to make sure that there are no people at risk – and if there are, to do everything possible to get them to safety.

In the eighteenth and nineteenth centuries the insurance brigades existed in order to protect property. People lost in burning buildings were simply a consequence of fire. There was no loss of profit involved, therefore the companies accepted no responsibility for lives lost. As late as the 1820s there was not one fire brigade in Britain equipped with an escape ladder capable of rescuing trapped people from upper floors of burning buildings. Yet throughout the nineteenth century new houses, factories, warehouses, schools and hospitals were being built higher. The inevitable result was that even when small fires occurred there was often multiple loss of life. But when fire broke out in a principal thoroughfare, such as happened in the case of the Kildare Street Club fire with large numbers of spectators present, and if deaths were the result questions were then bound to be asked about the competence of Dublin's new civic fire brigade.

One of the most fearful of conflagrations yet experienced was discovered at 9 pm on the evening of 7 June 1866 in Westmoreland Street, one of the widest streets in the city. The two four-storey premises, numbers 19 and 20, were occu-pied by a drapery, a hatter, a gents' outfitter, a solicitor's office and a photographer, all on the ground and first floors. The second and third floors were occupied by the Delaney family. The staircase in No. 20 had been removed up to second floor level to provide more space for the commercial offices, so that access to the top floor was by a staircase on the southern side of No. 19 with a landing leading across to No. 20. There was access to the roof by way of a ladder and skylight. The premises are now part of the Ballast Office.

The fire started in the ground floor shop of No. 19 where the shutters were down, and was first noticed when smoke issued from that area. Some passers-by proceeded to pull open the shutters and smash the shop window, causing an ingress of air. This rapidly caused the flames to spread and destroy the wooden wall between shop and the drapery on the ground floor next door. With the city full of people the fire quickly attracted hundreds of onlookers as the buildings became enveloped in dense black smoke. Mrs Delaney with her three daughters aged twelve to twenty-one, a Mr Strahan and the maid Bessy Kavanagh were spotted through the smoke at the third floor window in No. 20.

The Dublin Fire Brigade escape was summoned from Nelson's Pillar and the police were simultaneously notified. When the fire escape arrived it was placed against the wall of No. 20 but short of the window where the unfortunate Delaney girls were now to be seen crying out for help. The fireman mounted the ladder but only proceeded up beyond the first floor. At the subsequent coroner's inquest he maintained that he had intended to enter the building by the lower window but was driven back by the intense heat and dense smoke. He decided to extend the escape by attaching the front and back fly-ladders, and called for assistance.

In spite of the fact that some constables were at the scene none of them was known to have joined in the attempt to extend the escape. The front fly-ladder was erected without any trouble but before the back fly-ladder could be attached civilians began to pull on the ropes and press the extension levers. Two or three people, one reputed to be a plain clothes constable, then tried to climb the ladder.

With up to three people on the escape and many untrained hands trying to operate the pulley ropes, the mechanism jammed and the ladder almost toppled onto the pavement. By this time a second fire brigade escape arrived from Foster Place. The conductor pitched and extended this ladder outside No. 18 and it was quickly mounted by a fireman and some civilians who entered that premises. In the fear and confusion a disastrous attempt was made to transfer this escape, while fully extended, to the premises where the people were trapped. This foolish attempt to move such a large and heavy appliance resulted in it being smashed against the wall and rendered useless.

Within a very short period of time, certainly less than twenty minutes from the time that the shutters were removed from the burning building, the two premises were totally engulfed in fire and dense smoke billowed from the upstairs windows. No more was seen of the six people trapped in the house. Soon, fire engines began to arrive to tackle the blaze and to prevent the fire spreading to the Ballast Office. The brigade steamer and three manual engines were assisted by the engines of the Sun, National, Globe and Royal Exchange insurance companies. During the firefighting some hose was seen to burst, apparently because members of the public had driven the wheels of an engine over it. The steamer worked successfully from the time of its arrival at about 10 o'clock until it broke down just before midnight. The fire raged until both Westmoreland Street buildings had collapsed. Structural damage was sustained by the Ballast Office.

Fire engines were still on the scene the next day, damping down in preparation for a search for the bodies of the victims. The first attempt at this grim task had to be called off until the area was declared safe. The city engineer instructed that the chimneys and some of the remaining walls be pulled down before a search for bodies could commence.

There was a huge outcry throughout the city as the news of the terrible tragedy spread. Editorials in the *Irish Times* and the *Freeman's Journal*, coupled with their reports on the fire, made damning allegations against the corporation, the Dublin Fire Brigade and even the constabulary, and the call for a wide ranging inquiry quickly gathered strength. Many tons of rubble were removed before the bodies were finally recovered on 10 June. By then most of the city's population had taken time out to visit the site, make comment on the tragedy, and join in the demand for answers.

It was clear that the victims were on the third floor when they became aware of the fire. They were obviously unable to reach safety by way of the landing and stairs so they proceeded to the top floor. For some reason they could not gain access to the roof via the ladder and skylight, probably because that area was filled with black smoke. They were last seen alive at a window on the top floor. The bodies were eventually discovered, all close together, in an area directly below the room where they were last seen. A newspaper report surmised as follows.

> *It is supposed that they proceeded to the sitting room at the northern end of the third floor when the fearful shock burst upon them. The general impression is that the fire, which extended with extraordinary rapidity, took them completely by surprise and the terror stricken people fled to the top storey, where the faces of the two females were last visible.*

The editorial in the *Freeman's Journal* of 11 June summed up the public mood.

> *Never perhaps in the annals of our city did an event occur which created greater sympathy for the surviving relatives, greater horror at the contemplation of its own intrinsic terrors, or greater public indignation at the supposed or real errors of those who were entrusted with the duty of saving life and protecting property from fire .... It is the duty of the government to investigate the whole of the case – not only the events of the terrible Thursday evening, which cast such a gloom over the city – but the antecedents of the case – the provisions made by the civic authorities to guard against such casualties, and whether or not any reasonable and obvious precautionary arrangements had been omitted or neglected.*

The people were incensed by what happened. A fire engine returning to the William Street station on the day after the fire was followed by a large hostile crowd, such that the police had to escort the firemen back to their station. The newspaper columns were filled with irate letters demanding immediate action and a full public inquiry into the running of the corporation and the organisation of the fire brigade.

There is an interesting poetic composition relating to this fire among the old broadsheets in the Civic Museum. It is entitled "Elegaic Lines on the six persons who were burned in Westmoreland Street on the evening of 7th June 1866 in the broad daylight, and in sight of thousands". And it seems significant that there is no mention in the poem of firemen.

*Oh! sad and dismal is the tale, alas! I have to tell,*
*In tears of bitter anguish it will be remembered well,*
*An awful conflagration upon the seventh of June*
*In which six souls as we are told were burned in their bloom,*
*In Westmoreland Street on Thursday night, 'twas nearly eight o'clock,*
*The people of the neighbourhood they got a dreadful shock,*
*The awful flames they bursted forth, the smoke it reached on high,*
*While sighs and cries did loudly rise and seem to reach the sky.*

*Much thanks are due to those who strove their precious lives to save,*
*Some of our gallant seamen their kind assistance gave,*
*Any many of our citizens their endeavours they were great*
*Who kindly ventured life and limb to save them from their fate.*

(The time of the fire quoted in the poem is not correct.)

The lord mayor called a meeting of the waterworks committee on the Saturday following the fire to get a full report from Ingram and his officers. But because of the public outcry it was decided not to issue any statement until a decision was made on whether an inquiry would take place or whether public reaction would be abated by a full coroner's court investigation.

The newspapers carried many letters such as the following in the *Irish Times*.

*It is time for the citizens of Dublin to awake to the position they are left in by so-called Dublin Corporation. Is it not dreadful to think of the awful loss of life that has occurred by their gross neglect? Everything is done by them to stir up animosity and dissention, but after all the expense and heavy taxation the citizens are put to, a fire takes place early in the evening in the widest and most central part of the city, and though two fire escapes are within three minutes distance, yet six beings are allowed to be burnt to death. If those Corporation shams had not been there, no doubt many plans might have been adopted to rescue them from the flames.*

Strong words indeed! The letter was written by a Samuel McComas who was, after some objection, elected foreman of the coroner's court jury.

Another letter claimed that

*… as long as the services of the Fire Brigade are confined to show and parade, or to swell the ranks of a Lord Mayor's procession, they do and look very well. But when they are required – as last night – for the work for which they were instituted, they are worse than useless … after our Corporation had taken the control of those affairs into their hands, and spending a vast sum of money in the supposed perfection of everything pertaining to this vaunted Brigade, an occasion unfortunately arises to test their usefulness and they are found wanting. The public insist on a strict investigation into the whole business.*

The coroner's inquiry was to be wide ranging, and would uncover some issues regarding the fire brigade which, in its short existence, had been kept from public notice. Three years previously, in April 1863, a letter from James Hanlon, secretary of the Dublin Metropolitan Police, had uncovered a serious rift between Superintendent Ingram and the police.

> *The Commissioners of the Police having received a report of your having upon the occasion of the recent fire in Sackville Street called upon the superintendent of the C division to direct the police to work the fire engines, I am instructed to inform you that the duty of the police on such occasions is limited to the protection of life and property and keeping the place free from obstructions.*
>
> *I am to request that you will not in the future interfere with them in the discharge of these functions, as you may be sure that the police will fulfil all the duties properly belonging to them without any foreign interference.*

Bearing in mind this curt rebuke, one cannot help wondering what the police at the 1866 fire were doing to control the crowd whose well-intentioned interference undoubtedly contributed to the failure to rescue the trapped people. In the broadsheet already quoted, the line "many of our citizens their endeavors they were great" masks the awful consequences of the same citizens' misguided actions.

Under cross-examination Ingram said that he had passed along Westmoreland Street at 8.45 pm without seeing any sign of fire, and he denied arriving late on the fireground. He highlighted his difficulties with shortage of manpower and equipment: "I asked a police officer to give me some of his men to assist at the engines ... I cannot say he declined, but I did not get the assistance". He was obviously glad to have the assistance of the insurance company brigades, and admitted that he looked on the engines of the National and the Royal Exchange as part and parcel of his brigade.

The superintendent was questioned about his 1865 report in which he had stated that there were nineteen cases of poor supply of water, and three cases of very poor supply. He did not think there was any problem with the water supply at the Westmoreland Street fire. "The moment I arrived I sent for the steam engine. I directed the hand engines to work but they made but little impression on the fire". Asked if he had not ordered the steamer would it respond to the fire, he replied "It would come by the direction of the man in command of the Brigade. The steamer does not start for a fire without an order from me or whoever is in command. It takes two horses to pull the engine, and if we drove the steamer there first she might be useless". He admitted that he had no knowledge of fire escapes when in New York, but that he had gone to London to see how they worked, particularly at fires.

He was confident in his ability to drill on escapes even with his limited working knowledge of them. He even went on to claim "I wish to God I had been there [at the fire earlier] for if I had, you would have been holding an inquest on my body, or I would have had them out".

One of his firemen was arrested drunk in Camden Street while the fire was still blazing, and it seems that the only other officer in William Street fire station was sick on that night. So when Ingram left the station there was no officer on duty. Complaints were also made of a slow turnout even after the fire call was relayed to the station from the police.

The court inquiry ended after the fifteenth day with the following verdict read out by the foreman of the jury.

> *We find that the death of Anthony James Strahan and the other five individuals was caused by a fire in the houses … the origin of which fire we have no evidence to enable us to determine …. we believe said persons could have been saved, and there was full time to save them, had the fire escapes been manned by experienced men, or had the police been called to the assistance of the men who brought the two fire escapes to the houses. We also believe that, had there been no fire escapes at the premises on the occasion, the six persons could have been saved, either by ropes from the parapets or by opening up the roofs. There was no such effort made to save the people by the Fire Brigade. We believe that the present system adopted by the Fire Brigade and its escapes for the preservation of life and property is insufficient, and that it should be brought under the special supervision of the Board of Works, the Metropolitan Police, or other experienced parties.*

It would have been interesting to have had some intelligence as to who the "other experienced parties" might have been! The verdict continued.

> *We consider the escapes and their present construction very imperfect for the saving of life from fire, and we strongly recommend some alteration or an adoption of the telescopic escape, which we consider a great improvement to those at present in use.*
>
> *We are of the opinion from the evidence that came before us, that the Fire Brigade, who are for the saving of life and property in our city, and due to the arduous duty and danger to which they are often exposed, are insufficiently paid, and that the Board of Works and the Insurance Companies should contribute to their support, as is the practice in London.*

The coroner, in summing up all the evidence and the statements of the various legal representatives, praised the members of the jury for their diligence, consideration and forbearance during the proceedings, and then went on as follows.

> *You have had a long and arduous inquiry, but not longer than the circumstances demanded. It was due to the friends of the parties who suffered by this terrible calamity, it was due to the safety of the lives of yourselves and fellow citizens, and it was loudly demanded by the public that a full and ample investigation should be had into all the circumstances connected with the unfortunate event. There is one thing I will call your attention to, and that is the exceedingly short space of time it took to reduce two first class houses from a state of apparent security to absolute ruin. The evidence of the two fire brigade men who first got the intimation of this fire fixes the time at about eight minutes past nine, and the evidence of the gentleman whom I examined today confirms that with great accuracy. The next thing is the time the Lord Mayor arrived. That was at twenty-seven minutes past nine, and at that time the roof of the house had fallen, and there was no possibility of any person being able to exist in the blaze. That you must bear in mind in measuring what efforts were made, whether there was unnecessary delay, and what more could be done under the circumstances.*

It is of interest to note that among those who gave evidence at the inquiry was one of the partners in the London firm Shand Mason. James Shand had travelled to Dublin to examine the escapes and engines of the fire brigade and, although his firm had not supplied any of the escapes, he commented on the severe damage sustained by two of them. Shand Mason had supplied the manual fire engines in use in Dublin and also the steam fire engine.

Thus, after a public inquiry lasting fifteen days, the first major trial and investigation of the Dublin Fire Brigade came to an end. Unfortunately it would not be the last.

On 16 September 1866 the brigade was faced with another serious fire, this time in Great Britain Street. It was attended by the steamer, three manual engines, a hose jumper and a fire escape. The jets were operated from the steamer working at 85lbs pressure in the front of the building, while two of the manual engines tackled the blaze at the rear to prevent it spreading. The property contained much inflammable material: show boards, paper, light deal in large quantities and twenty bales of hay in a loft over the stables. All was destroyed along with most of the building. The fire brigade came in for some praise on this occasion because of the rescue of two men, a girl of twelve years and a nine-month old infant, whose mother was also rescued. An impression of the overcrowding in the city's housing areas is evident in the report on this fire. The house which was destroyed had the following listed as tenants: a man, his wife, two children and a servant girl in the front drawing room; the back section on the same floor housed a family of six; the second floor front and back housed eleven persons; the top floor had thirteen people in residence. At the time the fire broke out there were in fact thirty-five people, young and old, in the house. All were saved.

The house, typical of properties in that area of the city at the time and indeed for many years later, was of four storeys with a shop at ground level and stables with loft at the rear. It took the brigade over twenty-six hours using sixty-nine men to extinguish the blaze. The cost of the supernumeraries and others involved in working the manual pumps amounted to three pounds.

At its October meeting the waterworks committee discussed the coroner's report in the aftermath of the Westmoreland Street inquiry. It was decided to appoint a lieutenant to act as deputy to Superintendent Ingram. Captain Shaw was contacted and proved willing to recommend one of his officers to fill the post at a salary of £150 per annum. The meeting also agreed to purchase a second steamer and some telescopic fire escapes, as well as to extend the telegraphic system. Plans for a fire station on the north side of the city were discussed. All of these provisions to improve the overall efficiency of the fire brigade could not however be immediately implemented because of the limitation of the funds available. Meanwhile six of the escapes had major repairs carried out on them, and in four cases this amounted to almost an entire rebuild: new main shafts, fly ladder, canvas trough, copper gauze, galvanised wire stays and swivels. In addition, two of the manual fire engines were repaired, repainted and lettered.

The fire brigade currently available to protect the entire city of Dublin consisted of a chief officer and twenty-six firemen, with one steamer, four manual engines and seven fire escapes. There was however an interesting addition to the Brigade's equipment in 1866. Because of the number of calls being received relating to persons drowning, a life buoy, ladder and drop-line were purchased for Winetavern Street station, being close to the river.

> *On 11 June between 1.00 and 2.00 am an alarm was received at No. 2 Station, from police constable 112A, that a woman had thrown herself into the Liffey at Ormond Quay. Two men were on duty (Patrick Balfe and Patrick Kelly) at that station and immediately attended with the life buoy and succeeded in safely rescuing the woman.*

Another rescue was carried out in the early morning of 29 September by Stephen Foot, the conductor of the fire escape at Nelson's Pillar. Foot received an alarm that his machine was required to rescue some occupants of No.1 Liffey Street following the collapse of a portion of the house. He responded at once and by means of the escape he rescued a woman who had been asleep in the top front room. The Dublin Fire Brigade was clearly already a fully fledged rescue and emergency service, though even today at the beginning of the twenty-first century it rarely receives credit for being the all-round emergency organisation it actually is.

The chief officer's annual report for the year 1867 showed that the brigade turned out to 191 fire calls, of which 104 were chimney fires. It also responded to five fires outside the city boundary, where much damage was reported at Dundrum lunatic asylum, Baggot Street Hospital, paper mills on Clonliffe Road, and hay ricks at Ashtown and Seapoint. A medical report stated that "the health of the firemen during the year has been uninterruptedly good. But we had one death, that of Richard Byrne, in March, from pulmonary consumption of long duration". This is the first recorded death of a serving member of the Dublin Fire Brigade. The total wage bill for the year was £1,241-11-0, and the second steam fire engine, costing £400 and similar in size to the first one, arrived from Shand Mason. The brigade also purchased two new telescopic fire escapes made by a Mr Clayton in Dublin. One other matter of some significance was mentioned in this annual report.

> *On several occasions we were enabled to dispense with the use of engines owing to the high pressure water from the Vartry being on the mains, thereby saving a great deal of time and labour, and enabling the Brigade to use larger jets of water without the assistance of the engines.*

Throughout the early years in which the Dublin Fire Brigade was getting itself established, a major infrastructural project was being tackled by the corporation. This was the construction of the new water supply system for the city. It was a great undertaking and by the end of 1865 the waterworks committee was able to report a big improvement in the city mains; 5,260 yards of 27 inch piping, 181 yards of 9 inch piping, and 2,068 yards of smaller pipes had been laid. The new system not only increased the possibilities for more and cleaner water for the citizens, but also altered the existing system to bring it more in line with the changing configuration of the growing city.

By June 1867 the Vartry high pressure supply had begun to arrive in the city. By 31 December 5,389 houses with a population of 59,279 persons were linked to this new supply, while 49 water fountains (street pumps) were now being supplied from the same source.

When the Vartry scheme was finally completed in 1868 up to fourteen miles of old pipes had been excavated, of which six miles had been cleaned and re-laid. Nine hundred and eighty-six ball hydrants had been fixed on the mains, one hundred and thirty seven new water fountains had been installed in the city, sixty-eight of which were replacements of older units and the balance were completely new. A bright new era was dawning for the Dublin Fire Brigade. At last the city would be served by a water supply system capable of delivering adequate quantities for firefighting when required. Water carts could now become a thing of the past and the use of supernumeraries could be reduced and finally discontinued. A few insurance companies still had fire engines and assisted at large fires, but the parish engines had passed into history. For the next fifteen years the annual reports show that about 70% of all fires within the city were fought with hose connected directly to the high pressure mains because of the great amount of water now available for firefighting from that source.

The need for a senior officer to assist Superintendent Ingram was finally addressed early in 1868 when the post of lieutenant in the Dublin Fire Brigade was advertised. The following four men were short-listed: Mr Moorehead from Belfast Fire Brigade; Mr Lamb from Ross, Murray & Co, Engineers, Plumbers and Brass Founders of Middle Abbey Street; Inspector Byrne of the Dublin Fire Brigade; and Mr John Boyle, late superintendent in the Dublin Metropolitan Police.

Boyle was successful and took up his post in April at a salary of £100 per annum. He was born in Enniskillen in 1819 and had served for many years in the horse police at Kevin Street barracks. He attended most fires in Dublin when an escape was stationed at Kevin Street and he may well have been one of those involved in pushing the idea of a police fire service during the preceding controversial years. He had been a defender of the police when they were under criticism for their failure to turn out the fire escape when it was housed at Kevin Street and also when they were castigated for inactivity when dealing with fire situations. He was destined to become the fire brigade's second chief officer in 1882, and did not retire until he was seventy-three years of age in 1892.

In 1868 the brigade received some new equipment: a horse-drawn hose jumper, five jumping sheets for rescues, and forty leather helmets. Now, for the first time, every fireman wore proper protective headgear, and there was a sufficient stock of helmets to issue temporarily to supernumeraries if required when assisting the brigade. The following year there was a decrease of £57-14-0 in the cost of turning out because water carts were no longer in use and men were not required to operate the manual pumps during that particular year – this of course was thanks to the new water system. Two Dublin firemen left the brigade having been selected to take charge of brigades in the provinces. The total wage bill for the year were put at £1,440-6-6.

A much-needed new engine house and a stall for one horse were built at No. 2

Station. Annual fire calls had now reached 238, and it was now taking only two minutes for engines and hose jumpers to be horsed and fully manned in turn-outs.

The year 1869 saw the thorny matter of recoupment of expenses for attend-ing out-of-area fires really come to a head. The brigade, in spite of its legal rights under Section 9 of the 1862 Act, was only successful in recouping a small propor-tion of the money owed to it for out-of-area attendance. The position was aggravated by the failure of the townships outside the city to provide some form of fire service of their own, coupled with their continued opposition to making payment when the Dublin fire engines responded to their calls.

In compliance with an order from the waterworks committee the fire brigade superintendent issued a notice "withdrawing the attendance of the Fire Brigade from County districts on the occasions of fires". The full municipal council promptly requested the committee to explain its reasons for withdrawing the attendance. The committee, not surprisingly, stated with righteous indignation that "the wealthy residents of the various Townships are better enabled to contribute towards the support and maintenance of a brigade than are the popu-lation of Dublin, who are taxed in the annual sum of £3,000 per annum ... and these Townships have not organised such a corps". The two largest areas referred to were Rathmines to the south and Pembroke to the east. These townships did not establish their own fire brigades until some years later. The committee pointed out that the duty of the Dublin Fire Brigade should be to aid the town-ships energetically when needed, but not to relieve them of their own clear responsibilities. It so happened that year that the Dublin brigade was called to three fires outside the city – at a soap-boiling factory and a hay rick both in Rathmines, and at a house in Pembroke. The problem of non-payment was to drag on for many years before a solution was agreed.

Following a house fire in Clontarf, beyond the city limits, the corporation took the householder to court for failing to pay under Section 9 of the Act. This was the first such case to come before a magistrate, and the bench admitted having some difficulty in deciding what a "reasonable charge" might be. He made an order for £2-15-0 for actual expenditure and £13-18-6 for the brigade's attendance.

In the the report of the waterworks committee chairman, Sir John Gray, to the corporation in August 1871, after the completion of the Vartry water scheme, there was evidence of the extent of Gray's anger at the opposition he encoun-tered to the project. He referred to attempts to undermine the waterworks committee by a continuous lobby of conspiracy against the elected council.

*In connection with your water service your committee obtained powers to organise an effective Fire Brigade. The same active party which opposed your Water Bill attempted to deteriorate the status of the Corporation by promoting a Bill to authorise the Police authorities to form a Fire Brigade, and tax your citizens for its support ... your committee organised a Fire Brigade which for efficiency, bravery in circumstances of danger and good conduct cannot be surpassed ... out of their limited taxation, three half pence in the pound, nearly £3,000 a year to the maintenance and equipment of the*

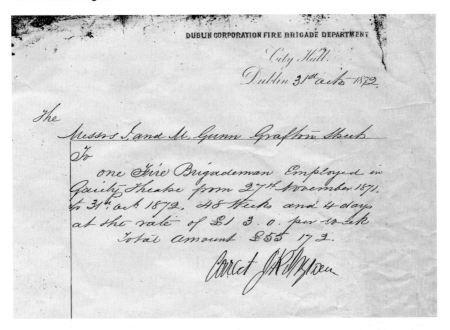

**Invoice from Dublin Fire Brigade to Messrs Gunn, Grafton Street, 31 Oct 1872:**
**for the attendance of one Fire Brigade man in Gaiety Theatre, 27 Nov 1872 to 31 Oct 1872**

*Fire Brigade, an amount not contemplated at all when the limit of taxation was fixed.*
*All these extra costs were carried without any extra burden on the rate-payers.*

The Dublin city street scene continued to change. The expanded omnibus system was now working in conjunction with the main railway companies which had opened up travel to outlying areas of the country. Bicycles had made their appearance and traffic flows were much altered. Private horse traffic, horse-drawn cabs and drays, carts, floats, coal cars and hand carts congested the principal streets, all adding extra obstacles to the fire brigade turn-outs, particularly during daylight hours. And in February 1872 the first tram line was opened between College Green and Rathgar.

At national level the most notable issues at this time were the activity of the Fenian organisation and the continuing decline in population. The Dublin population, however, continued its slow growth. The change in the water rate contribution now revealed a problem which was to adversely affect the Dublin Fire Brigade for some years to come. A reduction to one penny in the pound meant that the brigade received £3,139-13-6 in 1872 yet the expenditure had risen to £3,331-12-8. The following year matters became even worse, when the rate receipt came to £2,304-10-0 while expenditure was £2,712-18-0. Something needed to be done if the brigade was to maintain its level of service, purchase extra equipment, and continue to improve its contribution to the safety of the citizens.

The superintendent issued his eleventh annual report of the Dublin Fire Brigade at the end of 1873. At this milestone in the brigade's early history it is worth quoting from the report in some detail. Previously one saw the disastrous results of interference from members of the public who were trying to "help" the firemen, and the report highlights two more instances of this behaviour. In April at a fire in South Great Georges Street "through the interference of the crowd with the appliances the Brigade was prevented from getting to work as quickly as they otherwise would have done". Two months later more serious trouble surfaced during an extensive blaze at a timber merchants in Thomas Street. Ingram described the situation in the following terms.

*Worked 6 lines of hose from stand-pipes… the line of hose on left of gate entrance, Thomas Street, was pulled off by the people, the flanges of hydrant broken off, when the water escaped from hydrant. This cut the water off from me for nearly one hour. I immediately brought up the steam fire engine, made a hole in the street, and set the engine to work from the large column of water escaping from the hydrant; another line of hose was also pulled off hydrant by the people near Vicar Street, breaking flange; a fresh line had to be laid down from corner of New Row.*

During the ten years since the formation of the brigade its appliances and equipment had increased, but not its personnel. The shortage of staff and the need to increase the establishment, already referred to in the previous year's report, is mentioned again: "I am only able, during the time in which large fires generally occur – that is from 7.30 pm to 7.30 am – to turn out to a fire 13 men, having to leave one man in each station to receive alarms during the absence of the Brigade at a fire – the other portion being employed on escape duty from 7.30 pm until 7.30 am; and I have no reserve." The manpower of the brigade grew slowly and the hours of duty were consequently very long. The men who manned the escapes in the streets during the night were expected also to attend any fire calls during the day.

The equipment was distributed between the two fire stations as follows.

- No.1 William Street: 2 steam fire engines, 1 hose carriage, 1 tool cart, 4 hand pumps, 3 stand pipes, 2 sets of double harness, 2 single harnesses.
- No. 2 Winetavern Street: 3 manual fire engines, 1 hose jumper, 4 stand pipes, 4 hand pumps, 4 sets of double harness, 1 single harness.

The total amount of hose available was: leather 5,709 feet, canvas 617 feet. The fire escapes were stationed at the following locations.

- St Paul's Church (telescopic)           North King Street
- Nelson's Pillar (telescopic)            Sackville Street
- Kildare Street (telescopic)
- St George's Church (fly-ladder)         Hardwicke Place
- Bank of Ireland (fly-ladder)            Foster Place
- St Catherine's Church (fly-ladder)      Thomas Street
- Kevin Street (fly-ladder)

The annual inspection of the fire brigade by the lord mayor and the chairman of the waterworks committee, Sir John Gray, took place on 20 December. The steamers, hose reels and other appliances were inspected first, then the horses, stables, harness, etc. The men were paraded in drill and marching, after which the fly-ladder escapes and the telescopic escapes "were raised and lowered with great quickness, and men of the Brigade brought others down from a great height with the greatest ease and with perfect safety. The merits of Clayton's telescopic escape were exemplified on the occasion".

The telescopic fire escape was designed by William Clayton. He patented his invention in 1874, and became a successful manufacturer of these ladders at his premises in Camden Street, Dublin. His reputation was such that four of Clayton's escapes were ordered, at £85 each, for use in London by the Metropolitan Fire Brigade. Orders were also received from the brigades in Glasgow, Liverpool and Cardiff, as well as from Kilkenny and other brigades in Ireland. It is surprising to note that the firm of Merryweather, which designed and built its own escapes, acted as the sole agent in England for Clayton's escapes.

There were three specific technical elements which Clayton claimed as his own invention.
- constructing the ladders with sides of iron latticework trussing
- the application of steel wire ropes for raising the ladders
- utilising the axle of the running wheels as a windlass.

A significant difference from the older type of escape was the absence of a canvas chute or any inflammable material attached to the ladder. The most popular size which he manufactured could reach a height of sixty feet.

# 5    A growing emergency service

The brigade's report for 1873 lists six fires attended outside the city boundary: at Milltown, Booterstown, Rathgar, Northumberland Road, Drumcondra and Kingstown. The account of the latter incident is interesting.

*The fire originated in the ship "Nagpore". On receiving written directions from the Lord Mayor, I proceeded to Kingstown (a distance of seven miles from the fire station) with No. 3 and 4 hand engines, tool cart and nine men.*

On 10 November a severe storm struck the east coast causing many vessels to seek refuge in Kingstown (Dun Laoghaire) harbour. As so often happens on such occasions, many sightseers made their way down to the harbour area to watch the massive waves beat over the piers and crash against the shore and the ships at anchor. Shortly before midnight they witnessed an iron East Indiaman cutter, the *Nagpore,* with every inch of canvas set and oiled to catch the wind running direct for the harbour mouth with volumes of smoke pouring from her hatchways as the furious waters of the storm drove her forward, causing consternation and panic among the vessels already taking shelter. It was quite clear that the 1,500 ton *Nagpore* was on fire, and the crew were making no effort to shorten sail.

As the ship was driven into the harbour she let go both anchors but, owing to her speed with so much canvas aloft, she swung round dragging the anchors and fouling, smashing or sinking everything to left and right. The ship carried a cargo of jute, cotton, hemp and about fifty tons of saltpetre and had been on fire for three days. The crew had spent that time battling a most dreaded enemy, a fire on board, with the prospect of a fierce explosion if the saltpetre came in contact with the flames. They had become so exhausted from operating the pumps – spurred on by knowledge that continuous pumping was essential if they were to have any chance of survival – that no effort was made to shorten sail. The burning ship first came in contact with the schooner *Pilot*, a Wexford coal boat which sustained serious damage, but the crew were fortunate to be able to scramble on board the *Nagpore* as their boat sank. The captain of the *Pilot*, however, was tragically crushed to death between the two vessels. Then a collision occurred involving the *Manager of Wales* which soon became swamped, but here again the crew were lucky to be able to board the runaway ship that was causing the disaster. A trader, the *Diligence*, was then struck and de-masted and the trawler *Echo* had her moorings severed, causing her to be driven westwards where she was broken to pieces on the harbour barriers.

At last the *Nagpore* was brought-to as her anchors finally held in about twenty feet of water in the high tide opposite the Royal Irish Yacht Club. Those on board quickly scrambled ashore to inform the anxious harbour officials of the immediate danger to the area from the burning vessel and her saltpetre cargo. Because the Kingstown authorities had no means of their own to deal with such a disastrous situation, a member of the harbour master's crew was immediately despatched to the Dublin Fire Brigade headquarters to seek assistance from Ingram's fire crews in extinguishing the fire before damage was done to the harbour buildings. The brigade's report records the time of the call as 8 am. Following the usual delay in contacting the lord mayor to obtain his permission to travel outside the city boundary, the two Dublin fire engines proceeded to Kingstown. Ingram's report continued as follows.

> On arrival I had one of the engines placed on board a launch and brought alongside the burning ship, and worked two lines of hose on it. At 2 pm another engine was ordered out with two men but, finding that the fire could not be got under, in consultation with Captain Barnes (of the cruiser "Victoria"), the Lord Mayor and other gentlemen, I recommended the scuttling of the ship. This was done by the "Victoria" firing six rounds into the hull, striking her at the water line, after which she settled down till water covered her upper deck. The men and engines returned to town at 4.55 pm.

So ended this dramatic episode.

At the council meeting the following day the lord mayor, who had been present to see the drama in Kingstown harbour, justified his order for the Dublin Fire Brigade to attend by stating that

> a man who would not risk his life to save a fellow creature was not worthy of life. Captain Ingram at great personal risk had got on board the vessel and deciding that she was capable of blowing up part of Kingstown, had got the assistance of a gunboat to put holes in her bulk so as to sink her.

This report was received by "hear hear" from the assembled councillors. There was comment in the *London Morning Advertiser* that this disaster "should impress upon the authorities of the City the necessity of providing themselves with at least one floating fire engine". The authorities did not consider that this was a necessity, but the lord mayor concluded by saying that, after the example they had that day, he thought it was a matter for serious consideration whether they would not extend the fire brigade. He regretted to add that there appeared to be no facility in Kingstown to put out the fire.

When the ship fire was discussed at the November meeting of the water-works committee there was some controversy regarding the Dublin Fire Brigade's response. The superintendent had authority to allow up to two engines to respond outside the city boundary at any one time, but on this occasion a third engine was sent. This should have had prior authorisation from the waterworks committee. The controversy was ended when Sir John Gray stated that, as chairman, he would take responsibility for ordering the third engine.

In the early 1870s differences had arisen between the corporation and the commissioners of Pembroke, Kingstown, Dalkey, Killiney, Ballybrack, Bray and Clontarf. The corporation had been forced to concede to these wealthy areas which surrounded the city special terms for water supplies. Cheaper terms were given to these townships in order to get them to drop opposition to the Vartry scheme but the problem was that they were not honouring the agreements. Each township was consuming a much larger quantity of water than that to which it was entitled under its contract with the corporation, ie twenty gallons per day per head of the population. The cost of water had been set under the original agreement at 3½ pence per 1,000 gallons, with the amounts to be supplied fixed by the total sum. This stealing of the water, causing further subsidy from the Dublin ratepayers, was one of the major factors in the refusal of Dublin Corporation to allow the fire brigade to provide a service in those surrounding areas. The corporation thus sought a new Amended Waterworks Act for 1874.

The waterworks committee report to the municipal council on 28 October of that year once again dealt with the waterworks estimates and the continuing disagreement regarding the water supply. The corporation maintained that the townships owed it £2,612-17-7 for that year because excess water to that amount had been drawn off the Vartry supply. It had borrowed a total of £511,527-11-7 in order to complete the radical change in the water system for the city and to provide the citizens with a fire service, and now it needed to borrow more funds to extend filter beds and enlarge the Stillorgan reservoir. The report to the council continued as follows.

> *Your committee, in the interests of the citizens, voluntarily undertook the charge, equipment and maintenance of the Fire Brigade, which amounted to about £3,200 per annum … a gross sum paid out of your water income since its formation for fire purposes of £38,933…your committee would remind your Council that an effort was made to take the management and direction of the Fire Brigade, and the protection of the lives and property of the citizens of Dublin, out of the hands of the elected representatives of the Burgesses, and place it under the control of a Government force. For this purpose it was proposed to increase the police tax by 1½d in the pound, imposing the whole of the burden upon the inhabitant householders and not as a property rate. It was also proposed to give to a force unconnected with your Corporation, and in no way amenable to the representative power of the city, an absolute right to interfere with mains, valves and waterworks generally, leaving them entirely at the discretion of an unskilled body, and to place this expenditure of £3,200 in addition to your existing water rates upon the inhabitant householders alone. … The Council in the year 1862 applied to Parliament for protection on all of these grounds and from that period to this have voluntarily incurred the expenditure already referred to … to secure the best protection against fire, and the most efficient Fire Brigade which any similarly situated city in the United Kingdom at present maintains.*

The Dublin Fire Brigade had now been in existence for over ten years, and the Act of 1862 stated in Section 12 that the proportion of the public water rate to be applied to fire brigade purposes was, as we have seen, to be reduced. By this time it was only too apparent that a reduction in available funds would be disastrous. In July 1873 Sir John Gray proposed that the corporation seek an Amending Bill to the 1862 Act "to restore the Dublin Fire Brigade to the original efficient strength by empowering them to apply the same amount of the water rates to its maintenance as they were allowed to use up to 3rd June 1872". But despite the corporation's representation to Lord Hartington (the Lord Lieutenant) the reply was "there are technical objections to the introduction of any amended Private Bill at this period of the session so that the matter must be left over during the recess".

In 1874 an Amended Water Act did go through parliament. This enabled the corporation to stop water to townships or charge the commissioners for any excess taken; to apply out of Dublin domestic water rate an additional sum for the purpose of the fire brigade; and to compulsorily take lands for the purposes of the Act. The position of the corporation was greatly strengthened by this legislation in facing the problems that had arisen while operating under the 1862 Act.

The new Act came just in time to allow the corporation to deal with a serious case of pollution in the old city water course. The Old Bawn paper mill, above the city weir, was polluting the water supply by the discharge of chemicals and refuse. The mill owners on the water course paid no rent and refused to pay for repairs to the city weir, the water course or for cleansing the river. The new Act allowed the corporation to proceed against these polluters.

It was agreed at this time that a new hose jumper would be purchased for Winetavern Street; the existing one at No. 2 Station would then be converted to a tool cart. In addition 1,000 feet of leather hose and two telescopic escapes were to be purchased to replace two old London pattern escapes. The hose jumper and the leather hose were received in 1874, but before the new escapes arrived the brigade was confronted with the biggest fire of the entire nineteenth century in the city of Dublin.

Some time after 8 pm on 18 June 1875 vast sheets of flame were seen to rise above the houses in the area bounded by Cork Street, Ardee Street and Chamber Street. Not long after the fire was discovered in the bonded Malone warehouses off Cork Street the buildings were engulfed in flames. The stores were filled with an enormous quantity of whiskey, brandy and wine, and the burning liquid flowed down Ardee Street and Mill Street like streams of lava. The onlookers who had gathered at the scene were forced to run for their lives and the flames set fire to the homes of the unfortunates who lived in their path. The spectacle of the poor people rushing from their burning homes was a harrowing sight: "some of them only half-dressed, others with children clutched to their breasts, invalids tottering on some helping arm", as described by one reporter. Even the Carmelite convent in Ormond Street was threatened for a time, but the wind blowing the flames in the opposite direction saved that large building from destruction.

By midnight the fire was at the height of its intensity and could be seen for miles around. The police were praised for the speed with which they arrived on the shocking scene, as were the firemen, arriving as the fire spread to Reid's malt-house in Chamber Street. The brigade received the call at 8.30 pm and turned out with two steamers and two hose carriages. At first the firefighters tried to douse the flames with water, but this effort was halted because it was clear that it simply accelerated the flow of the burning whiskey. Fire engines were abandoned as the firemen, assisted by the military, began digging up roadway and baking rubble to form dams, particularly in those streets downhill from the escaping liquid. Luckily the tide of flames left Watkin's Brewery in Ardee Street untouched, but houses in Mill Street were engulfed in flames.

The lord mayor arrived at the fire as rubble and sand was being drawn from the local corporation yard for use in damming the burning flow and materials were being comandeered from adjacent premises for the same purpose. However, even these innovative measures proved futile as rivulets of flames continued to seep through. The scaffolding on the new extension to the Coombe maternity hospital was soon alight and had to be pulled down rapidly by firemen and soldiers to prevent damage to the building. The public house at No.1 Chamber Street was not so lucky; it disappeared in flames. At the time of the outbreak a wake was taking place in a house in the same street as the public house and the occupants were forced to flee with the corpse to mourn elsewhere, while the home of the bereaved and their belongings were totally destroyed. Many courageous rescues were carried out by people living in the affected area, as well as by soldiers and firemen. Horses and pigs in Mander's yard broke out and rushed headlong through the burning whiskey, scattering police and onlookers as they stampeded to safety. Soon the tannery in Mill Street was engulfed, pouring out palls of black acrid smoke over the area. It was estimated that up to thirty-five premises to the value of over £100,000 were destroyed along with £54,000 worth of spirits in the bonded warehouse.

In some of the adjoining streets people were seen to scoop up whiskey running in the gutters – with mugs, hats, shoes – and indulge themselves by drinking the contaminated, immature spirit. It is a fascinating experience to look at a newspaper photograph published eighty years later which records an identical occurence – a fireman and two onlookers drinking the whiskey that flowed down the street from a fire in Power's Distillery in July 1961.

At about one o'clock in the morning, when the danger from burning whiskey had passed, the firemen got to work with four jets,. The extensive bonded stores in Chamber Street/Cork Street were completely destroyed and more than twenty tenements, each housing ten to fifteen families, were either totally or partially destroyed. The cottages in Mill Street were burnt out and all their occupants were left on the verge of ruin. Looms, trade appliances, beds, clothing and furniture, all were lost in those terrible few hours, as were major businesses, warehouses, a tannery and bonded stores.

Over the next two days the site was visited by crowds of sightseers, some entering the Liberties area for the first time, who came to see perhaps how the poorest people lived and perhaps to support the appeal set up by the lord mayor to which the archbishop Dr Cullen had contributed £10. The firemen remained at the scene until 3.40 pm on 23 June. That fire was the calalyst for one of the fire brigade's finest "hours". The debacle of Westmoreland Street was finally buried. The brigade was highly praised for its "most commendable bravery and fearless activity. They worked as hard as men could to retard the progress of the fire". Unfortunately it was also reported that it was Mr Crofton (son of the former waterworks inspector) who had made the decision to block and dam the streets with clay, sand and rubble. Superintendent Ingram was swift to put the record straight.

*I deem it necessary to report that there is a feeling among the officers and men of the Brigade, owing to reports that have appeared, that on the occasion of the late disastrous fire they did not take proper means to extinguish it. I beg to state that that which was attributed to another corporation officer was done long before he appeared at the fire. I had given orders to have the fire earthed, and sent for the corporation carts to convey ashes, tan, manure, etc and previous to their arrive I had the streets dug up and sewers stopped.*

The final toll from this disastrous fire was thirteen dead from drinking dirty liquor as it ran along the gutter, and many more treated in hospitals for the same complaint.

The following year the Dublin Traffic Act and the Explosives Act were passed, both of which would involve the city's firemen. It was in that year also that a major campaign was launched by the corporation to persuade the fire insurance offices to contribute to the annual expense of maintaining the Dublin Fire Brigade "to its present efficiency" – a call which was to fall on deaf ears. During 1876 the brigade answered 306 calls, and although the establishment was increased by two men to a complement of twenty-seven the annual report again called attention to the shortage of manpower and the need for expansion.

*It is to be regretted that the funds at the disposal of your committee for Fire Brigade purposes do not admit the erection of a station on the north side of the river, where there is a vast amount of property unprotected. That portion of the quays extending from the Custom House to the point of the North Wall, which is increasing in valuation each year, would necessarily require at least one fire station. The very large fires which have broken out in this district warrant me in calling your attention to the necessity of a station.*

In a move to raise some of the funds required to expand and improve the brigade, Dublin Corporation took the unusual step of setting up a sub-committee to press the insurance companies to act as they did in other cities in the United Kingdom and make direct payments towards the upkeep of the city's fire service. John Gregg, chairman, wrote to all local offices of the insurance companies requesting them to suggest a plan by which "an equitable arrangement might be come to with regard to the extra support required for the corporation Fire Brigade". He also pointed out that since the establishment of the fire brigade the offices had made a considerable saving by disbanding their own staff of firemen. The offices also received much valuable assistance from time to time from fire brigade staff in their evidence as witnesses, whereby the offices were not only able to resist fraudulent claims but were able to punish crimes which would have escaped detection but for the vigilance of the firemen. The local offices of course had to refer the matter to their head offices but in the meanwhile were pleased to reply the "preservation of life and property from loss by fire is a subject of vast importance", and that they felt "certain it will receive that mature consideration which the striking facts detailed in your circular demand of it". The head offices in London thought differently. The insurance managers in Dublin had their attention called to the Act of 1862 and the power given therein to the corporation, and exercised by them, of charging owners and occupiers for the services of the brigade, and therefore the companies felt that they could not entertain the application now before them: "They hold strongly that fire extinction is a public matter, and should be provided for by the community through its executive Corporation, or other local authority, out of the funds raised for all public objects, by the method of rating only".

They were the companies who for 150 years had maintained fire brigades from their own funds to protect their insured properties; the same companies who had been fully supported by Dublin Corporation in their private lobby to parliament to be exempted from tax on their fire insurance policies. Finance corporations hardly change in essence over time. Even then they showed the unacceptable face of capitalism.

The city fathers concluded the correspondence, which had commenced in June 1876, with a compelling argument "…from the statistics carefully compiled for the last ten years, it is computed that the property lost by fire in Dublin amounted to an annual average of £25,000. 10% of this amount alone, if given to the Fire Brigade, would, the committee are convinced, save at least 25% of the gross loss". Despite this the fire insurance offices said No, and the special sub-committee was abandoned in August 1877.

The narrow street that constitutes Hammond Lane is the thoroughfare that runs westwards from Church Street towards Smithfield, opposite the side of the Four Courts. It was for much of the twentieth century associated with the factory of Maguire & Patterson, which stood on the site of one of Dublin's worst disasters. The highly populated area housed the foundry and ironworks of Messrs Strong, millwrights and engineers.

Just before 2 pm on Saturday 27 April 1878 the large twenty-foot boiler exploded, reducing a major part of the building to rubble, destroying some tenement houses on the opposite side of the lane, and causing the licensed premises of Duffy's which adjoined the foundry to collapse. The blast, which for some reason seems not to have been very loud because it was not heard by many of those attending the courts, scattered death and destruction in its immediate vicinity. Many people were buried in the debris of their collapsed homes, while others stumbled out with varying degrees of injury, trauma or bewilderment. The scene, which a few seconds before was one of life, activity and neighbourliness, was turned into a virtual battleground, with everywhere the pleading cries of people seeking immediate help for themselves or those they knew who were now buried or lying amidst the carnage.

The fire brigade, assisted by the local constabulary, was soon on the scene and a call was made for assistance from the military. At once the brigade and some locals set about the grim task of extricating bodies from the rubble or pulling seriously injured and shocked families from what was left of their homes. The boiler which had exploded was lodged in the tenement structure opposite the foundry. Doors, windows, bricks and slates littered the roadway while walls stood at dangerous angles, beds and the remnants of furniture dangled from open floors, and staircases stood without support, ready to collapse if disturbed. In these dangerous locations firemen worked with friends, relatives and neighbours of the trapped and dying until the police threw a cordon around the area.

Explosion at Hammond Lane foundry. Photograph, © National Library of Ireland

They expelled the "helpers" in spite of their protestations of wanting to assist family members or neighbours from the rubble. It was evident early on that many lives were lost, many families were devastated and large numbers were suddenly homeless. Duffy's public house was a shambles, having collapsed like a pack of cards, burying those inside. The firemen were joined by one hundred men from the 91st Highlanders who, with spades, shovels and pickaxes, heedless of their own safety, set about removing debris, which in places was piled up to fifty feet high, to try to extricate the victims. It was known that members of the publican's family as well as some customers were inside the public house when the explosion happened. That proved to be the case when in the pile of rubble which had been a three storey building the bodies of Mr Duffy and two of his daughters were recovered along with a number of seriously injured men.

Further bodies were recovered from the location as the evening progressed and many of the injured were removed to the Richmond Hospital. The borough engineer, Mr Parke Neville, then halted all operations because of the imminent collapse of remaining sections of the foundry and bar. Crowds watched corporation workers pull down the offending structures, while police and fire officers tried to ascertain the exact number of those who had been in the street when the explosion occurred. The sombre task had to be continued until darkness descended, by which time thirteen bodies had been recovered and sixteen injured had been removed to hospital where they were detained suffering from fractures to arms or legs, scalds from steam or serious cuts and abrasions. Many others were discharged after treatment and told to report to other hospitals where they could receive further treatment.

The foundry, the public house and two three-storey tenements were completely destroyed by the force of the explosion, and a further four tenements suffered severe damage. Throughout the following days the corporation carts removed rubble as the rescuers continued their gruesome duty of searching for further victims. Normally up to one hundred men worked in the foundry, but it being Saturday lunch time when the accident took place, only about thirty workers were on the premises.

All through Sunday the fire brigade and the military continued their dangerous but necessary work. Both groups received high praise from the newspapers and the city dignitaries for their relentless efforts. A seamstress was found dead with a scissors clutched in her hand; another had part of a sewing machine frame still attached to her foot. Many died from suffocation caused by the intensity of the blast, and apart from the circumstances of sixteen who were detained in hospital, forty-six people were left homeless and destitute.

The *Freeman's Journal* in its editorial of 29 April maintained that "of late years boiler explosion has followed boiler explosion with alarming and increasing frequency. This is most extraordinary… as a simple precaution of a safety valve and a fusible plug, which melts and allows the water to escape and quench a fire when the heat becomes dangerously great, will render the catastrophe of an explosion almost impossible". No one was listening to this sensible precautionary advice. So the newspaper called on the coroner's court for the amplest inquiry into the facts, circumstances and causes of this tragedy: "We trust the Coroner will do his duty by resisting any of those governmental attempts, now so common, to smother and obstruct the inquest to prevent the fullest evidence being given". Even in those days politicians obviously approached all calamities for which they shared some responsibility as matters to be dealt with on the basis of "maximum damage limitation". They were, of course, vocal in their support of the newspaper's call for the setting up of a charitable fund to assist the injured, bereaved and homeless.

The coroner's court convened on 30 April when identification of fourteen bodies took place. It then adjourned until after the funerals. The lord mayor called an emergency meeting on 2 May to set up a relief fund and issued the following statement.

> *A sad catastrophe has come to our doors, intensified by the suddenness with which the victims were crushed, worse than being immured in mines, for they had not time to make an act of contrition or an exclamation "The Lord have mercy on me".*

The area involved was one of the poorest in the city and there was an immediate response which produced almost £600 at the time of the meeting. A Mr V. D. Dillon, when seconding the motion for the fund however, clearly could not resist delivering one of those openly supercilious and patronising remarks which one still hears from time to time from complacent public figures. His fear was that the affected people would get "too much money, and there would be a danger of it being wasted".

The coroner's court reconvened on 4 May and the *Freeman's Journal* editorial writer seemed to know what would happen. The crown solicitor opening remarks included the statement: "No facts have been reported to the government justifying any charge of criminality against any person". The Board of Trade had sent over an engineer to examine and report on the boiler, and the coroner had invited a Manchester engineer to do likewise. Both engineers concurred on the cause of the tragedy.

The victims who worked in the foundry and those from the tenements were not represented, the members of the Duffy family being the only ones who could afford to have legal representation. Fourteen people died in the explosion, of whom twelve were suffocated and two killed by crushing, and over thirty were injured and received hospital treatment.

The engineer's report stated that the boiler was not properly maintained and was weakened by corrosion. No independent engineer had examined the boiler in the previous two years, that job being left to an employee engineer of the foundry who was still hospitalised and was not called to give evidence. The Board of Trade and the Manchester engineers both agreed that the boiler should not have been in use. There were no statutory regulations under the Factories Act 1875 for the inspection of boilers, although such provision had been demanded from parliament by engineers throughout the United Kingdom. Counsel for Strong's foundry tried to maintain that the explosion was caused by the collapse of Duffy's pub, which two years previously had developed a "bellied out" front wall, thus causing the tragedy. The jury however brought in a verdict that "the explosion was caused by the faulty boiler" and that Mr Strong had tried to keep it in reasonable order. No one was found negligent, criminally or otherwise. The unfortunate residents in the overcrowded tenements, those drinking in the pub and those in the foundry were simply victims of an accident and soon their plight was no longer covered in the newspapers and they passed quietly into history. On 10 May the relief fund was reported to have reached £972. The relief committee was of the opinion that about a further £1,000 was required to give aid to all those injured and to relatives of the deceased – so much for Dillon's rather stupid declaration.

In 1878 Ingram was still concerned at the state of the fire service. He expressed his concern in his annual report to the waterworks committee.

> *I do not consider the department at all as effective as it might be. I have 7 fire escapes and 2 engine stations (both unfit). I would require 2 more stations on the north side, and a repair yard. I consider the system of the telegraph alarm communication very uncompleted.*

He went on to deplore the fact that the Dublin Fire Brigade still had to attend fires outside the city boundary without any financial contribution from the townships or the insurance companies. The brigade surgeon also complained.

> *The Central Station is an unwholesome, foul smelling building, to which may probably be credited the pulmonary delicacy which more than once of late prostrated the*

*energetic and able Chief of the Brigade; whilst the married men around the Winetavern Street station live in filthy, decaying houses, damp in winter and most unsavoury at all times.*

Disturbing comments, but the waterworks committee was unable to attempt any major changes because of the shortage of funds. The cost of the brigade for that year was £3,881 while only £3,650 was transferred from the water rate. The salary of Superintendent Ingram had recently been raised to £400 per annum. Lieutenant Boyle was still paid £100. Wages for other ranks amounted to £1,332-3-11 and uniforms cost £264. Dr Thomas Nedley, the brigade's surgeon, received £52-10s-0d. One positive step forward was the contracting of maintenance and repair of the machinery and appliances to Hutton's, the well-known coach builders.

The end of the year showed an increase in the number of fires with the brigade attending 174 during the months of November, December and January 1879. The situation regarding out of area calls still existed – the Dublin Fire Brigade responded to a fire at St Patrick's College Maynooth, in November. One steamer, one hose jumper and three horses went by special train, and returned to Dublin the following day. The waterworks committee agreed to lay a fire main to Molyneaux Asylum in Leeson Street for use by the brigade even though it still declined to make an order for automatic response to townships beyond the city. It is interesting to note that the brigade's response to Maynooth received fulsome praise and no cash.

At this time the Dublin Metropolitan Police objected to the fact that large quantities of petrol were being stored in premises close to their barracks in Linenhall Street. They wrote to the corporation demanding that the Petroleum Act be enforced. In response to this Ingram inspected the premises and discovered that up to 325 barrels were stored about forty yards from the barracks, and another 400 barrels were kept by the same company in a building in Henrietta Lane. The corporation took immediate action to have most of this fuel removed to other areas of the city. Exploding boilers and suchlike were obviously making even the police and military aware of dangers, especially to their own premises and personnel.

An indication of the rate at which poverty was growing was revealed in January 1880. The South Dublin Union reported that it had 524 more inmates than in the previous January. The institution was not exactly popular with the poor in the city; the wretchedness in the rural areas was having the effect of forcing families to abandon their homes there, travel to the city and seek shelter in the workhouse. Further large fires and other human disasters affecting the inhabitants of the city continued to occur throughout 1880 and the first of them struck at the time when the report on the numbers in the workhouses was being published.

Jacob's factory stood large on the corner of Bishop Street and Peter's Row. The biscuit factory employed over 200 workers in its production area, stores, offices and yard. A fire was first noticed sometime after 9 pm on the dark evening of 29 January. It spread rapidly into the bakery, boiler house and stores. Ingram, who happened to be in the vicinity, arrived in advance of his fire crews. Firemen from William Street and Winetavern Street attended and immediately set to work laying out hose, forcing the front doors and getting water to the flames. They were hampered in gaining access other than through the front of the building because the factory was surrounded by tenement houses. The unfortunate families in many of these houses, terrified at the extent of the blaze and the speed with which it was spreading, began removing their meagre belongings to the streets outside, causing major problems for the firemen, their engines and ladders.

By 11 pm the entire factory complex, with the exception of the offices, was fully ablaze and sparks and burning debris, fanned by the high winds, were causing smaller fires in the tenements. The brigade had laid seven lines of hose into the building in an attempt to surround the conflagration.

For the first hour or so all the exertions of the firemen seemed to produce little progress as flames shot out of the warehouse roof illuminating the murky winter sky. The brigade however had in fact stopped the spread of fire to the dangerous boiler house and had formed a water curtain between most of the office area and the rest of the building. It was clear by 1.30 am that the seemingly uneven struggle had turned in favour of the firefighters, and although fires continued to burn in a number of sections of the premises they were being surrounded by water from the hoses. It came as a happy relief to the many frightened people in the neighbourhood that their homes would be saved and they would not be forced onto the parish. The possible explosion of the boilers, because they were kept steamed-up all night, had been a great cause of panic, but as the flames abated the confidence of the onlookers grew. Soon the extent of the damage became obvious; the warehouse was destroyed as was the manufacturing area, but the main office section was saved. Most of the stock and machinery was damaged by either fire, water or falling masonry, but there was unanimous agreement that the fire chief and his small complement of fifteen firefighters had once again tamed the fire devil.

Damage to the £10,000 building was extensive and although there were no fatalities the livelihoods of a large number of families were suddenly at risk in a city already in the throes of depression. As dawn broke, the gaunt walls and the smouldering debris inside gave testimony to the inferno that the fire crew had faced throughout the winter night. It would not be long before they faced even more formidable odds in the densely populated and growing city.

The Dublin Society (later Royal Dublin Society) was founded in 1731 and occupied various sites in the city. In 1821 it vacated its premises in Hawkins Street and moved to Leinster House in Kildare Street. The Hawkins Street building was totally remodelled by a leading English architect and it became the Theatre Royal. This magnificent theatre was the pride of actors, comedians, singers and musicians for the next sixty years. When it opened in late 1821, with its stage sixty feet wide by fifty feet deep, it was immediately classed with Drury Lane Theatre in London for its lavish scale and the comfort it provided for patrons. In 1823 it was fully lighted by gas, and over the years it listed most of the great actors and variety artistes of the day on its programmes. It was purchased and completely refurbished by the impresario Michael Gunn in 1874.

On Saturday 7 February 1880 a special performance of *Ali Baba and the Forty Thieves* was to take place at 2 pm in the presence of the Lord Lieutenant, the Duke of Marlborough, and his wife. Lady Spencer Churchill, the Marchioness of Blandford, Lord Crofton, Lord Wallacort and many other notable dignitaries and military men also attended. The show included a song which praised the Duchess of Marlborough for her charitable work on behalf of the famine victims on the west coast of Ireland. Death, starvation and disease were rife in the country at that time so much so that the Mansion House committee had met on that day to discuss the question of aid for those suffering from famine in the western counties.

The more modest Dublin Grocers Assistance Association was also holding a special meeting to agree "to take up a collection to help alleviate distress at present existing in the city". The plight of many in the country had been mentioned in the Queen's speech on the previous day, and this pressing matter was to be the subject of a full debate in the House of Commons on the following Monday. It was in this broad context that the Theatre Royal management was putting on a charity performance of the pantomime on that Saturday.

About an hour before the matinee was due to begin a catastrophe occurred. Most of the staff were still at lunch, and the crowds due to attend the performance had not yet arrived, when a serious fire broke out behind the richly upholstered Vice-Regal Box and quickly spread throughout the interior of the theatre. The fire brigade turned out with one steamer and three hose carriages, as well as the fire escape from Foster Place. It was later joined by some military fire engines. In all twenty-seven firemen fought the fire with nine jets from Hawkins Street, Burgh Quay, Townsend Street and Poolbeg Street.

The cause of the fire was not discovered but it was believed to have been started either by a young lad who was lighting the gas jets behind the Vice-Regal Box with the aid of a spirit lamp or that some jets had not been turned off and the use of a naked flame in the area caused the instant ignition of escaping gas.

With the brigade's arrival it was reported that the manager Mr Egerton, some cleaners, a barmaid and two boys – who were due to take up their positions inside "the donkey" which was to take part in the panto – were trapped inside. The firemen had problems gaining access to the burning building because all doors other than the main entrance were locked. Axes kept in the building were taken by the firemen and used to break down the doors, while rescues took place from upper windows by means of the escape ladder. One barmaid was lowered from a second floor window into Poolbeg Street by means of a rope.

The newspapers reported that "In far less time than it takes to tell the tale, the fire had got such a grip of the building that it was beyond human power to save it". Yet, as in all such instances, the fire brigade tried to do just that. On the arrival of the first fire engines the drapery and curtains were fully alight, and the seating and scenery was in flames "dancing like a harlequin through the great Thespian edifice, rushing out of the windows and eventually through the roof, as that great mass collapsed in portions with loud thuds". Within twenty minutes the roof was a flaming mass, groaning above its supports, until in one final roar it completely gave way, releasing the flames into the sky and showering the surrounding buildings with sparks. As the firemen battled away the crowds gathered, to be soon subjected to pandemonium when a large portion of the eighty foot high external wall suddenly collapsed into Poolbeg Street injuring several men. It was fortunate that the water supply was excellent and also that the theatre was detached on all sides, which meant less risk of the fire spreading.

Some two hours after the start of the fire heavy rain began to fall and that certainly helped in the struggle to extinguish the flames. Almost 400 police were drafted into the area but, despite this and the continuous rain, a nasty situation arose as evening fell. Crowds had gathered in Great Brunswick Street, between Trinity College and the entrance to Hawkins Street, and were pelted with squibs and stones by some college boys who had clambered onto the roof of the college. A riotous atmosphere developed as the incensed people retaliated by throwing back stones, slates and other projectiles into the college.

James Ingram, the chief superintendent of the fire brigade, was personally complimented for the way he had acted in rescuing a man from the second floor of the theatre with the help of another fireman, and all the fire crews were reported as having worked beyond what might reasonably be expected of them during the protracted struggle to contain the fire. By midnight, some ten hours after their arrival, most of the firemen were being stood-down with their engines and equipment and returned to stations, but three lines of hose were left to continue damping down pockets of fire still visible in the wet, smoke-filled gloom.

The great Theatre Royal, which had cost £55,000 to build, was a shell of fallen masonry and dangerous high walls – a part of Dublin's history gone for those visiting the site on the following morning. (See colour section, illustration numbers 2 & 3.) The great centre of entertainment, pageant, pomp and spectacle had vanished. It had had its notoriety as well as its fame: it was here that a bottle of vitriol was thrown into the Vice-Regal Box when it was occupied by the then Lord Lieutenant, the Marquis of Wellesly. The city authorities were saddened to see the destruction but relieved that the fire had not occurred during a performance when panic would surely have resulted in numerous deaths and serious injuries from fire, smoke, stampede and falling masonry. As it was, the theatre manager tragically died in the fire and four others were hospitalised with serious injuries. Lieutenant Boyle and two other firemen were also taken to hospital with injuries sustained while tackling the blaze. The brigade was still in attendance a week after the fire. Much debris had to be removed and, although Ingram approved the use of explosives to remove the dangerous seventy-foot wall in Poolbeg Street, he was overruled by the city engineer Parke Neville who insisted that the structure be removed by other means.

The destruction of the Theatre Royal was given great prominence in all the newspapers of the day, as was the dramatic week-long search for the remains of Mr Egerton, the theatre manager. Strange to say that the papers gave a great deal less coverage to another tragedy which happened on the same evening as the fire when five watermen from Ringsend were drowned in the Liffey estuary.

The Dublin Fire Brigade was held in such high esteem following the Theatre Royal fire that the Patriotic Assurance Company of Ireland, at its annual general meeting on 27 February, proposed a public testimonial for James Ingram, calling on all property owners and on other insurance companies to support their proposal. There was irony in the fact that those who had refused to help finance a brigade that would match the needs of the city and its people were now proposing a formal display of public recognition for its chief fire officer. An inexpensive way perhaps to overcome the contempt the insurance companies were held in by the same chief when they continually rejected his pleas for their assistance in maintaining and improving the city's fire service.

By 1880 the annual cost of the fire brigade had risen to £4,500. This prompted the city council to request the members of parliament from the city of Dublin constituencies to seek a grant of £1,000 from the government towards the expenses of the brigade. There was no response. Ingram's health was beginning to deteriorate from age and the constant battles with both fires and politicians though he still continued to press for much-needed improvements. A new escape was placed on the north side at Bolton Street in 1880. This brought the number of street escape stations to eight. As far back as his annual report for 1876 Ingram had recommended the introduction of a street fire alarm system and the building of a more central headquarters station, but progress on both these matters was sluggish, particularly the installation of fire alarms.

It was a devastating time for the poor of Dublin. A report by the King's and Queen's College of Physicians and the Council of the Royal College of Surgeons contained the following sobering details regarding the death rate in the city.

> *The average death rate in Dublin city for the past ten years was the highest, with three exceptions, among 17 large towns or town districts in Great Britain.*
>
> *Whereas the death rate of all such English towns, with one exception, has decreased within the same decennial period, the death rate in Dublin city has largely increased.*
>
> *The City of Dublin is, and has been, unhealthy in an excessive degree and its extraordinary death rate is attributable to long continued disregard of sanitation, and omissions to enforce the legal means provided for a remedy.*

Suptintendent Ingram and his firemen were part of these same statistics, and their daily work brought them into contact with the victims, because the men provided their life-saving assistance in all types of emergencies. The city medical officer, agreed with the report of his peers and went further:

> *the death rate is due to unsanitary arrangement, overcrowding of tenement houses, the defective structure of new dwellings, dissemination of poisonous vapours by chemical and other manufacturers. Also, the insufficient provision for conveyance of infected persons to hospitals.*

A growing problem for the fire brigade at this time was the storing of hay or the placing of hay ricks in locations where once they caught fire they caused damage to buildings and even loss of life. The corporation frequently had to go to court to get hay removed when it was reported as a serious fire hazard. In a typical case in 1881 the corporation proceeded against a merchant on Montpelier Hill "for having straw ricks erected to the serious danger of that neighbourhood". The owner was fined £1 plus ten shillings costs.

A subject first mooted almost twenty years earlier cropped up again in December 1881. A letter was received from the chief commissioner of the Dublin Metropolitan Police regarding the telegraph system.

> *I am to establish a system of telegraphic inter-communication between several police stations and the central station at Chancery Lane. It occurs to me that advantage might be taken of this opportunity to place Dublin Fire Depot, Winetavern Street, in connection by a single wire with that station. By this means the Dublin Fire Brigade could receive almost instantaneous notice of fires from the police.*

At last, at the end of his career, Ingram might see his long-cherished hope realised, that of a direct link between the fire service and the police stations. He was not however destined to live to see either a new headquarters station or the communication link. James Robert Ingram died on 15 May 1882 following a short illness. Referring to him following his death the lord mayor recalled

> *the zealous, active and intrepid Chief … this valuable public servant and widely lamented gentleman.*

Ingram had served the people of the city well by leading the brigade through its difficult formative years. Much was still to be done, but solid foundations were laid; respect from the public had been earned, professionalism had been established. The fire brigade could now proceed to greater heights with the right leadership and with support from the corporation.

# 6  Growth, death and reports

On 26 May 1882 Dublin Corporation adopted the waterworks committee report declaring the position of chief superintendent of Dublin Fire Brigade vacant. The post was to be advertised at an annual salary of £300 plus accommodation, and the essential qualifications included "experience and ability in connection with the organisation, management, and drill of a Fire Brigade". James Ingram had been paid £200 per annum on appointment. This had risen to £250 in 1869, £300 in 1871, £400 in December 1876, and £450 in November 1881. After much discussion it was agreed that the post should be filled at the October council meeting.

On 2 October the council considered the four candidates who had applied, two Irish and two English: John Boyle, lieutenant in the Dublin Fire Brigade since 1868; P. J. Graham, in charge of the recently established Pembroke Fire Brigade at Ballsbridge; H. E. Davis from Margate and W. Wheeler from Ipswich. Boyle received nineteen votes, Graham eight votes, and the other two nil. Boyle was therefore declared duly elected as chief of the Dublin Fire Brigade. (See colour section, illustration number 4.)

The post of lieutenant (established in 1868) was, in effect, assistant to the chief. Because this post was now vacant, James Byrne, chief inspector of the fire brigade, wrote to the corporation seeking to fill it. Thecorporation law agent advised that the position must be advertised. This was agreed, a salary of £100 per annum offered, and the duties specified. The list of duties included the following.

- *To be in the station at 6.00 am to see the Escape Conductors return from duty, see that they are sober, and take any reports that they may have.*

- *To see that all horses are exercised daily, all engines washed, cleansed and packed with hose or other equipment.*

- *To gather all particulars at fires, and make out assessment papers against the owners for expenses incurred.*

At the January 1883 meeting James Byrne contested the post against a fireman from Cork Fire Brigade and was duly elected by thirty-one votes to one. The brigade now had its two senior officers in place, officers who would certainly be needed to guide the brigade through difficult times.

In July 1882 the corporation adopted a report to take a 200 year lease on two plots of ground, formerly the Clarendon Market, at a rent of £50 per annum, for the purpose of erecting a new central fire station. A waterworks committee report in that year noted that No. 1 Station "has become practically useless as a fire station during the day time, as all heavy traffic to the south of O'Connell Bridge passes through William Street and completely blocks it". Traffic jams in the city – more than a hundred years ago! The committee report continued in the same vein.

> *William Street station has been considered unfit and unsuitable as a place of residence for the Chief and for the men of the Brigade for a considerable time past, but no site was procurable, or anything approaching the means at the disposal of the committee.*

In September the city architect submitted proposals and estimates for "a plain substantial building, plus two extra dwelling houses and a large range of Chief's apartments" at a cost of £2,000 to be completed on the site as a new fire head-quarters. In November 1883 the corporation agreed to deal with Connolly Builders for the construction of the station at Clarendon Row. But in January 1884 an attempt was made by some councillors to revoke the original decision. They lost the vote however and the work commenced. The new station was completed in January 1885 and insured for £6,000.

It soon became evident unfortunately that the new buildings were not of a quality that would last. While the telegraph instruments were being installed it was discovered that no provision had been made for drying hose, so in June 1885, at a cost of £16-10-0, a mast was erected in the yard to facilitate this necessary function. The year passed by with the firemen still in their South William Street headquarters, and the courtyard of the new station (not very far distant) was still unpaved. The next flaw to be discovered was that the engine room was not deep enough to admit the horses under shelter while attached to the fire engines. Tenders were therefore invited to "extend a veranda or shelter into the courtyard to protect the horses from rain or inclemency of the weather". This was not the only remaining problem. Before the brigade moved into the premises discol-oration began to appear on the walls, and there were problems with smells and blockages of the sewers.

The station remained vacant. In February 1886 an English architect carried out a study of the building to ascertain if it had been constructed in accordance with the specifications agreed by the committee. The survey highlighted many faults in construction, pointing out that the internal walls and partitions, which ought to have been brickwork, were in fact lath and plaster, "infringing the Corporation's own bye-laws, endangering the stability of the building and increasing risks in case of fire". The survey went on to claim that "the stairs were warped, insecurely upright", and that "the sewers were wrongly laid". This report of the survey was rejected by the city architect and the builders, but it led to a long drawn-out, acrimonious row. The premises were still empty in November 1886 when the waterworks committee requested the Royal Institute of Architects of Ireland to nominate three independent architects to survey the buildings and make a report on their findings.

It was not until the meeting in July 1887 that the whole sordid episode reached a determination. The final report found that

> *... having regard to the limited area of the site on which these buildings are erected, we are of opinion that they are well designed ... some deviations from the plans have been*

*made, but we do not see how this could be avoided … the building has been carried out in a workmanlike manner, and with good materials throughout.*

Thus ended the saga of the new headquarters and the firemen finally moved from South William Street to Clarendon Row, later to be known as Chatham Row.

During the lengthy interlude from the completion to the occupation of the new headquarters many interesting, important and sometimes tragic events occurred within the Brigade. Both fire stations, William Street and Winetavern Street, already had telegraph connection with the General Post Office, the Metropolitan Police in Castle Yard, and the City Hall. In 1883 communications were further improved by the installation of telephones in both fire stations and at the eight fire escape street stations which were now connected to the William Street headquarters.

In the years since the formation of the fire brigade the workload on the person in charge had increased considerably. Under the 1871 Petroleum Act and the 1875 Factories Act the inspections to be carried out by the brigade placed an extra burden on the chief. To this was added the inevitable increase in the number of serious fires in the city, and the deterioration in the living quarters of all brigade personnel. The Petroleum Act of 1879 meant even more work for the chief, because all petrol giving off fumes at 73°F was now included in the legislation as opposed to the 100°F level covered under the previous Act. John Boyle, now faced with all this additional pressure, was to have the added responsibility during his period in command of coping with the first fatalities within the brigade. The job of firefighting always entails danger, but few who serve expect to lose their lives in the fashion that befell John Kite.

Number 10 Trinity Street was a four-storey building erected without retaining walls, the floors being of light timber supported on wooden storey posts, not strongly seamed to the walls of adjacent properties. It was considered a dangerous building, although it had once housed the publishing offices of the *Dublin Evening Post*. It was now owned by a firm of stockbrokers who also owned an adjoining premises. A part of the firm's lease conditions was that it carry out major work to strengthen the roof supports and stabilise the interior. The house had no chimney but from the basement kitchen a connection had been cut through to the chimney next door. The building was let to a jeweller-watchmaker and was occupied by his brother, a porter and a youth. Housekeeping practice in the premises was a disgrace, the kitchen had packing cases, boxes and straw stored in large quantities; straw was also stored under the basement stairs, and oil for use in a heating stove was stored in the basement. In addition, a painter used the premises to store paint and overalls while carrying out decorating work.

The fire brigade received a call to this building just before 9.30 pm on 20 March 1884. Being so close to William Street fire station the crews were on the scene within a few minutes. On arrival they discovered the building on fire in the basement, at ground level and on the third floor. The building was empty at the time – the youth was seen to leave some twenty minutes before the flames were noticed by a passer-by. The brigade forced the front door, laying a line of hose into the hallway. The stairs to the basement and first floor were burning, so an order was given for a fire escape to be pitched to upstairs windows and that a further entry be made from the footpath into the basement. Two men entered the basement with a line of hose and immediately encountered a fierce blaze. Two men with another line entered the third floor level from the escape and discovered fire in the roof.

Within half an hour much of the burning had been extinguished or reduced to such an extent that Inspector Kavanagh was able to climb the stairs and join those tackling the fire in the roof void. Suddenly, without the slightest warning, the house collapsed and burried nine firemen in masonry, timber, slates and plaster. Luckily police and onlookers were at a sufficient distance not to be buried in the rubble which filled much of Trinity Street and portion of St Andrew's Street. The remaining eight firemen with help from the police, in spite of the danger, set about removing rubble in search of the trapped firemen. A contingent of soldiers was called in from the nearby Ship Street barracks to lend further assistance in the wretched task of extricating the unfortunate firemen. Doctors were summoned from Mercers Hospital as the first of the injured were removed. Darkness was now adding to the problems of the rescuers as, with the help of lanterns, they braved the dirt, smoke and danger to save those who were heard calling for help from the smoking pile of rubble. It took more than two hours before all were removed and rushed to hospital. The full extent of the tragedy became apparent when the lifeless body of fireman John Kite was lifted from the ruins of the house.

The firemen who survived had suffered in various ways: Inspector Kavanagh had a fractured skull and broken ribs, and was never to work again; Richard Cullen had serious head and internal injuries, and only returned to work a month later – but not as a full-time operational fireman; James Gildea suffered head injuries and a fractured leg; John Hines sustained head and pelvis injuries; Edward Forde broke his right thigh. Fortunately firemen James McEvoy and James Ryan, the two who had originally entered the basement and the last to be rescued, were discharged from hospital following treatment.

There was a public outcry over the incident and a heartfelt expression of sympathy from all sections of the community for the unfortunate firemen. Much concern was expressed at the serious injuries sustained by the courageous firemen, but the business community was quick to identify a primary consequence of the tragedy – that the brigade which had consisted of a total of twenty-eight men prior to the accident was now reduced to only twenty active bodies to cover the whole city in the event of fires. The brigade's annual report published at the beginning of the following year confirmed that as a result of the Trinity Street fire two officers and six firemen were incapacitated for periods ranging from ten to 200 days.

The members of the waterworks committee were only too well aware of the implications of the shortage of manpower: "having regard to the financial position of the Fire Brigade Department, resolved to take on additional men to the number of six, according as eligible men could be found".

The coroner's court called many witnesses, including those who owned the premises and those who resided there when the fire occurred, but no conclusions as to what caused the fire were reached. The general belief was that the outbreak was in the kitchen because packing cases with straw had been left too close to the kitchen fire. What became apparent was that the building's contents had been insured the previous year for £1,500. John Boyle, the brigade chief, when addressing the court had this to say: "My great difficulty is to keep the men out of danger, for the moment they get my back turned they are into danger in their anxiety to do service".

John Kite, a married man with six children, had been the recipient of three bravery awards during his brigade service. He died unnecessarily in carrying out his duties to the full, in a building that was empty and in a particularly dangerous condition even before fire made it a deathtrap for those venturing inside. Death was declared as due to "suffocation in the ruins of a house while carrying out his duties". The jury added the following to their findings: "We desire to express our hope that the Corporation may be able to increase the pay of the Fire Brigade who, in our opinion, are very much underpaid". A fund was launched for Kite's widow and children, who were without pension rights. That fund was then extended to help those who were seriously injured. One of the first organisations in the city to come to the aid of the firemen was the Irish Champion Boxing and Gymnastics Association. It held a benefit in the Round Room of the Mansion House on Easter Monday, all proceeds of which went to the family of the dead fireman and to his seriously injured colleagues.

In the meantime Dublin Corporation wrote to other major fire brigades in Britain seeking information on how they compensated their staff members in the event of death or serious injury while carrying out their fire duties. Answers were received from Liverpool, Manchester, Leeds, Glasgow and London, and the corporation decided to adopt the Manchester system. Full wages were to be paid to the injured men until further orders, and "instead of paying a weekly compensation to the representatives of John Kite, deceased fireman, we recommend that a sum in gross be raised". Kite was forty-four years of age with over six years service in the brigade at the time of his death, and his pay and emoluments at that time were calculated as thirty shillings per week. The corporation agreed to calculate the compensation payment at seven shillings and sixpence per week (quarter salary), the amount he would have received if he were retiring with that service. The actuaries using that figure authorised a payment of £270-12-0 to his widow as a lump sum final settlement, less the wages per week received by her since her husband's death. It was the price of a fireman's life in Dublin in 1884 – the first member of Dublin Fire Brigade to be killed on duty.

In March 1884 the Rathmines commissioners decided that the time had come to expand their fire protection arrangements. Rathmines and Rathgar township was a large and growing area which up to now had to rely on the city fire brigade for anything other than a small outbreak of fire. It appears that in the preceding nine months the Rathmines men had answered only three calls. Nevertheless, new premises were built providing accommodation for Chief Superintendent

Searles, Lieutenant Smith, five firemen and their families. Also built were stabling for horses and an engine house for the new fire engine. This was a steamer supplied by Shand Mason. A newspaper report described the engine's boiler as "of novel and handsome construction and, covered as it is with brass dome and funnel, looks bright and pretty". The township also had a hose cart complete with 500ft of hose, and an ambulance.

In 1880 the number of street fire escapes had been increased by one, on the north side of the city at Bolton Street. During 1886 a further increase brought the total number to nine. The extra fire escape, on the south side, was located at the new central fire station at Clarendon Row, even though this fire station still remained to be occupied.

At the same time a new escape capable of reaching a height of 70ft was stationed at Nelson's Pillar. This escape station covered the city centre, including the new hospital at Jervis Street. The escape was purchased from Jessop Brown of Great Brunswick Street, to whom the rights of the Clayton patent had passed.

In July 1887 the new fire brigade headquarters at Clarendon Row, having taken five years to become an operational reality, was finally occupied. It now housed one steam fire engine, one hose carriage, one hose jumper, one escape and one four-wheel tool cart. Three horses were stationed there. The late Chief Ingram's vision of progress for the brigade was now taking place as telephonic communication was established with all police stations and by the end of the year the tenth fire escape station had been set up at Wood Quay. The telegraph which in 1863 was considered extremely progressive was being rapidly superseded by the telephone. The strength of the brigade was increasing too, as demonstrated in following table.

|  | 1883 | 1886 | 1887 |
| --- | --- | --- | --- |
| Chief | 1 | 1 | 1 |
| Officers | 3 | 3 | 3 |
| 1st class firemen | 13 | 18 | 19 |
| 2nd class firemen | 3 | 8 | 2 |
| 3rd class firemen | 5 | 1 | 3 |
| 4th class firemen | 8 | 6 | 11 |
| TOTALS | 33 | 37 | 39 |

The class system for firemen related to the number of years service in the brigade: 1st class was attained following a minimum of three years service; 4th class applied during probation and for the first year of service. In January 1888 the death of Inspector Hines, one of the three brigade officers, led to the promotion of Christopher Doherty to that rank. In July of that year the township of Drumcondra and Glasnevin applied to have a number of their employees trained as firemen by the Dublin Fire Brigade. The waterworks committee approved of this application and a number of township staff were trained. Meanwhile the firm of Merryweather had introduced a hand-woven hose of Irish flax; the waterworks committee directed "that all future supplies be of this home manufactured material".

The fire service costs had now risen to almost £4,000 per annum. The city was still plagued with the problem of loss of life and serious injury being caused by the collapse of buildings. Following the tragic death of a child in one such incident in October 1888, the city engineer Spencer Harty produced a detailed report. He stated that according to the 1847 Towns Improvement Act the corporation had no authority whatever to take any proceedings with regard to ruinous or dangerous buildings, unless they were dangerous to passers-by or to the occupiers of neighbouring buildings. The corporation did however have power under the 1864 Dublin Improvement Amendment Act "to shore-up any building to prevent it falling into the street, and power to remove occupiers of dangerous premises". It did not have power to recoup any costs.

The corporation inspectors, under the Public Health Act, inspecting many of the city's tenements found – between October 1886 and October 1888–435 houses with various structural defects and in need of immediate work and 270 houses that were liable to collapse. They now sought to have amending legislation to recover their costs and extend their powers to take action against the offending owners. Buildings in these conditions were a nightmare for the firemen as they tackled even the smallest of fires; their Trinity Street experience was ever present in their minds.

The premises of 30 Westmoreland Street, next door to the *Irish Times*, were occupied by an apothecary and druggist on the first two floors, Lafayette – the well-known photographic artists – on the next two floors, and living accommodation on the top floor. At about 2 am on the morning of 20 May 1891 fire broke out on the third floor. Five people were in the building when the fire was discovered by one of two woman on the top floor who got the smell of smoke. As she proceeded to alert the other woman, two of the male occupants forced the door of the photographic room on the third floor and were confronted by billowing acrid smoke with flames visible inside the room. They left to raise the alarm. Soon some policemen arrived but were unable to get beyond the second floor level because of heat and smoke. Meantime the fire escape from Foster Place and an engine from Clarendon Row had arrived outside the premises where women's screams could be heard coming from the top floor. The sixty-nine foot escape was set on the roadway and fully extended to reach the top floor window.

Fireman Patrick Barry, the conductor, climbed the ladder and carried a woman, Miss McSweeney, to the ground where he notified Inspector Christopher Doherty that there were others trapped in the building. Fireman Peter Burke then climbed the escape and endeavoured to help a large woman out of the top window. He was soon joined by the inspector and they climbed slowly downwards through the billowing smoke. Some thirty-five feet from the ground they met two firemen with a charged hose ascending, and another fireman could be seen below them. Suddenly, without the least warning, the ladder snapped, hurtling the six on to the pavement below and bringing down signs on the outside of the building.

The sickening crash was witnessed by onlookers, police and firemen colleagues who at this stage included John Boyle the brigade chief. The injured were quickly removed to hospital and the firemen concentrated on tackling the

**Fatal fire in Westmoreland Street, Dublin, May 1891. Print, © Dermot Dowling**

inferno. This was a hazardous task because there was no street lighting and the blazing premises housed large stocks of chemicals, fuel and highly combustible materials. Nevertheless, the firemen succeeded in getting up the stairs and they also managed to direct a hose from the Nelson's Pillar escape which was now in position. In spite of the trauma they had witnessed, and the danger they continued to face, they battled on and within three hours they had the fire under control.

The brigade paid an extremely high price for its endeavours on that catastrophic night. Inspector Doherty, a man in his forties with over twenty-five years of meritorious service, the holder of three chevrons for rescues, died from a fractured skull; the young fireman Peter Burke, with just three months' service, also died in the fall. Three other members of the brigade were injured, one seriously. Miss Keys, the rescued woman, suffered shock and abrasions but no serious injuries as a result of her fall. All the occupants of the house were removed to hospital, but none had any real injury other than the trauma of what had occurred. History had repeated itself, because some twenty-five years earlier on the opposite side of the same street during a dramatic rescue attempt two fire escapes had been damaged and six deaths ensued.

The public asked many questions. The newspapers demanded answers. Disquiet was openly expressed regarding the safety of fire escapes. "These ladders were built for rescue. What if many were trapped, their means of escape totally cut off, could they trust these ladders if a human chain of victims were trying to escape?". Chief Boyle issued a detailed report, but more information was demanded. The coroner's hearing was eagerly awaited. It took place on 24 May attended by legal representatives for the corporation and fireman Burke. There was no representation for the late Inspector Doherty, nor for

Superintendent Boyle. Fireman Patrick Barry, who was to receive a chevron for his rescue of Miss McSweeney, stated that Inspector Doherty instructed that the escape be pitched on the roadway, the pathway being unsafe for full extension due to its width and granite blocks.

Boyle concurred with Inspector Doherty's decision regarding the setting of the escape, although it was refuted by Clayton the manufacturer and by some of the firemen. The chief agreed that drills or tests on the escape were carried out with usually two and certainly not more than four men on the ladder when extended to about forty feet and not more than ten feet from the base of the drill tower. At the time of the fire this particular telescopic escape was sixteen years old. The brigade carried out monthly tests on all fire escapes but this one had missed its last check because Inspector Byrne was off-duty owing to a serious injury sustained at a recent fire. Chief Boyle denied any negligence on his part through failure to prevent such a number of firemen from climbing the ladder at one time, or through failure to see that the angle of elevation was not acute enough to render rescue in safety. Clayton agreed that the ladder should have been pitched closer to the building, but he also pointed out that the modern tele-scopic design bore no relation to the escapes which were damaged at the time of the unfortunate death of the Delaney family in 1866.

The coroner, in summing up, told the jury that Chief Boyle was not present when the escape was pitched, the street was dark, the ladder was enveloped in dense smoke, and in the confusion "the men acting on generous impulse in the true traditions of the fire service took insufficient care for their own safety. The lives were lost and the injuries were sustained by the firemen in the conscientious and gallant discharge of their duty". The jury members were informed that it would be for them to consider whether they should attach blame to anyone, or whether they would wish to add an expression of admiration for the gallant way in which these poor fellows acted.

The verdict read out in court as follows.

*We find that Inspector Doherty and Fireman Burke came to their deaths through the collapse of the fire escape through too much weight on a particular part of the machine, and we hold that the discipline was not satisfactory. We recommend the wife and seven children of Doherty, as well as the mother of Burke, to the consideration of the Corporation.*

The funerals took place on 23 May. Firemen marched behind the fire engines carry-ing the flag-draped coffins. A large crowd including many civic dignitaries attended, but the Dublin Metropolitan Police Band, announced as attending, failed to turn up claiming "another engagement". The Council of United Labourers of Ireland Trade Union in its statement of condolence mentioned that fireman Burke and the fireman son of Inspector Doherty were members of that organisation.

It was a sad chapter in the history of the fire brigade – a tragedy which should not have happened. It reflected poorly on the officership of John Boyle, a man then in his seventy-second year, who should have been allowed to retire long before then.

The gravity of the brigade's situation was now so apparent that it could no longer be ignored. Unusually swift and appropriate action was taken by the corporation to identify and tackle the many problems. As a direct result of public pressure following the Westmoreland Street fire the waterworks committee directed the chairman, John O'Meara, and the city engineer, Spencer Harty, to visit cities in England and Scotland to study the organisation of their fire brigades. This was sanctioned by the full council, and arrangements were made to visit London, Birmingham, Manchester, Liverpool, Edinburgh and Glasgow. These two men faced the fundamental problems of the Dublin Fire Brigade with remarkable insight and their report, adopted by the corporation at the September meeting, was a most valuable document. It drew careful comparisons with the cities which had been visited and accurately summed up the defects and inadequacies in the Dublin brigade.

*Our brigade has had a laudable reputation and the frequent exhibitions of courage by its members have called forth renewed expressions of approval from the citizens. When we come, however, to examine the conditions under which these men live, the manner in which they work, and the appliances at their command, we can ascertain if their training, discipline and management is perfect, and their equipment satisfactory. The population is 279,896 and the number of houses 49,184. The protection of this large population is now entrusted to 47 men, and although the protection of the city is their first concern, they are accustomed in cases of emergency, on receiving sanction of the Lord Mayor, to extinguish fires in the surrounding townships, and have travelled for that purpose as far as Bray with successful results. The force comprises: 1 Chief; 3 Officers; 2 Engineers; 3 Drivers; 38 Firemen.*

*The Engineers maintain the engines and equipment. They are tradesmen who, in addition to their lodgings and clothing, receive 32 shillings per week. They take charge of the station with a fireman whilst the remainder of the force is away at work on a fire, and if another alarm is received the fireman is sent to the fireground to inform the officer. The Drivers and firemen are paid 21 shillings per week for their first year of service, 23 for the second, 24 for the third, and a maximum of 25 after three years satisfactory service.*

*Chevrons given for saving of life carry one shilling per week, and firemen can have up to a maximum of three chevrons for separate acts of bravery and life saving. The Brigade has 8 horses, 5 in the Central Station and 3 at Winetavern Street. The firemen are instructed in fire extinguishing, life saving and other drills, but in no way up to the complete system of drill practised in several important English towns.*

The central fire station is described as situated at Clarendon Row but "the approaches to it are narrow and drivers must exercise caution, proceeding at a comparatively slow rate until they reach open thoroughfares". Then follows a stinging comment on the competency of the station's architect: "It appears strange that the architect who selected the site and designed the buildings did not provide the essentials absolutely necessary for a Chief Station!". The report went on to criticise the absence of a proper drill area and dwelt at some length on other inadequacies.

1. Fire officer with the Royal Exchange Assurance Company, 1832.
Print, Johnson Collection, © National Library of Ireland

2. Theatre Royal fire. Coloured lithograph, " National Library of Ireland

3. Detail of above, Captain Robert Ingram (right) talking to civic officials.
© National Library of Ireland

4. Portrait of Captain Boyle. Watercolour, collection Old Dublin Society

5. "The Fireman's Polka" composed by Charles S. Macdona. Dedicated to Captain Boyle, Dublin Fire Brigade. Cover shows horse-drawn fire engine approaching O'Connell Bridge. Coloured lithograph, collection Old Dublin Society

6. Shand Mason steamer and crew, with Chief John Boyle on the right. Photograph, Collection Trevor Whitehead, © National Library of Ireland

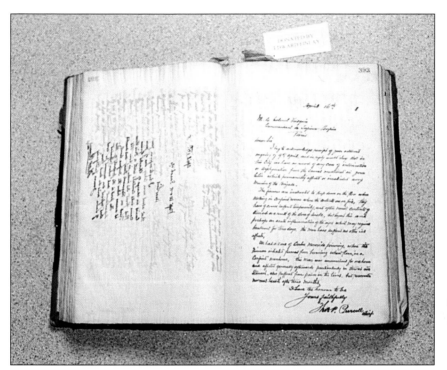

7. Captain Purcell's letter-book, mid 19th century.
Collection Dublin Fire Brigade Historical Society

8. The first motor fire engine, made by Leyland to design of CFO Purcell, 1909.
Coloured print, collection Trevor Whitehead

9. "Station Officer O'Brien" by James Conway.
Painting, collection Dublin Fire Brigade Training Centre, Photograph Gerry Allwell

10 "Burning of the Custom House, 1921", artist unknown.
Painting, collection Dublin Fire Brigade Training Centre, Photograph Gerry Allwell

11. Tara Street Fire Station, late 1920s. Back row: N. Bohan, J. Byrne, J. Gibney, J. Leetch, T. Smart, P. Cobbe, M. Murphy, L. Kiernan, T. Coyne, T. Walsh. Second row: P. Kelly, N. Seaver, P. O'Grady. N. Doyle, P. Bruton, T. Kavanagh, J. O'Hara, J. Dariton, R. Matthews. Front row: W. Carroll, W. French, J. Kinsella, C. McDonagh (Foreman), Lieutenant J. Connolly, Station Officer J. Howard, T. Curran, J. Markey, A. Kennedy. Photograph, collection of the Carroll Family (William Carroll, district officer, Dublin Fire Brigade, deceased)

12. The funeral of the victims of the Pearse Street fire, 1936 with the flag-draped coffin of Robert Malone in the lead as they pass Tara Street fire station, with full staff of Dublin Fire Brigade, led by Chief Fire Officer Connolly, forming a guard of honour.
Photograph, from Robert Malone, junior

13. Unveiling of memorial stone to firefighters who died in Pearse Street Fire, 1936.
Photograph, © Dublin Fire Brigade

*The impression gathered from an inspection of the station is that in every respect its dimensions are wholly inadequate for Brigade purposes ... the bedrooms are entirely too small, consequently, when fully occupied, the atmosphere must be very unhealthy. The enlargement of the station is a matter which must engage the attention of the committee AS IT WILL NOT BROOK DELAY. ... At present at the rear of the engine house there is a small square of one-storey tenements which might be obtained at a reasonable figure and which, in the interests of public health, it would be desirable to level ... there are 23 men at Clarendon Row and they are all unmarried.*

So at last the definite judgement was passed on the catastrophic buildings which constituted the recently built central fire station. Equally blunt comments were made about the Winetavern Street station. The twenty-two married men "do not live on the premises, and practically the only portions used for the Fire Brigade purposes are the engine and appliance room, the stables, office and bunk room. The remainder is occupied by the Cleansing Department". The firemen and their families occupied houses in Cook Street and numbers 11, 32 and 33 Winetavern Street, and

*a large number of them have only one room, though in several cases the number in family is large. This is highly objectionable and should be remedied without delay. There is no drying room provided at the station, so that men who get their clothes wet have to dry them in their solitary apartment.*

The report admitted that although the corporation had always condemned the evil of crowded tenements "in this respect some of their own employees have been neglected". It revealed how superior were the conditions for firemen and their families in the English brigades. In all these cases married men were allocated apartments, and single men lodged with them either on the stations or in superior accommodation adjoining.

It was acknowledged that Dublin was well provided for with its fire escape stations. The two latest, commissioned in 1891, were situated at Cardiff Lane on the south side and Wapping Street on the north. These were in fact sub-stations equipped with an escape, hose cart and hand pump. A large and ample wooden shelter with a stove was provided for the firemen who manned the sub-stations twenty-four hours a day, two men at night and one man during the day. The report mentioned that at all the other escape stations "the night boxes appear to be too small for the convenience of the men, and during the winter months should be provided with a small stove, otherwise if the weather is severe the men would be in danger from exposure".

An important difference between the brigades in England and that in Dublin was highlighted in the report. The former have

*powerful auxiliaries assisting them. In addition to the several fire brigade sub-stations they have the advantage of the assistance and co-operation of the police in extinguishing fire. The police, except in a portion of London, are under the control of the local authorities, and receive in several places training as firemen. The police stations thus answer the purpose of fire sub-stations, and are provided with the necessary appliances for rendering first aid. Each policeman receives a small gratuity for each fire he attends. In Dublin the role of the police is merely that of keeping back the populace at the scene of a fire.*

The report was highly critical of the training of Dublin firemen.

*The men are mere machines, instructed in the exercise of certain manual labour and in the manner in which the various appliances are used … no provision is afforded them for improving themselves, or of extending their education. Every member of a fire brigade can profitably study the principles governing the components of fire, the effect of heat, air and water on the different materials used in construction of build-ings, the science of fire extinguishment, and appliances used, and the saving of life and property. Knowledge such as this will raise the mental capacity of the men, and make them more proficient in their duty.*

John O'Meara recommended the provision of a library in the central station, as well as a billiard and smoke room and a gymnasium.

It is clear from reading their report that neither Harty nor O'Meara were happy with the exercise of discipline within the fire brigade. They were critical of the level and quality of officership, particularly when fire crews were attending fires where people were still inside the burning premises. Water supplies in Dublin were considered excellent, but problems in locating hydrants was often an issue, so firemen "should be acquainted with the position of all city hydrants by means of maps or by labels on the lamp posts". Other recommendations were that all houses should be provided with a trap door opening out onto the roof, and premises where petroleum or other inflammable oils were stored for sale should be frequently inspected: "In Birmingham the fire brigade look over the plans of new buildings lodged with the local authority so as to ascertain what provision is being made against fire".

*Perhaps the most obvious of all Dublin Fire Brigade's problems was the age of its two senior officers. The report was quite definite about what ought to be done. Chief John Boyle has rendered very valuable services to the committee during the period of his serv-ice, now extending over twenty years, and as his age has exceeded 70 years, the time has arrived when the committee ought to recognise his claims to an adequate pension, and relieve him from his present responsible and arduous duties. The like remarks apply to Lieutenant Byrne, now over 70 years of age, after an active service of over 28 years.*

The adoption of the report by the corporation was tantamount to retiring both Boyle and Byrne, leaving a major void at senior officer level. The section of the report dealing with the firemen's wages and hours of labour opened the way for a proper claim to be processed on their behalf to radically improve their condi-tions of employment. Working hours in the Dublin brigade were similar to cities in Britain, but the Dublin men had no annual leave or not even one full day's enti-tlement to be free of all duties. In English brigades each man had a full twenty-four hour leave on every thirteenth or fourteenth day plus one week's vacation in the year. Dublin wages at twenty-one to twenty-five shillings per week were lower than those in England.

The main recommendations of the O'Meara/Harty report may be summarised as follows.

- Superannuation of the chief and the lieutenant, and the appointment of younger men.
- The construction of a new central station, or the extension of Clarendon Row with adequate sleeping and recreation accommodation for the men, and sufficient yard space for drilling and technical instruction.
- The extension of Winetavern Street premises so as to concentrate on the homes of the married firemen and the provision of drying facilities and bathrooms similar to the central station.
- The provision of a more efficient system for call-out of married men.
- A more elaborate set of rules and regulations detailing generally the duties of the officers and firemen, and their rigid enforcement.
- Provision of a library of technical works, and leisure facilities.
- Construction of larger shelters at escape stations and provision of heating stoves.
- Inspection and identification of hydrants.
- Inspection by the chief of all plans of buildings in course of erection, and enforcement, where possible, of his suggested alterations.
- The immediate construction of the most approved system of electric alarms as agreed on 20 October 1890. [The implementation of this recommendation for street fire alarm pillars was to be delayed for a long period.]
- Provision of a small dispensary for treatment of cuts, burns and bruises.
- Alteration in a small way to the existing escapes. [This may have been due to Harty's concern regarding part of the city "now practically shut out from the use of a fire escape in consequence of overhead bridges crossing the thoroughfares". These bridges conveyed the railway loop line opened in 1891 connecting Westland Row and Amiens Street stations, and they created an obstruction for the escapes which had to be pushed in a vertical position.]
- The granting of a more liberal leave according to the practice in the English brigades.
- The Law Agent to examine the feasibility of insuring the lives of the firemen against accidents, and to pay premiums by a small deduction from the weekly wage of each member. The question of superannuation ought to be left over, and each case dealt with upon its merits.

The brigade had cost £4,350 during the year 1891, and there was no doubt in the minds of the members of the corporation that the carrying out of the recommendations of this most comprehensive report would entail the expenditure of a very large sum of money. But failure to carry out the proposed changes would lead to demoralisation among the firemen and, with it, a fall from the standards that had been painstakingly built up in the thirty years since the formation of the brigade.

The lord mayor, when speaking to the firemen on 4 May 1891, had said

*We have you engaged for a purpose which may at any time involve you in such consequences as may lead to loss of your lives; but if you do we will take care, as*

*representing the citizens of Dublin, that your families and those that you leave after you will be compensated to the full extent for the loss they sustain.*

What the lord mayor meant by these sentiments was made clear at the 23 June meeting of the corporation when the council authorised the payment of a gross sum of £450 to the personal representatives of the late Inspector Doherty, and the mother of fireman Burke was given £75. At its September meeting the corporation adopted what became known as the O'Meara Report, and also accepted what was to be John Boyle's last annual report of Dublin Fire Brigade.

The early years of the 1890s were an exceptionally busy period for the Dublin firemen with the city and the townships confronted by many large and dangerous fires. Following the disastrous occurrences at Westmoreland Street, the bad luck that had dogged Boyle's career in charge of the brigade continued. On 21 November 1891 a serious fire broke out in a wholesale stationary and printers at 53 Sackville Street. Large quantities of stored paper and ink facilitated the rapid spread of the blaze from the stores into the machine room. In the smoke-filled building four firemen were working from the stairs when it suddenly collapsed and they were thrown headlong into the basement among the wreckage of the staircase. Two of the firemen, Hogan and Cummins, sustained injuries which necessitated their removal to hospital where they were detained, while Lieutenant Byrne received an arm injury which also required medical attention. After more than three hours the fire was extinguished. This was the last serious fire in the distinguished career of the well-respected chief. Boyle's letter of resignation to the waterworks committee was both dignified and assertive.

*It is with feelings of great regret that I have to report that owing to failing health I find it will be impossible for me to resume active duty as Chief Fire Officer ... considering the long period of almost 24 years during which time I may say, without boasting, that I have worked incessantly in your service, both night and day. I therefore must respectfully place myself in the hands of your committee, trusting that you will deal generously with a faithful officer of long standing.*

He was to remain on the payroll as superintendent until the appointment of his successor in April of the following year. He spent the final months of his service on sick leave, during which period the administrative work of the brigade was dealt with by Mr Crofton, and Lieutenant James Byrne took control on the fireground. John Boyle retired on a pension of £250 per annum, and he died in 1898. He had seen death and injuries among his firemen during his years in charge, but he had also overseen substantial steps forward in the fire service for Dublin. He was commemorated even in music! In the 1880s Charles Macdona composed a piece of music entitled the fireman's polka and dedicated it to "Captain Boyle". A copy of this music is preserved in the Civic Museum. (See colour section, illustration numbers 5 & 6.)

An interesting development following the 1890 Dublin Corporation Act was the introduction of a fine for a chimney fire: "Occupiers of premises, the chimney of which may be on fire, are liable to a penalty of ten shillings". It was believed by the waterworks committee that the fines for chimney fires together with revised charges for attendance at out-of-area fires would be sufficient to cover the increased cost of the brigade. Attendance to calls in the surrounding townships was to be charged according to the following schedule.

|  | Day | Night |
|---|---|---|
| Attendance of chief | £1-11-6 | £2-2-0 |
| Attendance of lieutenant | 1-1-0 | 1-11-6 |
| Attendance of inspector | 0-15-0 | 1-1-0 |
| Attendance of engineer | 0-7-6 | 0-10-6 |
| Attendance of fireman | 0-5-0 | 0-7-6 |
| Use of steam engine | 5-5-0 | 10-10-0 |
| Use of manual engine | 3-3-0 | 5-5-0 |

Add one fourth to above charges for each hour above four,
and one fifth for each mile distance beyond five miles from the station.

At last a proper system of payment was put in place, but only when the townships agreed to pay these charges would the Dublin Fire Brigade respond automatically to the extra-municipal fire calls.

In the autumn of 1891 great havoc was caused in the city by an exceptional wind storm. Extensive damage was caused to all the electricity and telephone wires, which in turn interrupted communication between the two fire stations and all the escape stations. The incident of course resulted in extra charges on the rates to cover repairs and replacements.

Funeral of Charles Stewart Parnell, October 1891. Dublin Fire Brigade forming guard of honour at City Hall. Photograph, © National Library of Ireland

# 7 The end of the century

The most urgent priority for the corporation was to fill the vacancy created by the retirement of the brigade chief. Little did anyone realise that the year 1892 was to be the start of a completely new era for the fire service. The position of superintendent of the Dublin Fire Brigade was advertised in January, offering a salary of £300 per annum rising to £500. There were applications from fifteen candidates initially. It was then reported in the fire service journal *Fire and Water* that a member of the Dublin City Council "moved that the post be kept open to permit of applications being received from the United States. This was opposed on the ground of the expense of maintaining brigades in the American style". The number of applicants eventually reached forty-four and this included for the first time several civil engineers as well as superintendents and other senior officers from fire brigades in Britain. There were no applications from the United States.

A written examination, consisting of fifty questions to be answered within three hours, was taken by twenty-nine candidates, following which twenty-five were called for interview. Finally in March five names were presented to the council and the voting was as follows.

| J. Braidwood | Palatine Insurance Company, Manchester | 1 vote |
| P. Lambe | engineer, Liverpool | nil |
| G. Parker | superintendent, Bootle Fire Brigade | 23 votes |
| T. Purcell | civil engineer and volunteer fireman | 31 votes |
| T. Ward | chief, Datchet Fire Brigade, Berkshire | 2 votes |

Three candidates were eliminated after the first count, and on a second count Purcell received 31 votes and Parker 23, the former then being confirmed in the position.

There were some interesting names among the unsuccessful candidates. One of the short-listed candidates, George Parker, having failed in his bid for Dublin became Chief Officer of Belfast Fire Brigade later in the same year. Among the candidates not included on the short list J. Johnston, Superintendent of Southampton Fire Brigade will appear in our story again; John Myers, employed at the time by Dublin Corporation as a food inspector, joined the Dublin Fire Brigade five years later and eventually became its chief officer; Arthur Pordage, Superintendent of Portsmouth Fire Brigade, was appointed Firemaster of Edinburgh Fire Brigade in 1896 and became widely known and respected throughout the British fire service. The Chief of Newcastle-on-Tyne Fire Brigade withdrew his application when his local authority offered to increase his salary by £100 per annum. Alfred Hutson applied for the post in Dublin although only fire months previously he had moved from Brighton to become chief officer of the newly formed fire brigade in Cork. He did not, however, present himself for the examination in Dublin. Another candidate was C. G. Smith, Superintendent of Rathmines Fire Brigade.

Thomas Purcell, the man chosen to command the Dublin Fire Brigade, was an outstanding choice. Over the next twenty-five years he proved to be the most inspiring, innovative and forward-looking chief in the brigade's history. He was born in Kilkenny in 1851 and qualified as a civil engineer. For twelve years he had sole management of the contracts and works of an engineering contractor, and for two years prior to his Dublin appointment he was manager of the Shannon Foundry and Engineering Works at Limerick. He was a volunteer fireman and the members of Dublin Corporation were obviously impressed by his perform-ance in the written examination. He was also clearly a brave man because in 1876 he was awarded the silver medal of the Royal Society for the Protection of Life from Fire, an award not lightly given. Purcell had carried out a daring rescue of a woman from a fierce house fire and his citation was the following: "In testimony to the intrepid and valuable services rendered in the preservation of life at a fire at Messrs. Hennessy's, drapers, Kilkenny on December 19th 1875".

At a full parade of the Dublin Fire Brigade on 14 April 1892 presentations were made to the retiring chief and to the new chief. John Boyle received an illumi-nated address bearing a portrait of himself and some of his officers and men, and depicting a typical fire scene. He declared that the trust he had received from the late Captain Ingram he now handed over to his successor, and he was sure Thomas Purcell would feel, as he had felt, that it was a high honour to command the brigade. Purcell, replying to his address of welcome, expressed his desire to make the service one in which there would be few vacancies, long service, and a more permanent provision for those disabled by age or accident. So began the years of significant change for the fire brigade.

Following the publication of the O'Meara Report, the nascent trade union sentiment in the brigade finally came to the surface. The Dublin Fire Brigade Union, consisting of forty-two firemen, was founded on 5 June 1892 with its address at the Central Fire Station.

Although insignificant in national terms, because its membership never rose above fifty, it did however register on 28 December 1905 as an affiliate of the Dublin Council of Trade Unions, and it played a progressive role in the local labour movement. It was to affiliate with both the Irish Trade Union Congress and the Labour Party, regularly demonstrating a sense of solidarity by donating monies from its meagre income to help fellow workers in dispute. It gave £36 to the ITGWU in 1913 and a further £28 in 1914; Belfast barmen got £2 in 1917; Irish Seamen's Union £5 in 1933; and the Tramway Workers £5 in 1935. It also sent a wreath to the Blackpool firemen following a fatality in 1936.

Political contributions were common and were sometimes questioned by the Registrar of Trade Unions, as happened in 1920 when the union was obliged to state in its own defence "... regarding item £15 to Dublin Labour Party, this was decided by a vote of the Union, and according to paragraph 7 Rule 3 we are of the opinion that it is perfectly legal to do so, being for the benefit of Trades Unionism". The registrar agreed. Because of its small membership, the union retained the services of non-members to represent it professionally in its dealings with the municipal authorities. The list includes some famous names such as E.L. Richardson who served as General Secretary of the Irish Congress, and P. T. Daly, a long time colleague of Jim Larkin during the momentous years of the Irish Transport Union. The Firemen's Union finally dissolved on 26 April 1942, victim of the new demands of the Trade Union Act 1941.

On 24 August 1892 John Simmons, Secretary of the Dublin Trades Council, submitted a claim on behalf of the firemen and sought a meeting with the waterworks committee to process the claim. The firemen's position was simply and directly put.

Illuminated address presented to T.M. Healy from Dublin City firemen in gratitude to him for steering their pension scheme through the House of Commons. Illustration from *Fire and Water*, January 1892

*We are not making an unjust request, when it is taken into consideration the confined body of men we are, and the disadvantages we have in comparison to other prominent brigades in the United Kingdom ....A man is called at 6.30 am and comes on duty at 7.00 am. He remains on duty in the station all day until 6.00 pm, and if he is for chimney fires that night he remains until 10.00 pm .... The following will, we say, be part of his duty, providing he be the first man of the bunk: gate duty 6-8 pm; telephone 8-10 pm; he may rest from 10 to midnight; gate midnight to 1 am; telephone 2-4 am; he may rest 4-6 am; gate 6-7 am. This man gets but four hours rest out of 24 hours, providing there are no fires, or otherwise.*

This "memorial" from the firemen went on to describe the duties assigned to those at either Wapping Street or Cardiff Lane which often meant a day of "21 hours without resting". The disgusting condition of the married quarters was also referred to – "some rooms are only 15ft long and 9ft wide, scarcely fit for one to live in, much less a family". When on escape duty the men paraded at 7.20 pm in the summer and returned at 6.20 am. In winter the duty was 7.20 pm to 7.20 am. They had to remain outside their boxes "except if it is raining, but if we are tired we sit on the level of the escape". They told how they received only five hours leave about once a week and a day's leave now and then, but it had to be applied for. The memorial documented the position in London where working conditions were much more favourable, as was pay at thirty-seven shillings and sixpence per week for first class firemen.

The Dublin firemen made the following claims: a pay rise to thirty shillings per week for a first class fireman; leave from 5.00 pm to 11.00 pm once a week; seven days annual leave; the right of men to arrange changes with one another for a particular reason; special one-day leave for important business; continuous night duty to be confined to one month at a time; improvement in married quarters; escape duty hours to be reduced to 8.00 pm to 6.00 am all year round; a man to be allowed to call witnesses in his defence and be given a copy of the report when on a disciplinary charge; a change in the manner of testing escapes by weight only and not by men. The firemen also called for a pension to be granted after twenty-one years service: "a man deserves some recompense after devoting the best years of his life and all his energy in the service of the public".

> *Finally, Gentlemen, we earnestly request you to put a stop, once and for all, to the petty tyranny that we are continually subjected to … we do our duty, as the past can prove, therefore we strongly object to be continually haggled at.*

This document was signed by forty-two of the forty-four firemen.

The waterworks committee replied to the Trades Council on 12 November claiming that the memorial, and certain issues raised in discussion, were not entirely accurate particularly regarding dismissals that had taken place in recent years. The committee's view was expressed as follows.

> *All were given an opportunity of being brought before the committee by their superior officers. Some refused to appear, gave in their clothes and left the Brigade … The duty of a fireman is the saving of life and property, and the Corporation with that view have placed escape stations throughout the city, and experience has taught them the hours at which these escapes are most needed. The committee cannot see their way to alter the hours of attendance at present. … The Corporation are fully alive to the desirability of providing for the comfort of their aged and incapacitated servants, but the law as at present does not permit their doing so.*

However, the committee did agree to recommend to the council an increased pay scale, commencing at twenty-two shillings and sixpence per week, with annual increases of one shilling per week up to twenty-seven shillings and sixpence per week. It also announced that it was examining the question of "better accommodation for the married men and a scheme for the insurance of the men against accident".

At its meeting in December (the report of the meeting is dated Christmas Day!) the council agreed to the pay increases. It was also agreed to extend the hour of leave up to 11.00 pm and to sanction seven days leave every year, with permission to exchange the time for leave with each other. So ended the first confrontation between the corporation and trade union firemen.

Back in December 1891 Lieutenant Byrne had written to the waterworks committee seeking a pay increase. This might seem a little odd, because it was clearly indicated in the O'Meara Report that Byrne was to retire in 1892. He may have been prompted by the fact that since November he had taken on extra responsibilities during the absence of John Boyle on sick leave. In fact he did not retire, and in January 1893 he was given an increase of £50 per annum on a majority vote, an amendment to restrict the increase to £30 having been defeated.

Less than six months after Purcell became chief he was faced with his first major fire in Dublin. At about 1.30 am on 27 August fire was discovered in the extensive concourse area of the South City Market. The buildings formed a complete block covering an area of 82,500 square feet completely roofed in, having frontages on four streets: George's Street, Exchequer Street, Fade Street and Drury Street. The area consisted of over thirty shops, with basements and living accommodation in many of the upper storeys, while within the quadrangle there were numerous stalls in the arcades. An extensive furniture and carpet warehouse stood to one side of the concourse and the entire block was vaulted by brickwork arches used for the purpose of a bonded warehouse in which there were 7,000 hogsheads of whiskey and 2,000 cases ready for export on the night the fire broke out.

Because of the large open grilled entrances on the four sides, and the highly inflammable materials, within an hour the central core of the market was ablaze. Huge volumes of dense smoke billowed out into the streets and the roof trusses became a dangerous mass of flames.

On arrival the fire brigade set about preventing the fire getting to the whiskey stores, so the first two hose lines were laid down in that area. When the roof began to collapse, sparks and burning debris, aided by the high wind, were sent flying into Dame Street and South William Street. Eventually sixteen jets were brought into operation to try to extinguish the flames. The trojan work of the firemen was watched by the poor of the Liberties mixing freely with the Merrion Square gentry, all enthralled by the spectacle of bursting windows and crackling timbers falling to the ground spreading the blazing contents of the market stalls.

The police, fearing the collapse of walls or roofs, threw a cordon across the surrounding streets as the firemen attacked the fire through the premises of Pims and Eastmans. Twenty soldiers of the Royal Munster Fusiliers were called to assist in removing barrels and crates of spirits from the ground floor shop of Powers Distillers. While this work was proceeding the minaret above the entrance opposite Dame Court crashed into the street, damaging fire fighting equipment and threatening the lives of the firemen inside the burning buildings.

Three firemen were in fact injured, and the firemen's problems were exacerbated by a call to attend another fire, this time in Bridgefoot Street. This involved a tenement where two young children were lucky enough to be rescued by neighbours – their parents were not present when the fire broke out. The fire was confined to one room by the weary crew who had responded, but not before much of the contents was destroyed. By 5.00 am the brigade had the market concourse fire contained, and personnel were checking out all rooms in the adjoining premises. The fire certainly bore comparison with the great whiskey

**South City Markets fire, August 1892. Photograph, © Edward Chandler**

fire of 1875 and the rising smoke in the early dawn was an indication of the level of damage resulting from such a conflagration. The wreckage of burning debris blocked the central market arcade where the small stall holders, without insurance of any kind, had lost all. Most of these had stocked up on Friday in anticipation of their usual busy Saturday trade, and now everything was gone.

The tired fire crews spent all day Saturday dealing with pockets of flame in the highly dangerous buildings. Some of the men were up to fifteen hours on site continuously at work before the chief decided to reduce the attendance to two lines to be kept in operation until all was extinguished. The serious character of the fire, greatly threatening life and property, and especially the possibility of a catastrophic explosion of the large quantity of whiskey, had been faced so effectively by the courageous Dublin firemen that not only was the bonded store saved, but so also were most of the shops and houses adjoining, and no loss of life occurred. It was a tough time for the exhausted firemen because on the Sunday evening they faced another two hours of struggle while tackling a fire in Prussia Street, where a quantity of hay, thirty tons of it in fact, threatened to destroy two tenement houses as it blazed fiercely in their rear yards.

Monday's newspaper editorials were high in their praise of the way Chief Officer Purcell and his men had fought the market fire, but they questioned the sense of allowing bonded stores beneath commercial premises and chandlers shops in congested districts.

> Life and property should not be exposed to destructive carelessness on the part of those who manage or own premises …. We have a brave and well-managed Fire Service, a magnificent water supply, but whiskey stores or petroleum stores when once fired, mock all efforts to control the consequences.

Purcell had faced his first severe test as a fire chief. He had come through with the utmost credit and had begun to stamp an indelible mark on the Dublin Fire Brigade. By a strange coincidence the date of the south city market fire was also the date of two related tragedies outside Ireland. Fifteen firemen lost their lives fighting a huge fire at the Opera House in New York and 110 miners were killed in an explosion at Aberkenfig in Glamorgan: truly a night of terrible sacrifice for those who work in the forefront of dangerous professions.

The annual report for the first year of Purcell's leadership shows what a momentous one it was. Although the number of fires attended was thirty-three less than in the previous year, the number of serious fires showed a considerable increase. In 1892 the total amount of property at risk from fires attended by the brigade was assessed at £450,744, and the amount of damage caused or loss sustained was put at £82,475. Included in these figures were the relevant amounts for the south city market, which on its own had property at risk to the value of £242,235 and a total loss in the fire amounting to £63,263. Three firemen received chevrons for life-saving, and twelve extra-municipal fires were attended. The occupiers of forty-six premises in which chimney fires occurred were prosecuted under Section 97 of the recently enacted Dublin Corporation Act 1890, and fines totalled £16-2-6. (See colour section, illustration number 7.)

The brigade retired two old horses and purchased three new young ones. The central fire station was cleaned and painted and electric lighting installed, and a revised scale of pay and leave was introduced. Firemen in their first year were now paid twenty-two shillings and sixpence per week rising to twenty-seven shillings and sixpence after five years. The leave was now six hours per week and an annual holiday of seven days was granted to all firemen. The brigade staff at this time consisted of one chief, three officers and forty-two firemen. An interesting innovation introduced to the annual report by Purcell was a plan showing the south city market and the disposition of the appliances and jets used to fight the fire.

In February 1893 an extraordinary situation came to light when it was discovered by the corporation that the byelaws governing the brigade, which had been drafted in 1863, had never been signed into law by the lord lieutenant, and therefore had no validity. A special committee was set up to deal with this matter, and later in the year the byelaws came into force legally. The full title of the legal document was *Rules and Regulations for the government of the Dublin Corporation Fire Brigade, and for the more effectual prevention of disorder, neglect or abuse. Framed pursuant to the Dublin Corporation Fire Brigade Act 1862.*

The principal appliances in the Dublin Fire Brigade were two steam fire engines, one purchased in 1864 and the other in 1867. Both machines, supplied by the London fire engine makers, Shand, Mason & Company, had a single vertical pump capable of delivering 300 gallons of water per minute. Although they had been in use for over twenty-five years they were still serviceable owing to the fact that a new boiler was fitted to the older one in 1890 and the second one got a new boiler in 1893. However, it was decided that another steamer was needed and an order was placed with Shand, Mason & Company.

In the autumn Thomas Purcell visited America on vacation, which was in fact a working holiday at his own expense. He visited the fire departments in some of the chief cities to study their organisation, fire fighting methods and appliances. His future administration of the Dublin Brigade was partly influenced by the knowledge he gained on his travels: his technical mind was ever open to new ideas. During his absence Dublin experienced a serious drought and an outbreak of typhoid fever. Water supplies were cut drastically and firemen were utilised as waste water inspectors, assisting regular corporation staff from 6.00 am to 4.00 pm daily. It was a very busy time because Purcell had initiated a procedure requiring firemen to check and clean all the city fire hydrants and then to compile a record of their exact locations. This detailed survey of all 1,686 hydrants was completed in 1894. Printed copies of the survey were then produced and kept at each fire station.

During Purcell's absence in America the first serious problem in recent years of water shortage for fire fighting occurred. The fire was in a five-storey brick building on the corner of Beresford Place and Gardiner Street on 21 August 1893. The premises housed the company of J. Whitly, cork manufacturers, where supplies of bottle wax, shellac, resin and cork were stored. Although the brigade arrived promptly the building was a mass of flames before the firemen could get their hoses in place. It was then that the water shortage manifested itself, and an immediate call was sent out to the turncocks to rapidly increase the supply. Two escapes were pitched to the adjoining buildings from which the firemen fought the blaze in order to prevent its spread to the surrounding area. The fire took hold of the roof which eventually collapsed right through the top floors and led to the collapse of part of the front wall. The struggle continued until nothing was left but a huge pile of smouldering ash where a thriving business had been.

Lieutenant Byrne and his fire crews, hampered by shortage of water, showed exceptional determination in preventing the destruction of adjoining buildings despite the highly inflammable materials which they contained.

Not long after that fire the brigade took delivery of its new steamer. This was a Shand Mason Equilibrium engine and its official trial on 27 September 1893 was attended by Alderman Dillon, chairman of the waterworks committee, Sir Robert Sexton and other members of the corporation, Superintendent Purcell, and Spencer Harty, the city engineer. This new appliance was much more powerful than the two steamers already in the possession of the brigade; for example it could pump at the rate of 750 gallons per minute. Its usefulness and capability were certainly proved the following month when it was utilised to augment the city's water supply during the severe drought. The complimentary letter which Purcell wrote to the makers of the fire engine is worth quoting.

> The successful work which we did with your engine – when it was run continuously night and day for three weeks at high speed with a constant pressure of over 120 lbs at the pump, necessary to overcome the head of 2509 ft against which it forced water, and the friction due to a velocity of 10ft per second through 500 ft of 6 inch rising main – formed a test not often applied to a fire engine.

The steamer had been lowered by means of blocks and tackle down a 140 foot chasm to the Dargle river to pump to the waterworks reservoir.

In 1894 one of the largest and best known commercial premises in Dublin, Arnott and Company of Henry Street, wholesale and retail drapers and house furnishers, was destroyed in what was undoubtedly one of the most hazardous fires in the history of Dublin. It was a very windy night when shortly after 1.00 am a fire was sighted in the warehouse, and within a very short time the whole area – extensively stocked with goods – was involved. Flames spread across Princes Street, threatening the *Freeman's Journal* premises which was in fact slightly damaged, and down that roadway to Box's premises the company at the end of the block. On the arrival, the brigade was faced with a massive wall of fire

Large Shand Mason steamer commissioned by Dublin Fire Brigade, 1893. Print, © Trevor Whitehead

Destruction of Arnott's following fire, 1902. Photograph, © National Library of Ireland

in Princes Street and an ever increasing blaze consuming the stores and upstairs apartments of Arnott's premises in Henry Street.

At that time it was quite normal for drapery assistants to live on the premises where they worked. On the night of 4 May, when the fire broke out, some 200 male and forty female assistants were in residence in the Arnott building. Fortunately, although the spread of the fire was rapid, all were able to escape unhurt, though many were only partially dressed and the majority lost all of their personal belongings.

The usual quiet of the night-time city was disturbed by the roaring flames, breaking windows, collapsing floors and the noise of the wind through the tormented building. In spite of herculean efforts throughout that night, the intensity of the fire was such that it threatened to spread right through into Liffey Street and engulf the whole block.

Sometime after 3 am it began to rain heavily, cooling the remains of the collapsed warehouse and delivering some sort of respite to the hard-pressed firemen. By 4.30 am Purcell could announce that at last the fire was surrounded and firemen were able to advance inwards towards the core of the blaze. Much destruction of property had taken place; an estimated £100,000 worth of damage was the result. In the following days sightseers would look on the destruction and marvel at the work of the fire brigade in preventing such a conflagration from causing even more damage. Chief Purcell had shown again that his organised approach to fire fighting could be successful even in the most dangerous of situations, with fewer injuries being sustained by his better trained firemen.

The firemen were not entirely satisfied with the settlement received after their first claim backed by the trade union. So in 1895 they again presented a memorial to the corporation concerning wages and issues not dealt with by the committee on the previous occasion. The men now sought pay of thirty-two shillings and sixpence per week after five years service; the hours for escape duty to run from 10.00 pm to 6.00 am all year round; evening leave to be increased to allow twenty-four hours off, instead of 5.00 pm to 11.00pm; better watch boxes with some form of heating (already mentioned in the 1891 O'Meara Report); permission to wear old tunics when attending fires at night during winter months; improvement in the married quarters; and lastly that firemen be granted a pension after twenty-one years' service or, if disabled, that they be superannuated. As was the case with with the previous claim it was signed by forty-two men.

The memorial was referred to a subcommittee for detailed investigation and its report was issued in May, stating that "The wages of the firemen should not be altered pending projected alterations in the system, which the committee are considering". The projected alterations signified in fact a long-term radical reorganisation of the entire system of fire protection provided within the city, and they would involve great changes to every aspect of life in the fire brigade. Among these changes would be the abolition of street escape stations and the provision of escapes drawn by horses. No change was therefore recommended to the present hours of escape duty, and the wearing of red shirts without tunics had to continue.

The subcommittee's comments concerning the married men's quarters reiterated what had been said in 1891 by O'Meara.

> The rooms of the men are of the very worst description. We were utterly astonished at the primitive condition of affairs existing in Winetavern Street station and in the married men's quarters ... such a state of affairs, apart from any moral question, would be alike detrimental to the health of the men, injurious to the efficiency of the service and not creditable to the City and Council.

It was explained that on receipt of an alarm the men have to be called from various houses some distance apart, obviously causing delay. It was urged that "the existing state of affairs be grappled with and put an end as soon as possible". This urgent plea is particularly significant in view of a report issued at this time by Dr Charles Cameron, the city medical officer. He had recently visited London to equate life in tenements in both cities, and he commented: "The death rate in Dublin is very high due to the poverty of the population. One third of the people live in single room tenements in which they eat, drink, cook and sleep, and often carry on their work for a living". The firemen's lot was not any different to many of the city's poor people with regard to accommodation.

While considering the memorial from the firemen, the committee also had available to it an estimate from Purcell showing the gross capital outlay which would be required for the most urgent improvements to the fire brigade. The estimate included the cost of the construction of two new fire stations on the north side and the installation of the street fire alarm system "... so often previously reported on and recommended".

Purcell also outlined the probable cost of purchasing decaying property around Chatham Row to provide space for an adequate drill yard, alterations to the existing fire station and additional buildings. On the issue of pensions the subcommittee stated the following.

*The Law Agent advised that the Fire Brigade Act of 1862 gives no power to grant superannuation allowances to those employed under that Act. It may be well to point out that under the provisions of the Dublin Corporation Bill 1885 the Council could have obtained this power.*

Having discussed all the matters in the report at great length the council eventually decided that the maximum rate of wages on the tenth year of service should be thirty shillings, in other words as at present up the sixth year at twenty-seven shillings and sixpence, and in the four years after to reach thirty shillings. The council decided to press ahead with the chief's proposed fire stations on the north side.

In 1895 a number of the townships agreed to pay the charges required by the Dublin Fire Brigade giving them immediate response provided that fire cover for the city was not radically reduced. This development, which was welcomed by the corporation because of the income it would generate, nevertheless put an added burden on the officers and men of the brigade. On 13 May that year a fire occurred in the college chapel at All Hallows in the township of Drumcondra. The staff were unable to extinguish the blaze and called in the Dublin Fire Brigade who used two lines of hose to prevent the fire spreading into the residents' hall where 200 students and priests lived. Unfortunately the firemen were unable to save the old gothic chapel, which was over one hundred years old, or the new altar that had been installed only months before.

In October the Dublin firemen were again dealing with a serious outbreak of fire in a township. The location was the oil refinery and candle factory of Dunsinea near Blanchardstown. The property consisted of seven separate buildings arranged very close to each other, some five miles outside the north-west boundary of the city. A fire had been burning for nearly two hours before the arrival of the Dublin Fire Brigade. Police and locals had been unable to contain the spreading flames. The fire was tackled with the aid of the steamer in spite of a stiff gale which was blowing at the time. The brigade surrounded the refinery building, the danger area in the complex, which contained large quantities of whale oil, grease and barrels of wax. The crews worked for over six hours before the blaze was finally extinguished, but the refinery and some other buildings, all of light construction, were destroyed. Other properties, including the homes of both the owner and the manager, were saved. The owner informed a newspaper reporter when praising the endeavour and efficiency of the firemen that "were it not for the Fire Brigade not a stave of the property would have been left". He also commented on the "speed at which they came on the scene and the skill which they displayed in coping with the flames". Out of a possible total loss of property amounting to £20,000 only £4,600 worth of damage was caused, and the positive image of the fire brigade further enhanced.

The city of Dublin was now using about twenty-one million gallons of water per day and it had become quite clear that an increased supply was required. The city engineer requested finance to purchase additional land for a new catchment area and a new reservoir capable of holding 1,500 million gallons of water. With the period of severe drought being experienced at this time in Dublin getting worse and the brigade facing supply problems, two turncocks were appointed to the fire service, one to work a month on nights and the other a month on days, alternately. At this time also the fire brigade chief ordered that at least one steam fire engine must respond to all fire calls.

Thomas Purcell, in his annual report for 1895, paid a special compliment to Lieutenant Byrne on his retirement from the brigade. James Byrne had given sterling service to the people of Dublin since joining the fire brigade in 1863; he finally retired on a pension of £133-6-8. He was now aged seventy-four and had attended all the major fires in the city over a period of thirty-two years. He saw the Dublin Fire Brigade during its finest hours as well as during its lowest points. He had himself sustained many injuries. He had experienced both frustrations and acclaim, all, it might be argued, without adequate recompense – his salary at retirement was £200 per annum. Byrne's career epitomised what was good in the fire service and he left a worthy example for all the lower ranks who came after him.

The advertisement for Byrne's successor specified that the person appointed

*should be over 28 years of age and under 40 years, be examined by the Medical Officer, have active habits and be of good physique, educated to commercial standard and have practical experience with a fire brigade.*

Wages were to be £150 per annum, rising by £5 to a maximum of £200, uniform and accommodation supplied. There were twelve applicants and, following a written examination set by the chief and the city architect, five were deemed to have passed. The names were presented to the council and out of a total of fifty-seven votes cast, the main results were as follows: John Myers 31; Martin McCarthy 19; W.E. Cunniam 5.

Myers, who came fourth in the examination, was currently an Inspector in the Dublin Fire Brigade and was preferred over McCarthy who was first in the examination. One of the first duties of the newly appointed lieutenant was to train soldiers from the Curragh Camp who were sent to Dublin for fire fighting training. In 1896 the commissioners of Kingstown Township sought agreement with Dublin Corporation for their works foreman to attend all fire brigade practices. Permission was given and the man received basic training from Dublin Fire Brigade, presumably passing on his knowledge to some other employees of the township. It is known that at that time Kingstown had a wheeled escape and a horse-drawn manual fire engine.

Pile's timber yard and stores covered a large area stretching from No. 87 to 97 Great Brunswick Street, running east from Erne Street. The property extended even further at the rear, where storage yards occupied space behind six tenement houses. Immediately adjoining this was another potentially dangerous premises, the City Wheel Works, with huge supplies of timber, paints, oils and lacquers. Pile's property consisted of mills, showrooms and yards for the storage of vast quantities of timber. Shortly before 10.00 pm on the night of 15 January 1896, when there were gusting winds, smoke and flames was observed coming from one of the windows facing Erne Street. The fire brigade was immediately alerted and a full contingent of engines was dispatched.

On arrival the massive three-storey premises were well alight with flames coming through the shattered windows. Colossal tongues of fire could be seen shooting skywards from the open yards and sheds at the rear. It was soon apparent that Pile's buildings could not be saved so the brigade concentrated on trying to save the tenements immediately adjacent.

About seventy horses meanwhile had to be rescued from Tedcastle's stables which abutted the striken premises at the Erne Street end. Crowds gathered to watch their fellow citizens in the threatened tenements removing their meagre belongings onto the street, in fear of their lives and the destruction of their homes and livelihoods. In an incredibly short time the fire had spread into the upper portions of the main Pile building, seizing the floor timbers, reaching upwards until the roof was fully ablaze. The building became one vast furnace, lighting the winter sky with a violence and intensity that proclaimed to firemen, police, neighbours and bystanders that it was not going to be easily quelled.

The firemen, fortunately with water in plentiful supply from hydrants and from the Liffey, fought grimly in this unequal struggle. Great stacks of blazing timbers hissed and crackled as torrents of water poured down to arrest the fury of the flames. The solid and impenetrable inferno presented a weird and terrible spectacle. With the closing of the theatres and music halls those recently agape at the spectacle on the stage now hastened towards a greater spectacle by far – the hard pressed protectors of the city, with their puny engines but the will to win, in a death struggle with a ferocious and awe-inspiring enemy.

The flames and intense heat radiated across Great Brunswick Street lifting paint on shop fronts opposite and forcing fire crews to withdraw to the left and right. The massive roof soon collapsed releasing much of the destructive heat and increasing the height of the flames. On the following day it was said that people as far away as Howth and Rathfarnham could see the dark sky illuminated. Some four hours after the brigade arrived the tide of battle had turned and the firemen had succeeded in preventing the fire from spreading to other business properties and to the threatened tenements. Two steamers continued to pump water as the damping down was carried out, but by this time the dreaded "genie" was well on its way back into its box!

Great appreciation was again expressed in the days that followed to the chief and his firefighters for their gallant and successful action. Some newspapers, however, highlighted the dangers to life where such vast amounts of timber and other inflammables were massed in such close proximity to overcrowded tenements. It was stated that had the fire occurred at a later hour of the night, when few people would be about and the unfortunate tenement dwellers asleep, then deaths would surely have resulted. The papers called for some regulation so that the menace of such fires would not in the future threaten the young or old in their confined dwelling places.

A letter from Mr Pile, the owner of the yard and stores, thanked and complimented the firemen for their tremendous endeavours, and went on to outline how "the fire was a splendid, although expensive, test as to the qualities of breeze block concrete". When the strongroom, which was built of this material, was opened after the fire was extinguished, the books, papers and wooden shelves inside were found unscorched. This proved not only its fireproof quality but also its non-conductive properties – the strongroom walls were only nine inches thick.

The strongroom was in the centre of the three-storey building which was totally destroyed. Next door, at No. 167, the City Wheel Works of Messrs Brown also survived in spite of the intensity of the fire so close to the well-stocked premises. (The Brown firm was to figure in the construction of escapes, ambulances and fire appliances for the brigade.)

An example of the consequences of a very late fire call occurred in 1897. On 13 February just after 10.30 pm a small fire started in a four-storey tenement house in South King Street. For some unknown reason the central fire station, just around the corner, was not notified until over three hours later! By this time flames were shooting through the back windows on three floors, the stairs were well alight, and many rooms were ablaze. When the firemen arrived they found many of the inmates huddled in the hallway screaming that others were trapped upstairs. The escape was promptly extended and six women were rescued from a top-floor room, one with burns and unconscious. Firemen entered the building with a line of hose and fought their way up the burning stairs. They rescued two unconscious women in a smoke-filled room on the second floor. Five women were removed to hospital with various injuries. All recovered. The house, overcrowded like most of its kind in the city, was occupied by twenty-three adults, twenty of them women, all of whom had been in great danger.

In 1897 a deputation appointed by the waterworks committee visited fire stations in London, Manchester, Liverpool, Edinburgh, Glasgow and Belfast. The purpose of the trip was to study the construction of stations and the accommodation provided. The detailed report issued in November concluded as follows.

> *...we are fully convinced that the Dublin Fire Brigade organisation can never be considered satisfactory while housed under the existing defective conditions, and distributed according to the present system, and we unhesitatingly fully endorse the condemnation of our present stations. We go further, and say that the station in Chatham Row can never be properly adopted to the modern system ... even as a branch station ... it ought to be sold or converted to some other municipal purpose.*

The six recommendations in the report, obviously instigated by Chief Purcell, reveal the total reorganisation envisaged for the fire service in Dublin.

- *... that a suitable site be acquired on the south-east side, as near O'Connell Bridge as possible, and erect thereon a properly designed Central Station.*
- *That a convenient site be acquired in each of the other quarters of the city ... about*

Thomas Street, Bolton Street and Amiens Street ... compact branch stations to accommodate 1 officer, 7 married men, 1 horsed escape and 2 horses in each, be erected as soon as funds are available.

- That the Central Station be connected with each branch station, and the Police, by direct telephones.
- That a thorough system of electrical street fire alarms, with about 60 call points, ... be established.
- That all street fire escape stations be abolished, and a suitable escape to be run in a horizontal position by horses, carrying hose, appliances and a company of men, be provided for each station. The escape to be furnished with quick hitching harness, and the men to sleep in the room directly over the apparatus with sliding poles to the ground floor.
- That a properly equipped ambulance carriage be kept at the Central Station, and used for the speedy removal to hospital of all cases of street, fire building, and factory accidents.

The proposed reorganisation of the fire brigade would involve the addition of only three men and six horses, because the twelve men currently employed on street escape duties would then be available in the fire stations. In an interesting comment the committee claimed "... the expenditure will serve an important municipal purpose for the sites and buildings erected on them will be a permanent addition to the city property, the stations becoming an interesting and useful ornament to the streets on which they are erected". This report was adopted by the corporation on 29 March 1898, thereby taking the first step in the creation of the modern fire service.

At this time the area of Dublin City was 3,808 acres, or about six square miles, while the number of houses had increased from 27,500 in 1881 to 29,500 in 1891, and was estimated to be in excess of 30,500 in 1897. The population had grown to 245,000 in 1891 (the last census) with house values also rising. The demands on the fire brigade since its formation in 1862 had increased, and the proposed reorganisa-

**Dublin Fire Brigade at Stephen's Green West, 1897. Film still, © Lumiere Brothers**

tion, which would take the new situation fully into consideration, was an urgent necessity. Under Clause 31 of the 1897 Dublin Corporation Act the corporation was empowered to raise £12,500 for fire brigade purposes. Once these funds were raised they would be used to purchase sites for new fire stations and to pay for the installation of the proposed street fire alarm system. In addition to this, Clause 32 allowed for a greater portion of the water rate to be applied to the fire service.

Under the 1891 and 1895 Factory and Workshop Acts, the chief fire officer was responsible for inspecting premises and, where necessary, recommending means of escape in case of fire. The new Act of 1897 increased his workload, because he was now detailed to make recommendations relating to seventy-five specific factories as listed by factory inspectorate. The corporation agreed to increase his salary by £150 for this work in three annual payments of £50 each.

In 1895 letters had been appearing in the newspapers decrying the fact that Dublin did not have an ambulance and properly trained crews to operate it. In October of that year a Mr Johnston wrote: "In most important cities there is a well-organised ambulance system by means of which patients suffering from the effects of an accident are brought to hospital. In this city the first vehicle that turns up is pressed into service, and oftener than not a serious case is made a great deal worse by the treatment in transit." He went on to outline the practices in other cities where hospital authorities, mounted police or fire brigades carried out this function, and he ended by saying "…as projected changes are being brought on the constitution of the City Fire Brigade, it seems a fitting time to remedy such a serious deficiency in the city arrangements".

In the waterworks committee report in November 1897 it was made quite clear that the members were well aware of the need for an efficient ambulance service: "…the present method of removing persons with fractured limbs, dislocations, etc by unskilful people in a haphazard manner, on outside cars, or indeed any other vehicle which offers, being utterly unsuitable for the purpose, seriously aggravates the injury, and imperils the life of the patient". Training for ambulance work was commenced in 1898. Ten firemen from the central station were selected to be instructed by a Dr J. H. McAuley with the assistance of a sergeant-major from the army medical corps. By the time the new ambulance wagon arrived at the end of the year, the firemen had become very proficient in first aid and the use of splints and bandages.

The new ambulance was designed entirely by Chief Purcell and built in Dublin by Jessop Brown at the City Wheel Works. It was a four-wheeled covered vehicle with solid rubber tyres, drawn by two horses and crewed by three firemen. The space beneath the driver's seat held a battery to supply electric light, as well as bandages, splints, etc. The interior was 6ft 3in long, 4ft 6in wide and 5ft 10in high, with louvres on both sides and at the rear. There were two folding stretchers, one suspended above the other, the upper one supported by folding stays hanging from the roof, the lower one resting on upright supports fitted with joints enabling them to fold down level with the floor so that there was no obstruction when stretchers were being inserted into the wagon. In December the arrival of the brigade's new vehicle was given wide publicity in the newspapers and much favourable comment. The skill of Purcell in designing such an excellent ambulance was especially noted and acknowledged. In May 1897 the *Herald* had reported the plans for new fire stations and the introduction of an ambulance service. But in the autumn of 1898 when the new vehicle had still not arrived, the paper expressed outrage because Dublin "had the highest number of street accidents in proportion to its population of any city in the Kingdom". All criticism was now silenced!

The introduction of an ambulance service operated by the fire brigade was a tremendous boost for the people of the city. It was acknowledged by both the Dublin Council of Trade Unions – who had itself been advocating such a service – and the newspapers, who pointed out that the dreadful system of using "a jolt-ing float or an awkward outside car" to convey the injured to hospital should become a thing of the past: "anyone may be a victim of an accident and the provision of a few ambulances in the city is an expense no one will object to, not excepting the dreaded City Auditor".

On 10 January 1899 the corporation placed an advertisement in the newspapers officially announcing the arrival on the streets of Dublin of an accident ambu-lance service. It agreed to Purcell's proposal that private persons be charged £1 for the use of the ambulance, but that street accidents be dealt with free of charge.

The very first ambulance call was answered by the fire brigade five days before the newspaper advertisement appeared. In the early morning of 5 January 1899 a trawler, the *Curlew*, was fishing about twenty-five miles north-east of the Rockabill lighthouse when a serious accident occurred. The boat's fishing net became snagged in some underwater obstruction and this caused one of the steel hawsers to snap, fracturing the legs of one crewman and also injuring another. The captain immediately steered a course for Dublin and alerted the firemen on duty at the quay station as soon as the boat docked.

The central fire station was notified, and the ambulance, crewed by firemen Tom Dunphy, Joe Kiernan and William O'Brien, was despatched to the quay. The two stretchers were used to remove the injured fishermen to Sir Patrick Dun's hospital, where one man was detained and the other released after treatment. Unfortunately, John Smith, the seriously injured fisherman, who was a native of Fleetwood, died two days later.

A new, and much appreciated, service by the Dublin Fire Brigade had now begun – the long overdue, properly equipped ambulance crewed by trained first aiders was now a reality. In its first year of operation the service responded to no fewer than 537 calls.

The late 1890s saw moves by the Dublin United Tramways Company to intro-duce a rapid transit tram system powered by electricity. The issue was debated by the chamber of commerce as well as by the corporation, and the general opinion was "that an all out Dublin electric public transport system would be good for the few and bad for the many". The fire brigade chief maintained that

> it would be difficult to manoeuvre fire escapes into position and extend them to high roofs or windows in streets where tramway poles, trolley wires and span wires would

*form an obstruction. Also, the risk of fire spreading from one block of houses to another would be greatly increased. The introduction of electric traction with over-head trolley will compel us to alter our general system and apparatus.*

And indeed it was to do just that.

At that time the Bristol Tramways Company, which had taken control of Dublin's Southern Districts Tramway Company, was pressing for electrification and "direct and untrammelled access to the centre of the city, to cater for the growing traffic for Dalkey, Kingstown, Blackrock and other southern Dublin suburbs". An interesting report from a committee of American experts on electric tramways gave the opinion that "while the overhead single or double trolley system may be permitted on railroad lines running through rural districts and possibly in sparsely populated towns and cities, no overhead system should be tolerated in the mercantile, manufacturing or densely populated sections of the city". However, one of Dublin's most prominent businessmen, Martin Murphy, did not intend having a competitor extending a better transport system and putting him out of business.

He bought out Clifton Robinson's shares in the Southern Districts company and merged it with his Dublin United Tramways Company in 1896. (It was said of Robinson, the doyen of tramways: "what Edison Bell is to the electrical world, James Clifton Robinson is to tramways".) Murphy then got agreement to construct an electricity generating station in Clontarf and a mandate to operate an electric line between Annesley Bridge and Clontarf, a distance of four miles. Two great industrial advances were about to be made in Dublin – a rapid electrical transit system, and the corporation's electric lighting scheme for streets and private houses.

In 1898 the corporation withdrew its former opposition to the construction of an electric tramway for the city and made an agreement with Murphy. The deal would provide the corporation with a payment of £500 per annum per street mile of tramway for forty years from 31 December 1898, with a guaranteed minimum payment of £10,000 per annum when the electrical system was fully operational. Part of the deal was that "the Tramway Company shall not charge a higher fare from any termini to the city boundaries than one penny, provided the distance does not exceed 1½ miles". Thus arose the saying, well known at the time: "a penny to the Pillar".

Dublin was approaching the new century with electric trams for the city and suburbs, and the corporation was bringing electricity to both streets and houses. With the extra income available the council sought to alleviate some of the congestion and overcrowding in the city by seeking to extend the city boundary. This time they were to be successful.

In Victorian Dublin, firemen sometimes put their lives at risk to rescue those who had decided to take their own lives. One such case, which illustrates how the law looked on these situations, took place on the bitterly cold morning of 8 November 1897. A large crowd was gathered at the Liffey wall on Aston Quay watching a man struggling in the water. A police constable threw a rope and lifebuoy to the unfortunate individual, who seemed in imminent danger of drowning. But the man would not or could not take hold of the buoy. Fireman Henry Byrne, who happened to be passing, seeing the critical position of the drowning man at once removed his greatcoat and fire boots and plunged into the river.

After a long struggle, the drowning man was brought to the steps and taken from the river. Under police guard he was removed to Jervis Street hospital where he was detained for a period of observation. On his release from the hospital he was taken to the police courts where he was charged with attempting to commit suicide.

Henry Byrne had just a month previously carried out a similar rescue at Sir John Rogerson's Quay, where a man had fallen into the river just after midnight. Byrne was called from Cardiff Lane station where he was on duty. He threw a rope towards the man but, with the tide running out, he was unable to grasp the rope. The fireman then removed his coat and boots , took one end of the rope in his hand, dived into the river and reached the drowning man. With considerable effort he succeeded in tying the rope to the victim, who was then pulled ashore by those on the quayside. The rescued man was taken to Sir Patrick Dun's hospital where he recovered. Fireman Byrne received a chevron and a pay rise of one shilling per week for this courageous rescue.

At the beginning of the final decade of the nineteenth century the fire brigade faced a logistical problem presented by the extension of the railway system. Bridges had to be built to carry a new elevated railway line across many city streets between Westland Row and Amiens Street. This resulted in obstructions to the street fire escapes when they were being propelled in an upright position in response to fire calls. An additional problem, and much more widespread, was the electrification of the tramway system mentioned above, involving the erection of overhead live wires – not only a serious inconvenience to the brigade, but generally dangerous as well.

In 1898 four of the escapes stationed in the north-east and south-east areas were reconstructed, reducing their length to enable them to pass safely under trolley wires. This was not an ideal solution and a better one was obviously needed; the waterworks committee requested Purcell to design an escape which could be propelled in a horizontal position.

Using his engineering skill Purcell produced a detailed plan and specification for a horse-drawn vehicle, and advertisements inviting tenders for "one aerial telescopic extension ladder" were published. The response in May 1898 consisted of four tenders, three English and one Irish. Shand Mason of London quoted £575. Merryweather of London sent in a design of its own, but did not quote a price. William Rose of Manchester quoted £245. The Irish tender of £500 came from Jessop Browne of Dublin. The tender of Messrs William Rose and Co was accepted. The committee, "being greatly pleased with the ingenuity of the design submitted by the Chief, and the careful manner in which all the details are worked out", authorised Purcell to protect his invention by means of a patent, which he obtained in 1899.

The first of three Rose/Purcell aerial ladders went into service in 1899,

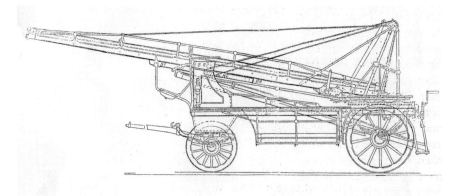

Purcell's aerial ladder, from patent, 1899. Drawing, © Trevor Whitehead

followed by one in 1904 and another in 1908. The new appliance consisted of a main ladder mounted on a turntable at the rear of a four-wheeled carriage. The ladder had two telescopic extensions giving a total height of sixty-six feet.

At the front of the carriage, below the ladder supports and double driving seats, was a large hose box. Space underneath the main escape ladder was designed for the stowage of one fifteen foot trussed ladder and one twenty-eight foot extension ladder. These ladders could be used independently or, if necessary, attached to the top of the main sliding ladder for extra height. The main ladder, when fully extended, had the great advantage of not requiring any additional supports (as some of the currently available ladders did) and could be used as a water tower. Another advantage was that all the ladder's movements – elevating,

Purcell horse-drawn ladder. Photograph, © Dublin City Council

106

extending, rotating – could be easily controlled by one man at the rear of the machine. Purcell's vehicle, equipped with a large escape ladder, two smaller ladders, hose, branches and nozzles, standpipe, crewed by five or six firemen, was a major addition to the fire fighting and rescue capabilities of the brigade. It also, of course, reflected great credit on its designer and was the forerunner of the modern turntable ladder.

**Mercer's Hospital, showing the Purcell Escape, first-ever 100-foot extension ladder in Europe. Photograph, *The Fire Call*, July 1902**

# 8     The years of planning

Sites for building new fire stations were purchased by the corporation at three locations: Great Brunswick Street/Tara Street for £7,632-14-0; Buckingham Street for £1,010-4-0; Thomas Street for £3,724-3-9. These costs were covered by the £12,500 loan which was such a significant part of the 1897 Corporation Act. The proposed central station in Great Brunswick Street was to be designed to accommodate the chief superintendent, inspector, twelve married men and twelve single men, eight horses, three steam fire engines, one hose tender, one horse-drawn fire escape, one tool cart and one ambulance.

The completion of the purchase of sites took place when the corporation acquired a plot of ground extending along Dorset Street to St Mary's Terrace, with a frontage of over sixty feet to a depth of twenty-two feet, for a sum just in excess of £1,000. The four new fire stations, as proposed in 1897, could now be built as soon as financial constraints allowed. The disastrous Chatham Row premises and the tenement accommodation would pass into history, as would the bleak, unhealthy escape stations. But all this change was going to take many years.

The first of the new fire stations, to serve the north-east area of Dublin, was completed in 1900. Situated in Buckingham Street, it was officially opened in August. It was built at a cost of £5,999 by Joseph Pemberton to a design similar to that used in the construction of such buildings in other cities. It was in fact Dublin's first purpose-built fire station, three stories high with a frontage of seventy-four feet and depth of sixty-four feet, containing living accommodation for an officer, seven married men and seven single men. The engine room, forty by twenty-eight feet, housed a steamer and the new aerial ladder, instantly ready with the suspended quick-hitch harness. At the rear was a stable for four horses. Sliding poles from the upper floors gave the firemen speedy access to the engine room. The officer's quarters consisted of five rooms, while the married men each had three rooms, with separate bathrooms, sculleries and toilets.

A spacious enclosed drill yard had a sixty foot high tower for hose drying and drills, and in the basement under the watch room a steam boiler supplied all the hot water required throughout the station. All floors and partitions were of "fire-resisting" construction, walls plastered in Portland cement or lined with glazed bricks. The front of this impressive building, divided into five bays, was constructed of red Portmarnock bricks with moulded granite bases to the piers. The brigade had certainly set off with an enlightened start to the new century.

All the personnel required to staff the newly opened Buckingham Street station transferred from the Winetavern Street station. They vacated the tenements they had endured while providing the city with such excellent service, and this removed some of the contempt which the corporation was held in for the conditions it had forced its employees to live in for so long. With the new station came the new rank of station officer. On 7 August James Markey, with eleven years service and the possessor of two chevrons, was given a wage of £2 per week plus free fuel and light when transferred to the new station as station officer.

The selection and acquisition of sites for fire stations was not without its difficulties. Most of the frontage of the site chosen for the city's new fire brigade headquarters, in Great Brunswick Street, was already owned by Dublin Corporation. The brigade chief considered it "the most suitable and best owing to its proximity to the quays, shipping, and important property, and having such fine open thoroughfares leading directly to every part of the city". However, although a large part of the site had been vacant and partly derelict for years, enclosed by a hoarding, and notwithstanding the fact that the owners and their predecessors had enjoyed the whole of Townsend Street from the city since 1702, at a nominal rent of £4-12-4, yet they refused to sell to the corporation.

The chairman of the waterworks committee decided that it had "no alternative but to obtain a Provisional Order under Sections 76 and 203 of the Public Health Act, 1878, for compulsory purchase under the Land Clauses Act". The land was thus purchased, providing a site on which

> *it was desirous of providing a building which will be satisfactory in every way, combining every facility for conducting the training and maintenance of the Brigade in the very best and most efficient manner. It should also be of substantial construction and harmonise architecturally with other buildings on the fine thoroughfare to which it will form a striking addition.*

It was to fulfil all of these expectations, but not without delays!

In order to pay for the new Dorset Street fire station the corporation agreed to take a loan of £7,500 to be repaid over thirty years at 2¾% from the Local Government Board. The building cost was estimated at £5,700 to £6,000 exclusive of special fittings such as sliding poles, trap doors, hose brackets, engine room and stable fittings, steam boiler, plumbing, gas fittings, electric lights, fire alarms, paved yard and water tanks – these would be covered by an extra £1,500. Under the 1862 Act the corporation was limited to 1½d on the water rate for funding the fire brigade. In spite of the fact that this amount was increased to 2½d as a result of the 1897 Act, the brigade was run on a deficit in both 1900 and 1901 when expenditure rose respectively to £10,200 and £9,182 against income of only £8,320 in each of those years. The corporation was then forced to announce that "owing to the large amount of expenditure on the extensive works being carried out, and the general desire to curtail further outlay of capital, the building of the much needed Central Station, for which plans are prepared … has been postponed for a short period".

The construction of the new station in Dorset Street had been much delayed when the builders went into liquidation, leaving the corporation to complete the work. The stonework on the front of the station bore the date 1901 but it was finally opened in November 1903. It was to cover the north-west area of the city. Inspector James Guildea, with twenty-one years service, was transferred from the central station to take charge of the new one as station officer. His place at Chatham Row was taken by Joe Kiernan, promoted to station officer there after seventeen years service. The rank of inspector was now being phased out and replaced by that of the more focused station officer. Dorset Street was staffed by one officer, six married and six single men, the firemen and their families coming from the condemned slums on Winetavern Street. The engine room accommodated two appliances, and there were four horses.

A few years previously the United Corporation Labourers Union of Ireland had made representations to the city council seeking the provisions of the Local Officers Superannuation Act 1869 "to be extended to labourers and other Servants in the employment of the Corporation" under the proposed Corporation Act, which eventually became law in 1897. The failure of the corporation to include these provisions led to much dissatisfaction among the staff in the brigade. In November 1899 a new claim relating to pay and conditions was served on the waterworks committee by the Firemen's Union. It followed the previous pattern, being signed by almost all the firemen. They called for an extension of annual leave to fourteen days, shorter escape duty periods, twenty-four hours leave every two weeks, a canteen in No. 1 station, a pay increase to thirty-two shillings and sixpence per week after six years service, oil coats for men on escape duty in wet weather, all charges to be cancelled after twelve months good conduct, and a pension after twenty years.

The chief opposed much of this claim, particularly any reduction in escape duty times or the provision of extra leave, which would reduce staffing for fire calls, or permission for a canteen, which would permit "storage of drink in the premises". He maintained that the rate of pay was a matter entirely for the committee and he concluded his comments by pointing out that

> ... if the scheme of reorganisation, as previously recommended in my reports to the committee and adopted by the Council, was carried to completion, most of the matters referred to would no longer have any existence.

The waterworks committee agreed to provide frieze overcoats for men on escape duty, and the removal of a record of misconduct after a reasonable time. They referred to Purcell's improvement scheme which was presently being implemented. This would eventually do away with escape duty, and "when the Brigade is better distributed and housed, night duties in the stations will be modified and further concessions made in periods of leave".

The firemen were not satisfied and re-submitted their claims, changing their call for a pension to one after twenty-five years service. The committee, having just replied to the claims in considerable detail, refused to reopen the matter. In 1898 the firemen's pay ranged from twenty-two shillings and sixpence in their first year to a maximum of thirty shillings after seven years service and four shillings per week lodging allowance. The corporation was empowered to compensate a man if injured in brigade service, but not otherwise. All those seeking retirement had to be certified as unfit to continue in the service through injury or ill-health attributed to the job before they could receive a pension.

In 1900 a special committee of Dublin Corporation visited a number of European cities, particularly in Germany, to examine electric lighting systems.

On its return, a contract was agreed with the British Insulated Wire Company for the provision of a new street lighting system for the city, at a cost of £109,447. This important development was to be a tremendous help to the firemen, especially when dealing with street accidents, river drownings or night-time fires. Unfortunately, it was also to become one of the greatest causes of fires throughout the twentieth century.

In his capacity as a factory inspector Purcell was responsible for fire prevention measures. A factory employing more than forty persons had to produce a certificate stating where the the management had provided adequate means of escape. Every floor had to have alternative exits, and all rooms occupied by more than ten employees had to have doors that opened outwards, and such doors "may not be locked in a manner not easily opened". Purcell made it quite clear that he intended enforcing this regulation, and over the years his annual reports give details of the actions he took against those who failed to comply.

The Dublin Fire Brigade was rapidly establishing itself as a service that was forward looking, professional and a credit to the city. The success of the new emergency ambulance service led to further training of the firemen. The number of qualified men was increased from eleven to sixteen, out of the total complement of forty-four. Some of the trained firemen were taking part successfully in ambulance competitions that were then held annually at the Earlsfort rink. The fire brigade team won first prize in 1900, a success which increased the standing of that part of the service. It seems, however, that not everyone was happy with the progress of the brigade. Newspapers reported that the members of Rathmines Fire Brigade were known to turn their hoses on the Dublin crews when the latter turned out to fires in that township. Rathmines had at that time a brigade consisting of a superintendent (whose main job was manager of the disinfecting chamber) and five men. The Dublin press, witnessing one of these "water wars", maintained that the Dublin men "… getting much annoyed, returned the compliment" but with much greater force and quantity of water.

In 1901 the Boundaries Extension Act of 1900 came into effect and Clontarf, Drumcondra, Glasnevin, Kilmainham, Inchicore, Chapelizod and Phoenix Park became part of the City of Dublin. The area of the city protected by the Dublin Fire Brigade was thereby increased from 3,733 acres to 7,894 acres, and the population from 245,000 to 289,108. The corporation had sought this extra space to build working-class housing estates, to help alleviate some of the dreadful overcrowding that continued to exist in the city tenements. The parliamentary committee of the Irish Trade Union Committee, at its delegate conference in the council chamber of the City Hall in 1900, had issued a statement regretting that "… the Joint Committee of both Houses of Parliament, while recognising the justice of the case made by the Corporation for the inclusion of the Townships of Rathmines and Pembroke, has not given effect to the practically unanimous desire of the citizens of the capital of Ireland". The 1900 Bill to extend the city limits had been bitterly opposed by Edward Carson and Colonel Saunderson, purely on the grounds that the city council had a majority of nationalists and that some members had suffered imprisonment for political offences. Their opposition together with that of Lord Pembroke, Lord Ardilaun and other peers led to a much reduced extension of the city boundary, and the continuation of the two townships of Rathmines and Pembroke.

The ambulance service, which in its first year received 537 calls, responded in 1900 to 660 calls. There was obviously a need for another vehicle, so a second ambulance was commissioned in 1901 and this was stationed at Buckingham Street. From this station it responded to calls from the docks where, almost every week, serious accidents occurred during the loading and unloading of ships.

A new type of breathing apparatus was purchased in 1901. This was known as the Vagen-Bader Head Protector, an American invention. It was a leather helmet, with a cylinder of compressed air fixed to the back, which completely enclosed the head and neck. It enabled the wearer to breathe freely in smoky or other poisonous atmospheres for up to three-quarters of an hour. It was said that the Dublin brigade was the first European one to use this type of helmet even though it had been tested satisfactorily by Southampton Fire Brigade in September of the previous year. It will be recalled that the superintendent of the Southampton brigade had applied unsuccessfully for the position of chief in Dublin in 1892.

Another piece of equipment purchased by the Dublin Fire Brigade about this time was the Foley hose branch. (The branch and nozzle at the delivery end of a line of hose determine the size of the jet.) The Foley was the first hand-controlled branch to be used in Ireland, and it was used to alter the flow of water from a jet to a spray, or to shut off the supply, so helping to reduce water damage.

The Vagen-Bader smoke helmet proved its worth at a tragic fire in the early morning of 13 May 1901. A constable Kelly, on duty outside Green Street police station, noticed smoke issuing from the premises of No. 9 Green Street. This three-storey building with a cellar had a saddler's shop on the ground floor, a basement used for storage which included half a ton of straw, and a total of sixteen tenants. When the policeman raised the alarm the stairs was burning and the saddler's shop was ablaze. Despite this the eight occupants on the ground and first floors were able to escape in their night attire. However, the saddler's apprentice along with a Mr and Mrs Doyle and their five young children were trapped on the top floor. The fire escape from Bolton Street was soon on the scene, but by then flames were rising from the shop window up the front wall and licking the windows on the first landing. The escape was immediately extended and fireman Tom Dunphy climbed to the top-floor window.

With courage and determination he succeeded in carrying down a young boy and an infant aged two years, both in a semi-conscious state. In doing so he received cuts and burns to both hands and had his face scorched. The brigade under Chief Purcell arrived at about this time and soon had water on the fire. By driving some of the flames back from the shop window on the ground floor another rescue could be attempted from the escape ladder. With the aid of the Vagen-Bader helmet Purcell was able to climb the damaged stairs and meet with Inspector Guildea in the top front room. Mrs Doyle and her three other children, in an unconscious state, were taken down the escape by the firemen and removed by ambulance to Jervis Street hospital. Mr Doyle was able to descend the ladder assisted by the rescuers.

The house was extensively damaged before the fire crews could extinguish the flames. The fire escape had also caught fire. The apprentice, James Devine, giving evidence at the coroner's inquest, stated that he was awakened by some-one shouting "fire". The smoke was so thick that he could not get down the stairs, so he went into the room of Mr and Mrs Doyle, awakened them and opened the window. He handed the two children to fireman Dunphy before he himself descended the ladder. The firemen who entered the room from the escape only discovered Mrs Doyle and the other children by searching the dense smoke.

Four children, aged from six to twelve years, died as a result of this tragic fire, in spite of the work and bravery of the firemen. The coroner commending Dunphy declared that it was "a great credit to the City of Dublin that we have such men ready to risk their lives for the purpose of saving the lives of others". Chief Purcell, who saw the rescue, had ordered the hose to be played on fireman Dunphy as he descended the ladder with the two children. Otherwise further deaths would certainly have occurred.

After years of attempts to obtain legal powers for the corporation to grant pensions to firemen, steps were at last being taken. The firemen's cause was supported by Tim Healy, a well-known politician, who procured an amendment to the Dublin Corporation Markets Bill of 1901. The following provisions were incorporated into the new Act.

> *clause 15. (1) It shall be lawful for the Corporation to allow any person employed under the provisions of the Dublin Corporation Fire Brigade Act 1862 – (a) if he has completed not less than twenty-five years' service to retire and receive a pension for life; and (b) if he has not completed a period of twenty-five years' service and is incapaci-tated by infirmity of mind or body occasioned in the execution of his duty without his own default, to retire and to receive a pension for life or a gratuity.*
>
> *The amount of the pension or gratuity must not exceed two-thirds of his salary and emoluments immediately prior to his retirement.*

The first fireman to benefit from this pension scheme was Richard Cullen. He had sustained serious injury at the Trinity Street fire in 1884, and finally retired with a pension of £1-6-0 per week for his thirty years of service. In 1902 Station Officer Markey was diagnosed as having cancer, and was granted extended sick leave.

The Dublin firemen were so impressed by Healy's efforts on their behalf that they expressed their gratitude by presenting him with a beautifully executed illu-minated address and two silver candlesticks. The address included pictures of an officer and a fireman, Dorset Street and Buckingham Street fire stations, and a steamer going to a fire. The presentation took place in the City Hall on 2 January 1902. It is interesting to note that Tim Healy KC, MP was appointed the first Governor-General of the Irish Free State in 1922.

Concerning the granting of pensions, the *Irish Times* commented that there was

*no class of men in the service of the city better entitled to be provided for in this way at the close of a long period of arduous labour, and none of whom the citizens are so justly proud.*

The year 1902 was ushered in by a fire the magnitude of which was even greater than the burning of Arnott's some eight years previously. The magnificent premises of Todd Burns, stocked out with all types of household goods, stood four storeys over basement at the corner of Mary Street and Jervis Street. The brigade headquarters received a call at 2.40 am on 1 January by telephone from No. 4 escape station at Bolton Street. The fireman was alerted by a passer-by that flames and smoke were issuing from windows of Todd Burns. He then headed with his escape towards the crackling and illumination which lit up the nearby streets.

The 200 staff in their dormitory accommodation on the upper floors were awakened by the roaring of the flames as they burst through the shop windows. The fire in the lower floors was raging out of control devouring furniture, domestic appliances, carpets and lino, candles, fuel oils and everything that could add to the inferno. When three sections of the brigade arrived most of the occupants were out on the streets in their night attire, running from the flames which were shooting across both streets and threatening to engulf other premises. Miraculously there were no victims of the voracious tide of flame.

The fire escapes from Bolton Street and Nelson's Pillar were soon pitched and fully charged hose was laid out for the struggle. The darkened stillness of the rest of the city contrasted with the thunderous noise of masonry falling and the shouts of orders from fire officers as the water jets steamed to the sky and cascaded on to the blaze. The staff of the enterprise, who had gone to bed with the usual anticipation of the next day's routine work, were now in the surrounding streets, homeless and frightened as the firemen struggled to gain control. Floors collapsed, and soon the roof was on fire. Not long afterwards the nurses' home of Jervis Street hospital across the street caught fire. Kelly's bookshop also began to burn as a massive fall of bricks crashed across the road smashing its windows. The fire crews redoubled their efforts to contain the fire in the original premises, at times having to retreat or shelter from the debris which continually fell from the building.

When daybreak came the extent of the damage was fully revealed. Two external walls and a chimney stack were all that was left of Todd Burns. The nurses' home had suffered severe fire damage, with most of its windows now open to the elements. Three other premises had structural and fire damage, while Jervis Street and Mary Street were impassable. The total loss was estimated at £170,000 but the plight of the shop assistants was far greater. Their belongings were destroyed, they were now out of work and dependent on the charity of their former employer or the Drapers Assistants' Association, or the good people of Dublin.

Throughout the day firemen worked to extinguish the last pockets of flame, and to assist corporation workers in pulling down the dangerous walls. Crowds looked on in awe at the destruction. The lord mayor revisited the scene accompanied by civic dignitaries and business people, all viewing the damage and contemplating how awful a major fire can be. Newspapers reported on the various funds being set up to help the homeless assistants. The hospital converted a large ward into a nurses' home until reconstruction could take place. Everywhere there was admiration for the work and conduct of the firemen, noting their bravery and fearlessness as they worked in such dangerous and exhausting conditions. The sympathy of the city was with Todd Burns though of course the firm was well insured.

The issue of providing direct funding for the fire service, either from the government or the insurance companies, was again raised in Dublin Corporation. But again, as always in the past, no monies were made available from either of these sources. While the debate went on, the removal of the debris of Todd Burns from the streets continued and the lengthy job of clearing the brick-filled cellars began. Purcell's aerial ladder proved very useful in the hazardous work of pulling down, by means of ropes, the most dangerous of the remaining structures following the fire.

Sir Charles Cameron, the city medical officer, continued to highlight the plight of the poor and the appalling conditions they were forced to live in. The tenement houses were a major source of danger to their inhabitants and to those in the emergency services who were called out when fires occurred or when these dilapidated premises collapsed. Cameron's report listed 786 houses south of the Liffey comprising of

> *2,982 rooms occupied by 2,149 families of 7,984 persons. All should be de-tenanted. On the north side 700 similar houses of 2,401 rooms with 1,496 families of 5,082 persons were also defective and should be pulled down.*

In June 1902 a semi-derelict house in Moore Lane collapsed without warning. Children were known to play in this dangerous structure, so when a local twelve year old boy was reported missing the fire brigade responded. The firemen searched in the rubble for seven hours in the hope of finding the boy, but to no avail. On the following day, as corporation workmen removed the debris they discovered the child's body under a pile of masonry in the laneway.

On an October night that year a tenement which housed sixteen people, owned by a prominent Dublin councillor, Gerald O'Reilly, collapsed killing one middle-aged tenant and injuring four others. The three-storey building fell into the back yard trapping nine of the tenants in the rubble. In the most dangerous conditions and guided only by the cries of those trapped, the firemen dug through the bricks, timbers and masonry. The victims, all of whom were in bed when the calamity occurred, were pulled from the debris and conveyed to hospital by fire brigade ambulance. In spite of the possibility of further collapse, the firemen stuck to their dangerous task until all involved were accounted for. One family, a husband and wife with two young daughters, were asleep on the top floor when the building came down. They were rescued unharmed when dug out from under timber joints and floor boards in the pile of rubble at ground level.

With premises like these throughout the inner city it was no surprise when, on 26 February 1903, a gale force storm swept across Dublin and caused the deaths of two women, injured many more and left massive destruction to property in its wake. Two fishermen died at Howth, and the North Wall lighthouse where two light-keepers worked was swept away in the huge tide after it was struck by a ship which dragged her moorings in the storm. Between 8 pm and 7 am the next morning the fire brigade attended to thirty-five calls to houses that sustained structural damage in the gales.

Hurricane force winds of up to ninety miles per hour were recorded as the firemen traversed the darkened city attending to those in need of assistance. Police reported almost 200 women and children seeking refuge in Green Street station because of damage to their homes, as shop windows were blown in and roofs or slates crashed into the streets. More than thirty people suffering from various injuries were removed to hospitals by the firemen and police. So bad was the destructive storm that the entire telegraphic system for the city went out of service.

Damage was caused to Dublin Castle, Findlater's Church, Dominick Street Church, the Land Registry Office, the Rotunda Hospital, and houses on Mountjoy Square. The Swiss Village Pavilion in Santry was completely destroyed. Despite the terrible conditions of wind-swept streets strewn with slates, broken glass and other hazardous debris, the fire crews responded everywhere assistance was required.

When the storm abated and the canals were still blocked by fallen trees, as were some of the main roads into the city, the city councillors met to discuss the extent of the damage. A proposal by one of the councillors in praise of the firemen for their selfless work on the dreadful night was ruled out-of-order by the lord mayor, who stated his view that it was not proper on the part of the council to move a resolution with regard to its employees. He felt sure however that he reflected the feelings of the council members when he said they were proud of the manner in which their fire brigade discharged its duties; the firemen were a credit to the city and might always be relied upon by the citizens to do their duty.

Thomas Dunphy joined the Dublin Fire Brigade in 1891, just before the introduction of the comprehensive changes to the brigade recommended by the O'Meara Report. With his family he was to live in the deplorable tenement conditions of those who served at the Winetavern Street fire station. He was one of the contingent who displayed great bravery and resilience at the Markets fire of 1892, and the traumatic blaze of Arnott and Company in 1894. We know he took part in the rescues in South King Street in 1897, and received a chevron and a shilling a week extra pay for this outstanding work.

On Wednesday 25 May 1898 the brigade was called to a fire in a house off Infirmary Road. The scene was described by a police constable who was on duty in Dame Street.

*I saw the Fire Brigade men with the hose carriage come along down Trinity Street. They were going at a fast rate. The carriage swung round the sharp corner from*

*Trinity Street to Dame Street at a rapid rate. The next thing I saw was the driver toppling over on his head on to the street. I picked him up at once and conveyed him in a cab to Mercer's Hospital. I picked up the helmet, which was broken.*

The unfortunate driver was Tom Dunphy, and he was found to have a fractured skull and a broken arm. The constable also described how another fireman managed to leap onto the back of one of the horses and stretching down while at full gallop "secured the reins and had the vehicle under control before the corner of George's Street was reached". The doctors described Dunphy's condition as serious. He was unconscious for many hours and was not discharged until almost two weeks later. When the broken arm had healed he returned to work and was soon among those called for ambulance training when the new service was set up.

Records show that he drove the ambulance on its historic first accident case to Sir John Rogerson's Quay in 1899. He showed initiative and courage when he rescued two children, while again sustaining personal injury, at the tragic Green Street fire in 1901, for which once more he received a chevron. Unfortunately this man, so praised for his selfless sense of duty on behalf of the people of Dublin, developed a personality complex in 1902. That year saw a colleague, Fireman Mansfield, receive a pension after sustaining a spinal injury which caused permanent paralysis. The injury was the result of falling from a horse when returning from an out-of-area fire. Another colleague, Turncock Clancy, was similarly treated when he retired following injury after a fall from a horse-drawn engine on the way to a fire.

Following medical examination, Tom Dunphy was declared insane in March 1902 but not entitled to a pension. The city medical officer reported that "Mr Dunphy was twice under my care suffering from dementia ... he was on sick leave for some time but came back quite sane". After a short time back on duty he was again examined and removed to the Asylum of St Brendan in Grangegorman.

A case was made for a pension because his wife and four young children were destitute. But the claim was rejected by the waterworks committee because the medical officer's report maintained that "the lunacy was hereditary, the infirmity of the mind or body was not occasioned in the execution of his duty ... there is, however, insanity in his family". No notice was taken of Tom's work details or the injuries he sustained in rescuing his fellow citizens.

An appeal was lodged when the family, supported by the chief fire officer, procured a specialist's report following a further examination of the patient. In April 1903 the waterworks committee finally relented, and under Section 15 (2) of the 1901 Act they agreed to allow this heroic fireman "a pension of eighteen shillings and sixpence per week for as long as Thomas Dunphy remained insane". But even this was withheld, and was not sanctioned by the corporation until its October meeting. The sad story ended when Dunphy died by his own hand in December 1903, leaving a widow with four school-going children and no income.

The recently introduced superannuation system was being used by the chief to allow some firemen to leave the brigade, so that he could recruit younger and more active men. Typical was the case of fireman Joe Sarsfield who had over twenty-five years of service. He was described by Purcell as follows: "57 years of age, and intellectually he has got very stupid, and is no longer fit to act as a capable member of the brigade. I recommend his application to the favourable consideration of your committee." Sarsfield was granted a pension by the waterworks committee on two-thirds of his wages, an amount of £1-6-0 per week.

At this time the general workers of Dublin Corporation, through their trade unions, and supported by the Dublin Trades Council, demanded similar pension rights to those achieved by the firemen in 1901. In response, the Dublin Corporation Superannuation Bill 1905, supported by the lord mayor and a majority of councillors, was presented to parliament. Although, as with the boundaries extension, this Bill was opposed by Rathmines and Pembroke councils, it did become law in that year. All 1,206 servants employed by Dublin Corporation thus became beneficiaries of the success of their fire brigade colleagues, whose union had opened the door to pensions for all grades of corporation workers. The Act provided for pensions to be paid to those in wholetime employment with at least twenty years service and over sixty years of age, "certified by the Medical Officer on incapacity to do their work due to old age".

The introduction of a system of public fire alarm boxes in the streets of Dublin was first recommended by Superintendent Ingram in the year 1876. His subsequent annual reports continued to refer to the need for such a system, but twenty-seven years went by before it was installed. In 1891 the O'Meara Report had called for "the immediate construction of the most approved system of electric alarms as agreed on 20 October 1890". In November 1889 the waterworks committee had issued a most comprehensive and detailed set of proposals, but nothing was done. The proposals were the result of a visit to London by the borough surveyor, at the invitation of Captain Shaw, to examine the system in use by the Metropolitan Fire Brigade. In 1895 Purcell urged the corporation to implement his improvement scheme which included the installation of the alarms.

Finally, at the beginning of 1903, Dublin saw its first street fire alarms. Twelve of these alarm pillars were erected on the footpaths at street corners on the north side of the city. The box at the top of the pillar contained a telephone as well as the automatic alarm handle. All the caller had to do to send an alarm was to pull the handle. This then immediately registered in the fire station and indicated the location of the fire alarm. Every constable on that beat had a key, to enable him to use the telephone, and a key was also usually kept in the house nearest to the alarm pillar. These new alarms extended out into the new city areas of Fairview, Clontarf, Drumcondra and Parkgate Street, greatly improving the fire cover by a modern system of quick notification of fires direct to the fire brigade.

Following the introduction of the street fire alarms, and the arrival of a second aerial ladder (stationed at Dorset Street), Purcell stated in his annual report for 1904 that all the "old form of vertical fire escapes, together with night watches, have been discontinued on the north side of the city and the new system of aerial horsed escapes is in use instead". Unfortunately he also records that a man was fined £1 for the first false call from a street fire alarm.

The huge fires at Arnott and Company and Todd Burns had let to the questioning of the practice in the drapery trade of compelling staff to live above the potentially dangerous premises. The fact that no deaths had occurred to the occupants on these two hazardous occasions had led to a false sense of safety. This was shattered in 18 July 1905 when the drapery premises of Aiden Grennell in Camden Street caught fire. Although these premises were much smaller than those of the two high street stores, this time three young female shop assistants lost their lives during the course of the fire. There were twenty-three young assistants sleeping in different parts of the building when the fire was discovered.

There was much recrimination in the aftermath of this tragic fire, as some evidence showed that a lengthy delay occurred from the time the fire was discovered to the alerting of the fire brigade. It was stated at the inquest that the deaths could have been prevented even though the girls were "living in an establishment which was practically a death trap if a fire broke out". The city architect told how there were elaborate byelaws for preventing the spread of fire, but there were no byelaws as to the provision of proper means of exit and proper appliances such as fire extinguishers.

Counsel for the next-of-kin of the victims maintained that human life was the most important thing, and if people could not afford to have these facilities (exits and extinguishers) "there should be byelaws in force by which assistants should not be allowed to live over a grave which may open at any moment". Chief Purcell said that "many old residential houses had been converted into business houses, for which purpose they were ill-suited and many of their main supports had been taken away". The Camden Street fire was in premises which were originally four residences. There was a projecting shop front, which meant that firemen were confronted with a sheet of flame before they could reach the residential area and there was no properly constructed passage leading into the main street. Purcell strongly advocated the use of an automatic fire detector, especially in all large buildings. This type of detector had been perfected in recent years and he described in detail how it worked. He agreed to give a practical demonstration by displaying some of these inventions at Chatham Row fire station.

The city council then adopted a number of fire prevention measures.

> That the Corporation do immediately establish a system of street fire alarms throughout the south side of the city … that factories and business establishments be afforded facilities to avail of automatic fire indicators as recommended by Captain Purcell … that the living-in system in business houses should be discontinued.

Despite an evident awareness of some urgency, it was in fact 1908 before the initial stage of the street fire alarm system was completed.

In the meantime one of the townships on the south side decided to build a fire station. The town council of Pembroke had built a town hall at Ballsbridge in 1880, and the following year the Pembroke Fire Brigade was established. In 1905 a fine new fire station was built adjoining the town hall. Over the three-bay engine room and stables was accommodation for seven firemen, and the building cost £3,000.

# 9 Motorisation and a full emergency service

There is no monument to a fireman in the city of Dublin. There is however an impressive memorial, at the junction of Burgh Quay and Hawkins Street, to Constable Sheahan of the Dublin Metropolitan Police who lost his life in a dramatic rescue in May 1905. The city main drainage was being extended and improved by contractors employed by the corporation. An inspection of the sewers was due to take place on 6 May by engineers from the gas company, the corporation and the contractors. A manhole cover was opened in Hawkins Street on the afternoon before, and two men employed by the contractors were left to open the flush gate to allow water to wash out the system prior to inspection.

Just before 4 pm on 5 May one of these workers, a Mr Fleming, entered the sewer which lay some twenty-four feet below the roadway. As he descended he was overcome by fumes and fell into the sewer pipe. The alarm was raised and constable Sheahan who was on duty at O'Connell Bridge arrived on the scene. By then the second worker had entered the dangerous atmosphere and he too was in difficulty. An onlooker with a length of rope tied around him, accompanied by constable Sheahan, then descended into the sewer.

The fire brigade ambulance was called and on arrival was met by a large number of confused onlookers. Some maintained that an explosion had taken place in the sewer where a number of people were trapped. The firemen immediately sought further assistance from their station and set about effecting a rescue. Fireman Lambert accompanied by a Mr Murphy entered the sewer but both men were soon in difficulty. Murphy however did succeed in dragging the now unconscious Constable Sheahan up to the street with the help of the firemen.

When the fire crew and another ambulance arrived there were still five people in the sewer to be rescued. The firemen, realising the dangers involved and acting with great courage, carried out a focused and sustained rescue operation. They were now risking their lives because others with wasted energy and reckless courage had put themselves in unnecessary peril. Effecting a number of entries using the Bader smoke helmet, and needing to travel a distance of a hundred feet into the sewer pipe, they finally removed all the victims. Both Constable Sheahan and Mr Fleming were pronounced dead on arrival at the hospital and three others spent some days under treatment before being eventually discharged.

There was much conflicting evidence at the inquest, but the autopsy showed that both victims had died as a result of sewer gas from the stagnant water. The contractor, a Mr Martin, gave evidence that there was no reason for his workers to enter the sewer, because the flush gate could be opened from street level. He maintained that the tragedy was the result of "people going down the manhole and keeping ventilation from coming out of it. The manhole was 2ft 6in square and when a man's body goes into it much ventilation cannot get out."

The *Evening Mail* reported that there was excitement and confusion in a section of the crowd that gathered at the incident because "the crowd was largely composed of 'hooligans' of the worst possible type, utterly indifferent to all considerations save their own contemptible amusement, and to them anything in the shape of interference with the police was enjoyable". This criticism arose primarily because of a claim that the manhole cover had been closed after constable Sheahan entered the sewer. The allegation however was refuted in the *Freeman's Journal*: "A hero, not a 'hooligan', is the word to describe John Murphy who at imminent risk of his own life rescued, though alas too late, the body of heroic constable Sheahan".

Following this tragic incident seven firemen were recognised for exceptional bravery in risking their own lives attempting to save others. Firemen M. Lambert and C. Kelly were each awarded a chevron. Firemen E. Blake and M. Jennings were each awarded £2, and Firemen R. O'Hara, M. Kelly and M. O'Farrell £1 each. In addition, Martin Jennings received the bronze medal of the Royal Humane Society and the life-saving medal (bronze) of St John of Jerusalem. The brigade's one and only smoke helmet had been used in the rescue attempts by Blake and Kelly. The awards were recommended to the waterworks committee by the fire brigade chief.

Constable Sheahan was given a civic funeral and his colleagues in the police some years later received permission from Dublin Corporation to erect the substantial monument which now stands to his memory in Hawkins Street.

In 1901 Purcell, bearing in mind the considerable expansion of the city area, had expressed his anxiety about the water supply to the waterworks committee. Owing to the increasing demand for water for manufacturing and other purposes, the mains were no longer sufficiently large to afford a satisfactory supply and sufficient pressure for fire extinguishing purposes in some districts. A new reservoir at Roundwood was opened in 1905 and water from this extra source was used to improve the supply to areas on the north side of the city.

Work on the long awaited new central fire station was finally completed and the building was occupied by the brigade on 23 April 1907. It was not officially opened until September. The site comprised a series of buildings fronting on to Great Brunswick Street, Tara Street and Townsend Street. The main fire station building, a three-storey structure, fronted on to Great Brunswick Street, now called Pearse Street. The engine room, with five bays and glazed teak exit doors, was sixty-five by thirty-six feet and directly behind was an area sixty by twenty feet with stabling for seven horses. The ground floor was completed by a watch room, an office, harness room and lavatory.

Accommodation for twenty single men was provided on the upper floors, together with a clothing store, mess room, kitchen, two lavatories, bathroom, and a gymnasium sixty-five by thirty-six feet with an open roof supported by steel trusses. Three brass sliding poles from the upper floors to the engine room enabled rapid response by the firemen and the horses, trained to react to an alarm, could release themselves from the stalls and run forward to their places under the quick-hitch harness.

The watch room was the nerve-centre of the communication system. It contained telephones and the fire alarm terminals and annunciator. It co-ordinated as well all the services of the station including heating, lighting and water supply, and of course the room was continuously manned. Although all the wires and piping relating to the central station, together with those from the district stations and the street fire alarms, terminated in the watch room, not a single wire or connection was visible in the room.

In the basement was the battery room, coal store and boiler house which distributed heat throughout the establishment. Also included was an efficient laundry with drying closet for the married quarters.

The block of buildings fronting Townsend Street, three-storeys high, comprised the residence of the second officer and quarters for nine married firemen and their families. Each set of quarters consisted of two or three bedroom apartments, with a bathroom on each floor, and entrances from open balconies. The enclosed flat roof over the block provided a safe playground for the children.

The Tara Street frontage included the most impressive architectural feature of the site. This was a 125ft clock tower with an open balcony near the top. This is the only part of the 1907 building which has been incorporated without alteration into the present-day buildings on this site. Below the clock tower was the main entrance to the fire station. Internally the tower was divided into two sections, one fitted with hose hoist, racks and heating pipes for drying hose, the other housing a stairway up to the clock chamber and the balcony, from which there was an excellent view of the city and suburbs.

The official opening of the new fire brigade headquarters on 13 September was an historic day for the brigade. It marked the end of the wholly unsuitable Chatham Row building, with its poor location and lack of space. It also allowed the corporation to relocate the firemen and their families who had endured atrocious living conditions in the cottages opposite the old station. Contemporary newspapers gave extensive coverage to the pageant which accompanied the opening of the new station.

*The opening of the new station yesterday marked the end of the Chatham Row premises as the headquarters of the Brigade. The old headquarters will not be used for fire extinguishing purposes any longer and, indeed, their awkward situation in the matter of speedy egress always made them a most unsuitable spot for the work of the Brigade, while their limited space prevented adequate accommodation being provided, especially for married men. However, in the new Central Station the Brigade will possess housing accommodation of the most ample, comfortable and up-to-date character.*

*The opening ceremony took place at four o'clock and the guests, including the Lord Mayor, were received in the station by Mr. J. M. C. Briscoe on behalf of the Waterworks and Fire Brigade Committee of which he is the chairman. Captain Purcell, the efficient Chief of the Brigade, and Mr. C. J. McCarthy, City Architect, who were jointly responsible for the arrangement of the building, were also present.*

Mr Briscoe's speech of welcome was reported extensively in terms similar to the following account.

> *Having visited the fire stations in the chief towns in England and in many continental centres, he was glad to be able to say that only nine of them surpassed the Dublin station. In fact, in many respects, but especially in that of electrical and time-saving equipment, they in Dublin were far ahead of other places. This was largely due to the*

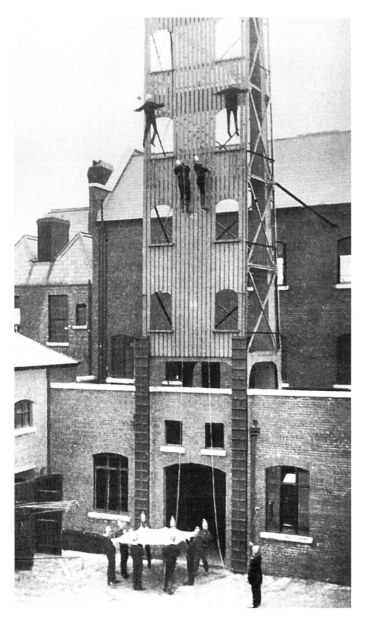

**Display of hook-ladder drill and jump-sheet, Tara Street Station, c. 1910. Photograph, © National Library of Ireland**

*Chief of the Brigade who was a practical engineer and electrician, and the Corporation in having the benefit of his experience and advice had been able in working out the details of the station to save the ratepayers a very large sum of money.*

*The building had been carried out entirely by Irish hands, and he might mention that the electrical and other fittings, manufactured in Dublin under the direct supervision of Captain Purcell, were a credit both to the contractors and to the workmen .... it was now his very great honour and privilege to declare the Central Fire Station of Dublin open, and to announce what would interest them as ratepayers that they, the much abused Corporation of Dublin, had been able to erect this magnificent structure at a cost of £21,840 without the addition of a single penny to the rates.*

*The company was then treated to a turn-out by portion of the brigade. When the alarm sounded the men dropped down a sliding pole into the engine room from the dormitory overhead, where the first turn-out staff are on duty. The horses at the rear of the engine room were at the same time automatically freed from their stalls, and showed their careful training by rushing up to their places in front of the carriages. To drop the harness hanging overhead on their backs and adjust the patent couplings was the work of a moment, and when the men took their seats on the carriages an overhead lever was pulled, as a result of which the doors swung open and the carriages, one of them belonging to the ambulance department, dashed into the street. Demonstrations with the aerial ladder followed, after which the men went through a course of life saving exercises on the drill tower, a steel skeleton erection 67ft high, with front and floors representing a six-storey building. Then the visitors were shown how the men are trained to work in smoke, in an underground room especially devoted to that purpose, from which dense fumes were issuing.*

The overall reorganisation scheme of the brigade, of which the new central fire station was the most important aspect, provided for modern district stations at Buckingham Street, Dorset Street and Thomas Street. The last named, yet to be built, would replace the makeshift station at Winetavern Street. Completion of the scheme would finally see the end of the street escape duty so often complained about by the firemen, and by using man's latest discovery, electricity, the central station could be automatically in touch with every part of the city. All that was said about the new headquarters building before it was built proved to be true: it added to the architectural grandeur of the city and it gave the south side of the city a focal point in the four-faced clock tower which was illuminated the following year. But most of all, it provided a worthy headquarters which was to serve the Dublin Fire Brigade for the following ninety years.

The year 1907 was such a milestone in the brigade's history that the chief's annual report deserves a close look. It is full of interesting statistics and gives the reader a clear picture of the progress being made and the problems faced.

The brigade received 250 fire calls during the year. The estimated value of property at risk was £971,092 and the value of property destroyed was £57,607. Twenty-two people received injuries requiring removal and detention in hospital, three of whom died from shock or burns. These figures did not include the many injuries to – and deaths of – infants and children as a result of burning from exposed or open fires in the tenements, in spite of Chief Purcell's continuing call for these to be included in the statistics.

The Brigade now had a total wholetime staff of forty-eight.

- One chief; one deputy (lieutenant); three station officers; one inspector of escapes; two foremen; forty firemen. There were also two turncocks attached to the brigade.

The staff was deployed as follows.

- Winetavern Street, 2 officers, 6 men, 1 horse
- Tara Street, 2 officers, 23 men, 8 horses
- Buckingham Street, 1 officer, 7 men, 2 horses
- Dorset Street, 1 officer, 6 men, 2 horses

The mobile horse-drawn appliances were as follows.

- 800 gallons per minute steam fire engine
- 300 gallons per minute steam fire engine, plus one similar in reserve
- 66ft aerial ladders
- hose carriages with folding pompier ladders
- hose wagon
- ambulances
- trap for the chief

The other main items of equipment were as follows

- telescopic fire escapes
- 3 telescopic ladders
- jumping sheets and life lines
- hand pumps
- Bader smoke helmet
- 1 72ft Clayton fire escape with fore-carriage and shafts, kept in reserve

The brigade received 1,780 ambulance calls in 1907, in contrast to the 537 calls in the first year of the ambulance service.

The annual report revealed other interesting items. Purcell was granted permission "to obtain an occasional suit of private clothes in lieu of uniform" because he often had to attend fires when wearing private clothes, which were then damaged. Lieutenant Myers, the deputy chief, was given a wage increase amounting to annual increments of £20 per year over the next five years.

The Irish Cap Factory of South Great George's Street, failing to comply with the chief's factory notice, was taken to court and a closing order obtained until it complied with the corporation requirements. Two new horses were purchased at £40 and £55 respectively, and one horse, Roundwood, was destroyed after twenty years service. The average cost of forage and bedding was eight shillings and eleven pence per week per horse. Two men were jailed for arson in the city.

In 1907 the brigade had three large outbreaks of fire to deal with, as well as a very strange occurrence. On 17 July a fire tender responded to a call to Berkeley Road church. In the early hours of the morning the firemen entered the smoke-filled mortuary chapel where they discovered the remnants of a coffin, its stand and black cloth cover burning away on the timber flooring. Lying in the burning debris were the charred remains of a female corpse whose coffin had been brought to the church the previous evening. The fire had evidently started in the coffin, but no one could give an explanation of the original cause of the fire. A macabre situation!

The first of three large fires in 1907 broke out at 11.25 am on 6 January in the gas works in Great Brunswick Street. It followed an explosion in a gas purifier which injured eighteen of the staff – two of them died later in hospital. Two chambers in the gas works were completely wrecked and a large heap of oxide under the purifiers caused serious problems for the firemen before the massive blaze was finally extinguished. It took over four hours to come to grips with the fire, but it was confined to the two chambers, where everything was destroyed with a loss of £2,170. Most of the injured were released from hospital later in the day and those detained were eventually discharged and were able to return to work.

The second big fire, in a building occupied by a silk merchant and costumer, was discovered on 2 November at 7.05 am following a gas explosion in the basement. It was a five-storey building over basement with a forty-five foot frontage in Grafton Street and four storeys in Wicklow Street with a thirty-six foot frontage which had only recently been totally refurbished. The explosion blew out the plate glass windows and shutters at ground floor level setting the contents of the 7,000 square foot warehouse totally alight. One servant on the premises was injured and taken to hospital and the other eleven occupants who had been sleeping on the upper floors made their way to safety through emergency exits recently constructed on the advice of Chief Purcell. It was fortunate that the explosion did not take place an hour later when the streets would have had a sizeable number of pedestrians passing by – serious injuries might well have been sustained by many as the glass exploded across the footpaths and streets.

The firemen, on arrival, found that the fire had spread from the basement through openings in the ground floor, setting alight the shop's stock and warehouse contents and quickly engulfing the roof. Firemen entered the sleeping accommodation to search for any occupants who might be trapped and rescued one injured person, a Miss Murphy, who suffered a broken collarbone and burns to her face and head.

Throughout the day the crews struggled to contain the growing inferno, but with such large quantities of highly inflammable material in store it was not possible to arrest the spread of fire. When the struggle ended bare walls and a mass of contorted ironwork gave a grim reminder of the ferocity of the fire. Stands, which the day before showed off the most expensive costumes, now stood in strange positions, charred and ugly. Glass and woodwork were strewn all over the place as plaster and furnishing gave a spectre of extraordinary chaos.

The firemen had done excellent work, as letters to the papers attested. One such from a prominent resident of Grafton Street read as follows: "nothing could exceed the systematic way in which the whole staff went to work, and they certainly deserve the best thanks of the citizens for their wonderful achievement". The total loss was put down at more than £11,000 but it could have been much greater if the attack on the fire had not been so well organised and persistent.

Only twenty-four days later a destructive and prolonged fire broke out at the

Dublin Granaries on Hanover Quay. It was discovered about 11 am on the ground floor of the two-storey building, which had a frontage of 180ft in a block of warehouses stretching back over 120ft. The premises were stocked with malt, maize and flour. On the arrival of the first fire engine the warehouse was an inferno and the timber roof was almost totally ablaze. Six lines of hose were laid down and the steamers, drawing huge water supplies from the canal basin, threw up to five tons of water per minute on to the flames. But even with this massive amount of water it took eleven hours solid work before the fire was under control.

For the next seven days exhausted fire crews damped down the site before the fire was completely extinguished. A loss of £31,817 was sustained by the owners as a result of this lengthy and spectacular fire. November had been a tough hard month for the fire brigade. The firemen were certainly earning their recent pay increase which brought the maximum on the pay scale up to thirty-five shillings per week. The station officers' increase of ten shillings brought their wages to fifty shillings per week.

Arson continued to be a problem in the city. Following a fire at Bridgefoot Street where two tons of hay and the roof of a storage shed were destroyed, a Michael O'Brien was convicted of arson and sentenced to one year in prison. After a fire in a stable in Westland Row, a Michael Brady, who had previous convictions for arson, was sentenced to three years penal servitude. Even these harsh sentences seemed not to deter the city's arsonists, because they continued their destructive way in various parts of the city.

Since the introduction of the internal combustion engine at the turn of the century the Dublin street scene had changed dramatically. Motor driven vehicles, both private and commercial, were steadily increasing in numbers. This, and the finalisation of the electrification of the tram system, now led the corporation to set up a meeting to decide on a speed limit for the trams in the city.

*The police were not prepared with their views on the matter, so the Corporation went ahead and introduced a speed limit of 8 mph in the city centre, 10 mph from Amiens Street to Dollymount or Whitehall, and 4 mph at curves, crossings and gradients.*

Fire brigades had been experimenting with motor vehicles for a number of years, and these vehicles were already in use in many places including Tottenham, Finchley, London, Birmingham and Glasgow. Chief Purcell was very conscious of the urgent need to find a replacement for one of his steamers which had been in use for forty-three years. Ever on the lookout for ways to increase the efficiency of his brigade, he began to seriously consider purchasing a motor engine. Over the years improvements had been made to steamers, the most significant of which was the design of a self-propelled steamer by Merryweather in 1899. The firm of Shand Mason some time later produced a similar type of steamer using the steam to drive the road wheels. Purcell contacted them to obtain full technical details of their machine for consideration. In September 1907 he wrote to Henry Simonis and Company in London requesting detailed information on their electric fire engine. He specified the kind of motor tender in which he was interested: "electric front wheel drive from storage batter, with a

127

discharging capacity of 20 miles. To carry six men and driver; one hook ladder; one 35 gallon first aid with 150ft of hose. Top speed 25 mph." Shortly afterwards he attended an exhibition of motor fire engines in Glasgow, meeting the different manufacturers and discussing the merits of their designs.

While Purcell worked on his plans for modernising his fire engine fleet, the firemen were pressing for a shorter working week. Their claim was for either twelve hours (11.30 am to 11.30 pm) leave every eighth day or twenty-four hours leave (8.00am to 8.00am) every tenth day, instead of the current ten and a half hours (1pm to 11.30pm) every eighth day. The claim, which was referred to Chief Purcell by the waterworks committee, was answered on 14 September.

> Our men are allowed to remain out until 11.30 pm every night. They come in without any inspection as to their fitness for duty, and if a man happens to be unfit and a fire call comes later in the night, he may not be noticed until an accident happens to himself or someone else as a consequence of his condition. The present system is not conducive to the efficiency of the service, and is not to the men's personal advantage.

Purcell went on to recommend the system then pertaining in Manchester.

> To grant every member of the Brigade, including Officers and men on night duty, one whole day and night of 24 hours off on every thirteenth day – in addition to the 14 days annual leave which is presently allowed them.

This proposal was adopted by the corporation before the end of 1908.

A fire brigade, consisting of sixteen men, was specially formed for the Irish International Exhibition which was held in Herbert Park in 1907. One morning in November it gave a public display. The turn-out was at 11 am. At that hour the brigade mustered, smart and business-like, under the command of Mr Harris. They lined up at the end of the Grand Central Avenue, near the Donnybrook entrance, and at a given signal – an imaginary fire alarm having been set off – they shot off into action. Each man bore a reel of hose, and this was unfurled, length after length, so as to connect up the line with the hydrant. In the first instance the line was connected to a distance "as far as the Vegetarian Restaurant from the Morehampton Road entrance".

The ladders were quickly run up against the building, and the hose being connected with the nearest hydrant, a liberal stream of water was directed onto the building.

> Having copiously douched the building to the satisfaction of themselves and the onlookers, the brigade was now in readiness to respond to another call. This time the scene of the "outbreak" was the Art Gallery, and once more the lusty men of the brigade sped off to extinguish the flames. Three lines of hose were rapidly laid down in sections, then connected and fixed to the nearest hydrant, and soon copious jets of water were poured from the roof of the Art Gallery. This operation, like the others, was carried out with marked rapidity and success. Mr Hutson, Chief of the Pembroke Fire Brigade, was present on the ground; and his valuable advice was freely placed at the disposal of the Exhibition force. The Exhibition Brigade is, of course, intended

*chiefly for first-aid work. In the event of the serious spread of an outbreak the services of the Dublin and Pembroke Brigades would be requisitioned. Incidentally, Mr Harris, who commanded the brigade, had been in charge of the Carlow fire brigade and was previously in the Melbourne fire service.*

When the exhibition was concluded, the large steam engine of the Dublin Fire Brigade was used to pump out the artificial lake which had been created in Herbert Park.

In September 1908 Chief Purcell reported that the street fire alarm system was now complete, following the installation of eighteen alarms on the south side of the city. The total number throughout the city was now thirty-three, although the original recommendation was for sixty alarms. All night duty with escapes, except that at St Catherine's Church, Thomas Street, was now discontinued on the south side. This one street escape station would remain until the proposed new fire station was built in Thomas Street.

In November of that year the council chamber in the City Hall was badly damaged by fire, with much of the original woodwork destroyed. There were differing opinions as to the cause of this fire, but it seems to have been the first serious fire in a public building in the city caused by an electrical fault. Following the fire Purcell decided to write to the Dublin Corporation estates and finance committee recommending a fire alarm system for both the Mansion House and the City Hall.

*In view of the serious consequences resulting from the recent outbreak of fire in the City Hall, and the irreparable loss of pictures … when it is remembered what valuable furniture, pictures and priceless works of art are in the Mansion House, the many historical and other important records in City Hall …. A modern invention providing a simple system of automatic detection, by means of which any sudden outbreak of fire will actually ring a warning bell … this "May Oatway" detector has been installed in many public and commercial buildings in England and Scotland … it has successfully saved many buildings including the Irish Times printing works.*

The fire chief's recommendation was adopted by the corporation and work on installing the system commenced in 1909.

Whether these extra costs had any bearing on the matter or not, things were not always going Purcell's way in his dealings with the council. Early in 1909 an attempt was made by a number of councillors to shelve the plans for the Thomas Street fire station. They demanded that the money allocated be used to pay off some of the loans incurred by the corporation "as the city seems abundantly supplied with fire brigade stations". Fortunately, after a lengthy debate, this retrograde motion was voted down. But it was not until November that a final decision was made to press ahead with the building of the station. (The original decision dated back to 1898.) The councillors agreed that Winetavern Street was no longer acceptable as a fire station.

*The firemen and their families are living away in Cook Street under bad conditions in the worst description of old tenement houses. The station is badly located with regard to lives and property in Thomas Street and large districts south and west of it. The horses have to crawl with the men dismounted, and the firemen walk up St Michael's Hill every time they have to proceed to a fire. It is not intended to make any addition to the staff or annual expense but merely to transport it to a more suitable centre.*

One can well image that this ludicrous situation must have made a laughing stock of the firemen every time they turned out!

Chief Purcell continued to study the different types of motor vehicle being brought into use by other fire brigades. He made himself familiar with the mechanical construction of the various commercial chassis being produced. In March 1908 he wrote to Argyle Motors in Scotland telling them that he was "interested in procuring a motor pump", and that he was favourably impressed by the first fire engine built by Dennis Brothers of Guildford for the City of Bradford in that year. He was not, however, entirely sure that any of the available motor fire engines would suit his particular requirements So he decided to use his engineering skills to plan his own design of fire engine. An advertisement inviting tenders to supply one motor fire engine for Dublin was inserted in the press in March 1908. After due consideration of the response, Purcell was convinced that Leyland of Lancashire was the company he should go to. In one way this was a strange choice because the company had never yet built a fire engine. But after considering Purcell's request and examining his detailed plans the company agreed to do the work. It had gained a very good reputation for both steam-engined and petrol-engined commercial vehicles and, because of its lack of experience with fire engines, it offered to take back the engine and cancel the order after a full trial within one month if it was not found satisfactory in every way.

At the June meeting of the full council the waterworks committee was recommending the purchase of not one, but two, new motors. Purcell pointed out that some of the present equipment had been in service for over forty years and was now obsolete and unfit for modern fire fighting. Also, water mains in some districts had become internally corroded so that they were no longer capable of delivering sufficient water to cope with a fire of any magnitude. The recommendation was to no avail, and the council rejected it by twenty-three votes to seven.

Purcell was determined to persevere. He appealed against the council's decision and requested that he be allowed to attend the next full meeting to explain in detail the savings which would arise and the major advantages that would accrue from a motor-driven vehicle. His request was rejected, but he then submitted a document listing the motors in use in Liverpool, Glasgow and Edinburgh, all cities which he said had better water supplies than Dublin. He went on to show that the replacement of two obsolete appliances would cost a total amount of £725 – a new steamer would cost £500, a new hose tender £105, and £120 for two replacement horses. A Leyland motor fire engine at £875 would cost only £150 more, yet it would mark a significant step towards bringing the Dublin Fire Brigade into the twentieth century.

The motor vehicle would mean an operational saving of £153 per annum for horse fodder and veterinary services, while "reducing the average time of extinguishing fires, including travelling to the scene and returning to the station, by anything up to 24 minutes per alarm".

The saga of the Dublin brigade's motorisation was followed in the widely read journal *The Commercial Motor* and its issue of 15 July contained the following comment.

> *The Dublin City Council, as we anticipated, has decided not to make itself foolish in the eyes of the public, and some 45-year-old engines are at last to be replaced with up-to-date plant. We congratulate the Chief Officer of the brigade, Mr Thos. P. Purcell, on the fact that his committee has carried the vote of the Council by 30 to 12. It is a Leyland engine that will be supplied.*

So Purcell's detailed submission had finally persuaded the city council to sanction the purchase. A Dublin newspaper's comment was the following: "Dublin will now be able to hold its own in reputation with Liverpool, Edinburgh and Glasgow which have 13 motor fire engines between them".

Chief Purcell's plans and technical drawings for the construction of the new appliance were obviously examined and discussed in great detail when he visited the factory in the town of Leyland. The company followed his instructions most accurately, with a few modifications mutually agreed. The original design called for one very large central headlight, but Leyland had a problem with this, and on 25 November Purcell agreed to the fitting of two headlights, one on each side of the machine. Another interesting point was the change of registration number. On 12 November Purcell notified Leyland that it would be RI 1080, but progress was slower than anticipated and by the time the appliance was actually completed and undergoing its final tests the number had to be altered to RI 1090 on 7 December.

At last the great day dawned. The first Leyland motor fire engine was shipped to Dublin, arriving on 16 December 1909. From the docks it was driven by Henry Spurrier, one of the two partners of Leyland Motors Ltd, to Grattan Bridge (Capel Street). There its pumping capacity was demonstrated to the delight and satisfaction of Chief Purcell, council officials and many members of the public. The motor had an internal combustion engine of fifty horse power, four cylinders having dual ignition, and forced lubrication to all bearings. Four speed gear was provided, with the highest on direct worm drive to back axle being thirty miles per hour. The Mather & Platt pump, situated at the rear, could deliver 350 gallons per minute at a pressure of 120 pounds per square inch. Behind the water pump was a large copper cylinder and air pump which could create an immediate vacuum when required to lift water from river or canal. The fire engine was designed to carry a driver, officer and eight men, with thirty-two feet of five-inch suction hose, 2,000 feet of delivery hose standpipes and hose fittings plus twenty-six gallons of fuel. (See colour section, illustration number 8.)

To firemen accustomed to handling reins and controlling horses, the new motor must have seemed a strange beast. It had "the contractor's special method of steering, by which the front wheels pivot in the centre of hubs, thus relieving the gear from shocks due to roughness or obstruction, and transmitting them direct to the axle. Solid rubber tyres of ample section are provided." Over the following months intensive driver training was carried out, and it is not surprising that there were accidents.

In January 1910 Purcell wrote to the waterworks committee apologising for accidental damage to railings in Phoenix Park caused by drivers practising with the motor. £4-3-0 worth of damage was paid to the public works committee as drivers gained experience. The engine was insured for accidental damage or injuries to the public for a sum of £3,500.

Belfast, with more access to revenue than their nationalist counterparts in Dublin, and probably in more immediate contact with modern industrial development, had set itself on a course of doing away totally with the use of horses in the fire service. Between 1911 and the end of 1913, Belfast Fire Brigade was completely motorised. The new fire appliances were supplied by John Morris and Sons Ltd of Manchester and the fire fighting fleet consisted of eight motor pumps, one turntable ladder and one salvage tender plus one car and three ambulances. The brigade's steamers were bought by another brigade in Ireland and the last remaining horse was sold in January 1914. The whole re-organisation scheme was an incredibly bold and forward-looking venture at this early stage in the history of motor fire engines.

Purcell undoubtedly followed the same logic as his colleagues in Belfast and elsewhere in the UK, but lack of financial resources and backward thinking by the councillors of Dublin held back such progress. It is interesting to know that William Cosgrave, who was to become the leader of the Free State, was one of those who opposed the motion to buy the first motor fire engine, claiming that "he had no complaints of the Fire Brigade as it at present exists being late for fires".

When fire broke out at the extensive business premises of Edward Lee in Georges Street, Kingstown, on the afternoon of 17 December 1909, the brigades of Blackrock and Pembroke were called to assist the local brigade. When the ground floor area erupted in flames the shop was full of drapery items for the Christmas shopping and its windows displayed the finest of materials. With the upper floors supported by cast iron columns on the cantilever principle, the firemen tried to confine the raging fire without playing large amounts of water on the columns for fear that if they fractured the whole three-storey building would collapse.

Mr Lee, the owner, was not convinced that the attending brigades could contain the fire without damage to adjacent properties, so he personally requested the attendance of the Dublin Fire Brigade. Purcell and his men responded without delay and their new Leyland, again driven by Mr Spurrier, reached the scene of the fire within seventeen minutes of the receipt of the call. With a total of eleven lines of hose from the four brigades present, the fire was surrounded and extinguished. The motor, at its very first fire, impressed all those present with its powerful engine and capacity to deliver such huge quantities of water at such pressure. Within half an hour of Purcell's arrival Pembroke and Blackrock firemen were gathering up and soon heading back to their stations.

Two Kingstown firemen were injured by exploding glass from windows and required hospitalisation, and a number of onlookers were similarly injured when the main ground floor windows of the Lee premises crashed out into the street followed by flames which actually shattered the windows of Hamilton and Long on the opposite side of the road. Damage was estimated at more than £10,000 but the new motor fire engine had shown its worth in the speed with which it could reach a location some seven miles away and the work it could perform. The Dublin firemen, and their motor, were cheered as they moved off and it was said that they arrived back in Dublin in time for Spurrier to catch the evening boat to Liverpool!

This assistance from the Dublin Fire Brigade was not always welcomed by the chiefs or crews of the brigades in the townships. Some years earlier when Purcell and his men attended to assist the Pembroke Brigade at a large fire opposite the Tramway Company power station on Ringsend road, having been called by the Metropolitan Police, their assistance was declined by Captain Hutson, the Chief Officer of Pembroke. Later, when the owner of the stricken premises personally requested the Dublin firemen to attend, Purcell advised that he could not respond unless requested by Captain Hutson. The *Freeman's Journal* reporting on the scene, watching the arrival and departure of the Dublin brigade while the fire raged out of control, questioned Captain Hutson on these strange happenings and received the following cryptic response: "I told Captain Purcell I did not need his services and as I did not trouble him, he should not trouble me". The rivalry of the insurance brigades had continued between the municipal brigades of the twentieth century, and it was to exist right up to the 1981 Fire Services Act.

Perhaps out of shame or embarrassment, when the extensive premises of Crampton's Hammersmith Works in Ballsbridge caught fire on 15 June 1910 Captain Hutson did call for assistance from Dublin Fire Brigade. It was clear once the fire had got a hold, that Pembroke on its own could not cope. Purcell responded on the Leyland motor with fourteen men and set about tackling the blaze from the Shelbourne Road side of the premises to stop the spread to the Veterinary College next door. Again with a plentiful supply of water present the motor pump was put to good use, first containing the fire and then playing the major role in finally extinguishing it. The premises were almost totally destroyed, as were the contents of building plant materials, huge amounts of timber, felt and fuel. However, although the attached stables were gutted, the firemen assisted by civilians saved all the horses, and prevented the destruction of a cottage adjacent, which suffered some fire damage. The scene drew crowds of the usual night onlookers as the timber and fuels threw massive flames skyward to be transformed to great clouds of steam and smoke by the water torrents poured on by the firefighters. Newspapers were again full of praise for the power of the motor engine, but even this failed to change the minds of the city councillors on the need to fully motorise the Dublin Fire Brigade in the shortest time possible.

Chief Officer Hutson, however, knew that he would soon be on an equal footing with the Dublin brigade, because he had already ordered a motor pump from a well-established manufacturer. The new appliance for Pembroke was supplied by Merryweather, the company that had built their its first motor fire engine in 1903 for the Borough of Tottenham in Middlesex.

A public trial of the new motor machine took place at the Tramways Depot at Donnybrook on 4 August 1910. The large gathering included the members of Pembroke Urban District Council, their chief fire officer, J. C. Hutson, and his father Alfred Hutson who was Chief of the Cork Fire Brigade.

The new appliance was of particular interest for two reasons – it was the first Merryweather motor fire engine supplied to Ireland, and it was painted an unusual colour. Instead of the traditional red, it was painted white and lined blue and gold. This was adopted as the standard livery for Pembroke's fire and ambulance vehicles. Enquiries regarding motors had been made to thirty brigades in England and Scotland before the Merryweather "Hatfield" motor pump was selected although, costing £1,000, it was more expensive than the Dublin Leyland. The Pembroke machine, registration number IK686, gave good service to the township, and eventually passed to the Dublin Fire Brigade after the city boundary extension of 1930.

As well as dealing with fires in the surrounding townships, the Dublin motor fire engine was also showing its prowess at fires in the city. On the cold murky morning of 28 October the biggest fire in the city that year required the commitment of all the firemen, the leadership of Purcell and the power of the Leyland to contain the blaze that destroyed Archer's huge timber stores and saw-mills on the North Wall. On the arrival of the brigade the interior of the premises was a sheet of flame shooting up in glowing showers of burning timber, illuminating the entire dock area.

A strong breeze fanned the flames, engulfing the 200 x 250ft two-storey main building and spreading to the surrounding sheds and yard, stacked with timber. From the far side of the river the steeple of St Barnabas Church seemed to be rising from the flames, and firemen were valiantly trying to save the cottage homes of twenty or so families which stood barely twenty feet from the raging inferno on the other side of the road. With the proximity of the Liffey, water was in abundant supply, and was certainly needed. It was later reported that some of the city councillors were on hand to see the trojan work of the fire crews and their engines as they came to terms with the flames at close quarters.

Cattle pens at the railway yard nearby had to be emptied because burning embers carried by the high wind terrified the cattle awaiting shipping to England. The efforts of the brigade at length put an end to further damage although the fire took another three hours to burn itself out. Dawn broke on a smouldering mass covering an area of two acres. Stacked piles of timber still glowed into the daytime as the exhausted firemen, blackened by the grit and grime, cursed the cold that enters the bones as the body, for so many weary hours heated by its exertions and the flames, rapidly returns to normal.

# 10    Civil unrest and revolution

By 1910 the number of street fire alarms had been increased to thirty-eight, but their appearance on the street corners unfortunately led to further malicious false alarms. Two youths were charged with sending bogus fire calls and one of them, because of a previous conviction, was given three months hard labour. But this type of antisocial behaviour was not confined to youths from the lower classes. A Harold Lancaster Rowells, with the rank of inspector in the colonial police, was arrested and found guilty of deliberately breaking the glass of a fire alarm. He was fined five pounds. John Deehan, an Inniskilling Fusilier, was convicted of sending in a false call and was sentenced to a fine or two months imprisonment.

The passing in 1909 of the Cinematograph Act after the opening of Dublin's first cinema led to extra fire prevention work for the chief fire officer. His request for a further pay increase, however, was turned down.

In October 1911 Inspector William Myers, with over thirty-nine years service, was retired on medical grounds, and the rank of inspector was suppressed. In the previous year following a rail crash at Roscrea, Co Tipperary, the Dublin Fire Brigade was requested to provide both of their ambulances to assist in removing some of the injured to hospital from a train that had conveyed them to Dublin. The Great Southern & Western Railway praised the brigade for "the promptitude with which the ambulances were placed at the disposal of the company's officials … and warm appreciation of the skill and dispatch by the men in charge of the ambulances in removing the injured from the train to the different hospitals". At the same time the secretary of the fire offices committee was again rejecting any idea of a payment towards the upkeep of the Dublin Fire Brigade from the insurance companies. At this time also Purcell was disposing of some old equipment, selling fire escapes to Lord Longford and to the Spa Hotel at Lucan for £25 each.

It was also a year when four more tenement houses collapsed, tragically killing two tenants, injuring others and of course putting work and strain on the firefighters. During 1912 the busy Chief Purcell travelled to Limerick to advise on the organisation of a fire service for that city.

The Brigade's annual report for 1911 stated that five people had died in fires, though Purcell commented that the overall fire loss in the city was reduced. He claimed that this was due to the new fire stations and the efficient communication system provided by the street fire alarms. To indicate a fire the caller simply pulled a handle. If an ambulance was required the caller could use the telephone to explain the situation.

The final contracts for the building of a new fire station in Thomas Street were now agreed and, as this was to be built without provision for horses, the purchase of a second motor fire engine was approved. The engine was manufactured by Leyland at a cost of £850. Basically it was similar to the first Leyland of

First motor fire engine, towing the steamer.
Photograph, © National Library of Ireland

1909, but there were some differences. The original large copper vacuum drum was omitted and consequently the rear bodywork was re-designed; the engine was slightly more powerful being fifty-five horse power; brackets were fitted to carry a twenty-eight foot extension ladder and a "Pompier" climbing ladder. The latter was a type of hook ladder first used in France which came into general use in English brigades at the end of the nineteenth century.

Although Purcell was well pleased with the performance of his first motor fire engine, and had now ordered a second, he warned against overlooking the value of steamers. He described them as "the most powerful and compact pumping unit for a given weight ever put at the service of Fire Brigades". Some of the English brigades, being now motorised, were selling off their old steamers to smaller brigades which had never been able to afford a "major" appliance. Purcell pointed out, quite correctly, that few cities actually had enough pumps to deal with very large fires, and it would be wise to retain steamers. He had a special attachment fitted to the Leyland motor to enable it to tow the large steamer, which it did successfully, and both steam-driven and petrol-driven pumps worked at all the largest fires in Dublin right up to the 1920s.

Meanwhile, the firemen's union was seeking, and getting, the backing of their employer, "for the right to attend at coroners' inquests to examine witnesses if the interest of their members was involved". Some serious labour disputes were however worrying the councillors and they adopted the following resolution.

*That this council considers that legislation, somewhat similar to that in New Zealand, should be enacted for the purpose of securing the settlement of labour disputes by arbitration and avoidance where possible of lockouts and strikes … this council is also of the opinion that legislation providing minimum subsistence wages for all adult workers, fair conditions of employment, including a 48 hour week and the right to work, would be fruitful of peace in industrial concerns.*

As an indication of their own concern in these matters the councillors approved a motion "that all labouring grades in the Corporation whose wages were less than 29 shillings per week should receive a rise of 2 shillings per week". The firemen claimed two shillings and sixpence per week, and this was conceded in November.

A meeting of the Association of Professional Fire Brigade Officers had been arranged to take place in Bristol in July 1912, and it was attended by Chief Purcell along with two members of the waterworks committee. One of those attending may have been William Cosgrave, the chairman. He had obviously been "converted" since 1909 because it was he who now proposed the purchase of the second motor fire engine.

The value of motor engines was again shown when on 5 April a call was received for assistance in dealing with a serious fire at Gibbstown House, Navan, Co Meath. The fire had originated between roof and ceiling some time after 10 pm the previous evening, and had now spread to all the rooms in the top floor of the three-storey mansion. The Dublin appliance covered the thirty-six miles in one hour and twenty minutes, and got to work with two jets from a pond. After seven and a half hours the fire was extinguished. There were thirty-two rooms on the upper two floors, all entered from marble balconies surrounding a grand central hall surmounted by a glazed dome and a clock tower. All of these rooms were destroyed, but the brigade succeeded in saving all of the ground floor, basement and extension at the rear. The building was valued at £100,000 but in spite of the hours of burning the loss was put at £40,300.

On 21 January 1913 the final part of the Purcell fire station building programme was completed when the station in Thomas Street was officially opened. It had a frontage of seventy-four feet and a depth of one hundred and thirty-four feet, with a two-bay engine room twenty-five feet wide wide and seventeen feet high with brass poles in recesses for descent from upper floors. The folding exit doors were fitted with special quick-opening gear. Officers' quarters had five

**Thomas Street fire station, 1913.**
**Photograph, © Dublin City Council**

rooms plus bathroom, and storage for petrol and coal was provided in an asphalted yard to be used for drill practices. The two upper floors were approached by outside iron stairs and ferro-concrete balconies giving access to quarters for eight married men and their families. Each of these quarters contained three rooms of good size with cooking store, lockers, scullery, hot and cold water and a WC. A bathroom was provided on each floor. These would have appeared luxurious residences to the families who had spent years in the Cook Street slums.

The frontage of the station was of classic design, with five arched openings all finely chiselled in brown Mount Charles sandstone. The total cost of this building, including all the special fittings, drainage and paving, was £7,800. It was an imposing structure that added greatly to the appearance of the locality, and was the first Dublin fire station designed for motor engines. All of the four new stations added much to the architectural heritage of the city at a time when many buildings from the Georgian period had become decaying tenements. Following the opening of the fine new station in Thomas Street the premises in Whitehorse Yard, off Winetavern Street, which had served as a fire station for fifty years, finally closed. At the same time the last remaining street fire escape station, at St Catherine's Church in Thomas Street, was removed.

In addition to a second motor fire engine Purcell had been given permission to purchase an oxygen breathing apparatus to enable a man to work safely in noxious fumes. The "Proto" set plus one spare cylinder was bought at a cost of £27-6-0 and was to remain in service until after the Second World War. Purcell had also received sanction to purchase a motor driven ambulance, and he now set about designing the vehicle and placing advertisements in the papers for the contract. He recommended that Leyland should get the contract and they were to construct, to his working drawing, a chassis with very long easy springs and compact wheelbase.

He had originally provided for an engine of fourteen horse power at 1,000 revolutions per minute, but after consultation with several expert motorists Purcell decided on a more powerful engine. Eighteen to twenty horse power

Dublin's first motor-driven ambulance. Photograph, © Dublin Fire Brigade Historical Society

would mean less wear and tear, the machine not having to be forced when hill climbing. Leyland agreed to supply the heavier engine for £20 extra so, their tender being acceptable, the motor ambulance was purchased for £540.

Contrary to the practice in many other brigades of obtaining loans for the purchase of modern appliances, Dublin Fire Brigade had always paid for its vehicles out of revenue. It was this case which won the majority of councillors over to support the purchase of the new ambulance. The crimson finish, with on each side a device consisting of a red cross surrounded by a scroll with the lettering "Corporation Ambulance, Dublin" and also the city coat of arms, attracted much favourable comment for the new ambulance when it came into general service in January 1914. It incorporated many new features recommended by Purcell and had ample space to accommodate a driver, two assistants and up to four patients. Its normal speed was twenty miles per hour.

The year 1913 is remembered in Dublin because of the infamous "lock out" which occurred that year. It was the nearest thing Ireland has ever had to a socialist revolution, and it could not but impinge on the brigade and its members. Firemen through the course of their work entered people's homes on a day-to-day basis, and they could see the terrible poverty that the working class were forced to live in. Most of the crews, coming from that class, saw at first hand the great divide that existed in urban society. Although the corporation workers, including the firemen, were not directly involved in any of the 1913 industrial action, nor were they locked out by their employer, their sympathy was with those in the front line of the dispute.

The first week of September was probably the most confrontational week of the whole 1913 strike. Two strikers were killed in police charges and hundreds were injured, as were more than forty police constables. The Dublin Fire Brigade ambulances had been working continually on the 1 and 2 September ferrying those injured in the bitterest of conflicts to hospitals around the city. Often the vehicles were packed full of seriously injured tenement dwellers including many women. It was the period in the strike most examined in the Report of the Dublin Disturbances Commission into the riots. With the ambulance crews stretched way beyond their capacity, they were forced to confront a particularly trying episode that tested the organisation of the brigade to the full as well as the quality of its firemen.

Just before 9 pm on the evening of 2 September, perhaps the single most difficult day in the great dispute, without any warning two four-storey tenement houses collapsed in Church Street. At that time many of the adult occupants were outside on the steps of the buildings or in groups on the pavement discussing the terrible events of the day. The roof and upper storeys suddenly collapsed after portion of an upper front wall fell into the narrow street burying those outside and trapping those within. The number of unfortunate tenement dwellers involved in the disaster was over fifty – men, women and children.

Soon the firemen were in attendance at the harrowing scene and, in spite of the danger of further collapse, they entered the dust-filled and smoke-filled rubble in the gathering darkness as the cries of pain and fear from those trapped rose from the huge pile of rubble and debris which had recently been their over-crowded dwellings. Lanterns were brought to assist in the distressing search operation. As victims were dug out and conveyed by ambulance to hospital, serious fire broke out caused by the scattering of domestic fires which were now part of the piles of rubble. The rescuing had to cease as the firemen began to extinguish the blaze while frantic cries arose from those trapped fearing their suffering would now be exacerbated by burning. Firemen must have felt they were witnessing a scene from hell!

The brigade was assisted by corporation workers as the police threw a cordon around the immediate location to control the gathering crowds. Access was hampered owing to the terrified occupants of adjoining slums removing their meagre belongings onto the roadway for fear of loss in any further collapse. Such scenes and incidents stay with firemen throughout their lives, though the trauma is partly erased by a kind of merciful black humour among fire crews.

Search and rescue operations continued throughout the night and the following day as the ritual visits of the city dignitaries and prominent business people took place. Seven souls perished in the disaster, many were seriously injured and over one hundred were left homeless and devoid of all possessions. Such was the poverty of some of those involved that a newspaper reporter noticed that "every bit of coal in the debris was sought for and carefully collected in sacks by the survivors" when they were allowed to return to the scene to repossess their remaining belongings. Most of the homeless and destitute were looked after by the Capuchin priests in the Father Matthew Hall.

The site of the disaster was part of an area under demolition, condemned as unsafe by the corporation. However, red tape and procrastination by the Local Government Board had prevented the clearing of the area in sufficient time to prevent this appalling tragedy. So when the coroner's court on the death of two victims was convened it was attended by many dignitaries including Keir Hardy, the then leader of the British Labour Party, in Dublin at that time to investigate the widespread rioting. The coroner's verdict eventually maintained that "the cause of the collapse was unforeseen" and mentioned the noble work done by the fire brigade in the rescue operations. At the city council meeting on 18 September the waterworks committee tendered

> their thanks to Lieutenant Myers and the men of the Dublin Fire Brigade for the noble and courageous manner in which they rendered their services in connection with the unfortunate occurrence in Church Street, and their approval of the prompt way in which the Brigade attended to render assistance to the injured.

At this time some of the brigade were operating as far away from Dublin as the Ring of Kerry in pursuence of their vocation of helping others. A gorse and woodland fire had been raging out of control for some days. When the fire spread towards the historic Parknasilla Hotel, Dublin Fire Brigade had been contacted for assistance. On 17 September one steamer and crew had travelled by rail to Kenmare where a local coach was procured to tow the fire engine to the scene some twenty miles away. This approach failed, so the crew waited until horses were obtained, and the Dublin firemen then travelled probably the longest distance the brigade ever travelled to tackle a fire. Pumping water from a nearby lake they succeeded in preventing the fire from engulfing the hotel. After two days' absence they returned to Dublin, appropriately remunerated for their efforts.

Up to this time many firemen had broken service. Typical of these would have been William Byrne who joined the brigade in 1888, resigned in 1894, was re-appointed in 1897 and went on to give another twenty years service. After the introduction of the Pension Act there is no record of firemen being re-employed if they left the brigade for a number of years.

Chief Purcell's annual report for the year 1913 lists seventeen fires outside the city to which the Dublin brigade was called. Practically all of these fires were related to the industrial conflict which had engulfed the city and county. They represented by far the greatest number of calls for assistance from outside the city boundary since the brigade was formed in 1862. The brigade also received twenty-six false alarms – almost treble the number normally received. The chief reported that owing to "abnormal labour disturbances in the city and surrounding county during the latter months of the year incendiarism was very prevalent, particularly in farm produce, and many requests for assistance were received". The losses sustained by many prominent business people in this respect ran into many thousands of pounds. Farm labourers were particularly vulnerable in the industrial struggle, often removed from their tied cottages with "imported" labour doing their work. They were faced with the prospect of a hungry winter for themselves and their families, so in many cases they resorted to the practices of an earlier era by setting fire to crops or other property. Typical of these fires were those at Hazelhatch where hay and straw worth over £600 was destroyed and a pub owned by a landowner had fire damage of £400. In Santry a shed containing about 300 tons of hay, 50 tons of straw, 300 barrels of oats, a quantity of potatoes, carts and other farm implements owned by Richard O'Malley JP was destroyed. But for the intervention of the Dublin Fire Brigade his offices and dwelling house would also have perished. Losses of over £2,000 were recorded for this blaze. In Artane on 21 November twenty tons of straw in a shed were destroyed, and similar fires took place in Raheny, Swords, Crumlin and Kilbarrack. Many other fires occurred which did not receive assistance from the city brigade. In the following year claims for this malicious damage cost the ratepayers thousands of pounds.

Life carried on in the fire service with three older firemen retiring and three others being promoted. Two cellar pipes were purchased at a cost of £15-12-6 for dealing with fires in basements, and eight factory premises were served notices by the corporation for failing to provide sufficient means of escape in case of fire.

The brigade had a visit from Harry B. Lee, Chief of the Melbourne Fire Brigade, who inspected three of the fire stations. He was very taken by what he described to the Dublin authorities as "the way you provide for the men" – one wonders what he would have said if he had seen the slum tenements which the men were forced to endure for so many years! He also remarked on the "street alarms, and motor service engines with pumping appliances which I have not seen elsewhere". The latter comment is quite inexplicable, considering that he had completed a tour of America and England before coming to Ireland. He left Dublin to continue his fact-finding tour in Belfast and Paris before returning to Australia.

Although the work of the fire service had expanded to deal with all kinds of human situations including street accidents, collapse of buildings, rescue of animals, inspecting places of entertainment, or simply forcing entry on behalf of the occasional person who forget the keys, the trade union-organised firemen maintained their neutrality when the city was faced with civil disturbance. In 1912 London firemen were used to spray water in breaking up a demonstration by strikers in the London Docks dispute. The ongoing friction between the corporation and the metropolitan police probably worked in favour of Dublin Fire Brigade.

Whatever about involvement in civil strife, the brigade was involved in many other non-life threatening incidents. On 14 July 1913 the *Slaney*, a Guinness lighter, was loaded with a cargo of export stout while moored at low tide at the Victoria Wharf opposite the Guinness brewery. When it was realised that the vessel was not rising with the tide it was found to be buried too deep in the mud bank of the wharf. Within an hour of the tide rising, water was flowing into the hatches and flooding the hold. Soon hogsheads were floating down river to be hauled from the water by cranes on the North Wall. When the Guinness employees with their small pumps were unable to stop the flooding of the vessel the fire brigade was called.

Using the motor pump from Thomas Street and a steamer from Dorset Street, water was pumped out against the running tide. Within a few hours, with the aid of cranes the *Slaney* was once more floating on the river. An interesting sequel to this incident was that one barrel rescued from the water by bystanders was rolled back into the river when police arrived to seize it! Later there were disturbances at Sir John Rogerson's Quay when the police tried to take two barrels from groups of people who had assembled and were indulging themselves with quantities of porter.

Charges for the call-out of the Dublin Fire Brigade to surrounding townships or the county had not been changed for over twenty-five years, but still the corporation had problems collecting the money owed. It had to process a claim for £79-10-0 before the southern police court against the owner of Springfield, Dolphins Barn, for a two-day fire starting on 2 December 1913. The claim for the brigade's items of expenditure make fascinating reading.

*52 gallons of petrol £2-18-0; one length of hose destroyed £3-15-0; wear and tear on 3,000 feet of hose used £15; motor fire engine and horses £31-4-0; damaged clothing and boots £14-8-0; for wages for six firemen who were absent from duty for one week while under medical treatment for injuries to their eyes £10-10-0.*

The eye injuries resulted in the men having to be conveyed blind-folded to the Eye and Ear Hospital twice daily.

The fire had been malicious, so the owner had received £650 compensation. It was an exceptionally large fire and it was accepted by the owner that but for the outstanding work of the fire brigade a much greater amount of his property would have been destroyed. However, he maintained that he should not pay more than £15 because the rest of the costs were covered by payment of water rates. The judge hearing the case, in accordance with what seems to be the general position when dealing with bodies funded by the public, gave his judgement against the corporation.

Whether as a result of this type of judgement, or because of the extraordinary number of calls to fires in the country, the corporation reviewed its scale of charges for the use of the fire brigade outside the municipal boundary. On 24 February 1914 the following revised regulation came into force.

> *Before turning out, the payment of expenses according to the scale below must be guaranteed to the satisfaction of the Chief Officer, who will exercise his discretion as to giving assistance, also to the number or type of fire appliances to be despatched. Payment of account for attendance must be made to the City Treasurer within ten days from the delivery of the account for services per hour.*

| | |
|---|---|
| *Chief Fire Officer* | *£3-3-9 for turn out + first hour* |
| | *10 shillings for each extra hour.* |
| *Other Officer* | *£1, and 5 shillings each extra hour.* |
| *Engineer or Motorman* | *4 shillings, and 2 shillings per hour.* |
| *Firemen (each)* | *3 shillings, and 2 shillings per hour.* |
| *Motor Engine and Hose* | *£6-10-0 and £2 per hour* |
| | *+ 5 shillings per mile from station.* |
| *Steam Engine (Large)* | *£3-10-0 and £1 per hour.* |
| *Steam Engine (Small)* | *£2-10-0 and 15 shillings per hour.* |
| *Horses (each)* | *Per mile from station, two shillings and sixpence.* |

> *Add 50% for night service (sunset to 7 am).*

On 14 January 1914 the new motor ambulance, while travelling to Jervis Street Hospital with a patient, was involved in a collision with a horse and dray. Damage totalling £31-1-2 was caused to the ambulance. In the accident the shaft of the dray rammed the front of the ambulance leaving a large dent; the horse was injured and his driver thrown onto the road. Before the ambulance stopped it struck a lamp pole causing further damage. One of the ambulance men received facial injuries, which required stitching, but the crew were able to remove the patient on the stretcher to the hospital close by. Because the ambulance was damaged the crew requested assistance from Tara Street headquarters and a motor engine was required to tow the damaged vehicle back to the station.

The recorder gave judgement against the owner of the horse and dray, but on appeal to the King's Bench on 30 July the judge changed that decision stating that "he had visited the scene and come to the conclusion that the driver of the ambulance saw the horse coming out of Denmark Street when he was thirty yards away, and he had time to pull up … negligence on the part of the defendant had not been proved".

Chief Purcell, having received a refusal from the waterworks committee to purchase further motor engines, bought two replacement horses at a cost of £50

Fireman and two horses, Tara Street, 1911.
Photograph, © National Library of Ireland

and £52-10-0. At the same time the corporation employed 119 men on the construction of the new Roundwood reservoir which was now needed to improve the city water supply, and the valuation of the Tramway Company's lines was increased to provide extra revenue. The city councillors were informed by one of their advisors, Dr McWalter, that "the effects of vibration caused by heavy motor traffic was causing the collapse of old buildings in the city". McWalter went on to propose that "vehicles capable of carrying five tons or upwards should be prohibited passage through the streets of Dublin". But other more ominous happenings were about to completely change the city.

In July 1914 the fire brigade ambulances were called to Bachelors Walk where a troop of soldiers from a Scottish regiment had opened fire on a jeering crowd, killing and wounding many. With their usual professional efficiency the brigade crews removed all the injured to hospital as tension mounted in the city. Shortly after that incident the motor ambulance began to have mechanical problems and when on a trial run to Inchicore, following cleaning and adjustment, the engine broke down. It had to be sent to the Wembley workshops of Leyland for repair. The half-crank cases, rods and one of the cylinders were so badly damaged that a new engine costing £88 had to be installed.

August saw the beginning of the first world war, and immediately the call went out for troops. The cost of living had already been rising and that, along with the growing workload, led the firemen to make a claim for an extra two shillings and sixpence per week plus twenty-four hours leave every sixth day. In September the waterworks committee agreed to recommend sanction of the pay increase, but deferred any decision on shorter working hours. The increase was agreed by the council at its November meeting.

In October, following John Redmond's speech as leader of the Nationalist Party calling for volunteers to serve in the British army, Dublin Corporation adopted a resolution entitling employees who joined up to be paid their wages during the army service and to be re-employed when discharged. This decision was approved by the Local Government Board and a letter of approval was sent to all other local authorities. The following January the original motion was amended to "All employees' families are entitled to half-pay, providing that they have twelve months service within two years under any committee or department of the Corporation". This amendment was to provide temporary or seasonal staff the same "perks" as permanent workers if they were to join the army. It was also agreed that all employees earning less than thirty-five shillings per week were to be paid a "war bonus". The firemen, not to be outdone, entered a claim for a larger coal allowance for their quarters, and as a result they wee granted four tons of coal each year instead of their current three tons, from the beginning of the next financial year.

The brigade now purchased another horse and also 1,400 feet of hand-woven hose, the latter costing one shilling and four and a half pence per foot. At this time there was some trouble between the corporation and the government authorities which took some time to cool down. When the city fathers decided to stop payment of £15,580 of the city police tax this amount was immediately deducted from sums payable to the corporation by the Local Government Board, and objections made to the military authorities on the question of commandeering hay were ignored. Despite these financial problems however, the fire brigade continued its humane work. On 15 October all three ambulances were involved in removing 643 wounded soldiers from the *SS Oxfordshire* to various hospitals over a period of almost four hours. They were assisted in this work by the Irish Automobile Club which provided cars and drivers to remove the non-stretcher cases.

In May 1915 the city witnessed another huge fire when Armstrong's paper and book factory of Lower Abbey Street was seriously damaged. The building was part of the Northumberland Works, separated from the Scottish Free Church by a four-foot lane, with Old Abbey Street at the rear the only protection against the fire spreading to the houses on Eden Quay. The building was alight on three floors when the brigade arrived. Hose lines had to be laid on three fronts to prevent the fire from engulfing surrounding properties, and firemen had to climb onto the roof of the church to pour water into the inferno after a large part of the roof collapsed. A heavy downpour of rain drenched the firefighters but assisted in the struggle to contain the blaze. Such were the crowds gathering in the area that the police requested help from a company of Royal Munster Fusiliers which was returning from rifle practice at Dollymount.

The factory employed over 200 people and had a huge store of paper, which had been purchased in anticipation of major shortages because of the war. All this paper was destroyed, adding to the firemen's problems in extinguishing the fire and preventing it from spreading. The citizens had the excitement of watching the fire illuminate the city centre before dense clouds of smoke began to cover the whole area as the deluge of water from the hoses and the continuing heavy rain began to extinguish the flames. Purcell, speaking on the following morning described the outbreak as the worst fire the city had endured in the past two years. Plant worth over £12,000 was destroyed as well as most of the other contents and, in addition, the structural damage amounted to nearly £20,000. Much worse off, however, were the two hundred mainly female workers who were now out of work. With the exception of Jacob's biscuit works and Goodbody's tobacco factory, Armstrong's factory employed more women than any other firm in the city.

On 13 November 1915 Kingstown Pavilion, one of the most attractive features of the township and a resort for many Dubliners, holidaymakers and music lovers, was consumed by fire leaving a mere skeletal building and a loss of £10,000. The fire was discovered during lunchtime by a passer-by who saw smoke issuing from the back of the premises. When Captain Carroll and the Kingstown brigade arrived the huge stage and part of the galleries were well alight. Assistance was requested from both Dublin and Pembroke. Purcell arrived with two motor pumps to be followed later by a steamer. Water was pumped from the sea by one of the Dublin engines, watched by many onlookers who were held back from the immediate area by a company of Irish Rifles with fixed bayonets.

The untiring efforts of all the firemen present, even with the pumping capacity at hand, could not prevent the destruction of this outstanding landmark. The building being mainly of timber construction burned fiercely and rapidly with tremendous shattering of glass as the windows exploded outwards. The picturesque turrets blazed like chimneys, eventually collapsing inwards. The fire crews could only retreat out of harm's way, taking with them furniture, paintings and other artefacts to be stored in the town hall for safety.

The Pavilion had been built in 1902 at a cost of £10,000. It was a handsome structure standing in beautifully landscaped gardens and containing a concert hall with seating for 1,000, promenades capable of accommodating 3,000 more, tea rooms and cloak rooms. Standing as it did overlooking the seafront it presented a stunning picture from both the sea and the harbour piers.

It was a bad weekend for the township because, apart from the Pavilion fire, a storm and floods damaged the Kingstown Gas Company and the east pier of the harbour. Roads were flooded, the Boyd monument collapsed, the band stand was badly damaged and the glass shelter of the west pier collapsed. The three-masted *Iveresk* ran aground at the "Forty Foot" and many small boats were smashed and sunk in the harbour along with a trawler and another ship the *Emerald Ray*.

The weekend was exceptionally busy and hazardous for the Dublin firemen. In addition to pumping out the Kingstown Gas Company and fighting the Pavilion fire, Dublin crews had to wade through waist-high water from the flooded Camac river at Old Kilmainham to rescue forty people trapped upstairs in their small houses and remove them by ambulance to temporary shelter in the South Dublin Union. They also pumped water from basements of two hotels, a well-known charitable institution, and houses on the North Strand. It was proving a memorable year for the brigade, soon to be overshadowed by the momentous events of 1916.

A harbinger of what was to come, perhaps, was an explosion which took place at an electricity sub-station in Forbes Street on the south docks on 27 November 1915. Such was the ferocity of the blast that the building was blown from its foundations, bricks and mortar scattered over a hundred square yards, and the heavy steel door flung into the middle of the river some eighty yards away.

Smith's public house on the far side of the road had all its windows blown in and slates torn off the roof, presenting the appearance of premises that had been under heavy bombardment. Many in the city who heard the tremendous blast believed that, following the recent bombing of London, Zeppelins had arrived over Dublin. The three transformers housed in the thick-walled sub-station lay twisted around the iron girder on which they rested. When the brigade arrived to extinguish the fire the scene was reminiscent of a gas explosion; no one had ever witnessed such damage at an electrical incident.

Since the outbreak of war the brigade ambulance service was continually called to remove wounded servicemen from ships to hospitals or from hospitals to convalescent homes. To assist in this work, the Cinematograph Association, the owners of the city cinemas, set up a fund to raise money to purchase an additional motor ambulance.

**Shooting incident near Custom House, 1920, showing Cinema Ambulance.**
**Photograph, © Edward Chandler**

In November 1915 the sponsors of the Cinematograph Trade Ambulance Fund presented a fully equipped ambulance to Dublin Corporation. It had been built in Dublin to Chief Purcell's specification. The new vehicle was to be used for Red Cross purposes during the continuation of the war, and to revert to the corporation at the end of the war. It was certainly a welcome addition to the Dublin Fire Brigade, and was to see continuous duty during the Easter Rising of 1916. Some time earlier, under the general auspices of the Catholic Church, a Dublin vigilance committee composed mainly of clerics had been set up. The members visited the cinemas regularly and complained about "objectionable films shown at certain picture houses". Whether through their intercession or that of others, the cinema owners were rewarded for their generous donation of the ambulance by the establishment of a censorship of films office on the direction of the Lord Lieutenant in October 1916.

The chief officer's report for 1916 showed that the brigade received 145 calls to normal fires, a reduction of 76 on the previous year. This figure, however, did not include the calls to 93 fires, 41 in day time and 51 at night, attended during the week of the Easter Rising. Of the ordinary fires tackled that year, loss was estimated at £41,200 whereas the 196 establishments destroyed during the Rising had a direct loss estimate of over £2,500,000. This figure is exclusive of the direct business loss during the last eight days of April. The locations destroyed by fires, looters or bombardment by the military were given as: 68,900 sqare yards in the O'Connell Street area; 29,140 in Linenhall; 1,020 in Lower Bridge Street; 225 in Mount Street; 135 in Stephens Green – a total of 99,420 sqare yards of the most prosperous areas of the city destroyed.

The fire engines were practically on continuous duty for six days dealing with this ravaging of the city centre, often under fire or held at gun point by the military, the crews carrying on their professional duties in probably the brigade's finest week. In addition to their perilous duties the firemen carried out an incredible amount of ambulance work. Over a period of eight days each ambulance attended an average of fifty calls daily, picking up wounded, injured or dead, many times actually under fire from both snipers and the military. Bullets damaged each of the ambulances on several occasions, and in one incident a horse was shot dead. On another occasion a civilian engaged in assisting a fireman to place a stretcher carrying a seriously injured woman into an ambulance was shot and severely wounded. Dublin Fire Brigade was awarded £54 compensation for the damage caused in these incidents. In addition to this work, crews had to crawl over piles of debris into dangerous areas to remove invalids or injured people trapped in their homes. With only three ambulances available there were hundreds of cases that the brigade could not attend during this terrible period.

The first ambulance calls came in just after noon on Easter Monday as the first casualties began to occur on the streets. The first fire calls were from the North Earl Street area where a fire was started by looters. By Tuesday morning looting of shops had spread and further fires were started. Commandant Brennan Whitmore of the Volunteers relates in his memoirs being sent by James Connolly to try to stop the looting in the Talbot and North Earl streets area. He

Dublin Fire Brigade fire fighting in O'Connell Street, 1916.
Photograph, © National Library of Ireland

maintained that he called the fire brigade who responded immediately on the first two days but were prevented from turning out by the military from then on.

When shelling began, houses and business premises were set alight by the bombardment, making fire fighting an altogether impossible task. However, during lulls, the fire engines responded and attacked the fires that were then occurring on an ever more widespread basis. When shelling recommenced the firemen withdrew to adjacent streets awaiting the next lull so that they could get to work again. As the week went on so the fighting, shooting and shelling increased. Fires caused by incendiary shells were relatively small but, if they were not tackled quickly, they spread until they merged with other fires, finally becoming a conflagration which would radiate to any inflammables in the vicinity and eventually shower burning debris over a wide area. On occasions when the military informed the brigade that they had ceased to bombard an area the fire engines were able to get in and try to extinguish the fire.

The British Fire Service journal *Fire* when reporting on the 1916 Rising had this to say:

> *The troubled days of the week-long rebellion proved how nobly Irish firemen, and more particularly Dublin and Pembroke firemen, answered the call to them as Warriors of Peace.*

During the week most of the brigade's work was carried out under gunfire and several fireman were hit and injured. More than once operations had to be suspended at a particular incident as the firemen retreated to the nearest cover.

On the afternoon of Easter Monday a fire broke out in the Magazine Fort in Phoenix Park. At Queen Street Bridge the Tara Street section of the brigade was held up at a barricade and the Republican officers present compelled the crew to return to their station. The crew from Thomas Street station, however, who had taken a different route, was able to reach the scene of the fire and saved the Magazine by working right through to midnight. Other crews dealt with two serious fires in the North Earl Street and Talbot Street area on the same night, eventually succeeding in extinguishing the flames and saving the affected buildings.

Next morning a major fire at 32 Sackville Street was fought, to be followed in the evening by three fires in the same street. As already indicated, these early fires were caused by looters who had ransacked the premises. By Wednesday the looting and the firing had extended to more of Sackville Street and into Henry Street where firemen were actually accosted by looters as they tried to tackle the growing flames. More fires occurred as the intensity of the fighting grew worse and, because all of Sackville Street was now under gunfire, the brigade ceased to attend fires within that street. Similarly, the brigade was prevented from attacking a large blaze in Clanwilliam Place owing to the fact that the military were shelling houses in the immediate vicinity.

On Thursday a number of fires broke out in Harcourt Street. These were quickly extinguished in spite of bouts of sniper fire in the area. A bomb exploded at the rear of the Linenhall Barracks and started a fire in the theatre, a temporary structure which survived from the Civic Exhibition held a few years before. The brigade was unable to attend this fire because of conditions prevailing in that part of the city and, despite the actions of the resident caretaker and thirty-two clerks, the barracks was completely burned down and the fire extended to five adjacent houses and stores.

Just after noon on the same day an urgent call came from Lower Abbey Street where a huge fire was raging, but because this area, so close to the General Post Office, was under continuous rifle fire and bombardment by both field guns and the gunboat *Helga* on the Liffey, the brigade could not attend. It was this blaze which consequently started the major conflagration which wrought such havoc in the Sackville Street area. Within some twenty minutes of a street barricade consisting of boxes, timber, bicycles and rolls of paper being shelled by the army, buildings on both sides of the street were alight. The printing works on the north side was soon alight and the flames spread into Sackville Place and Sackville Street. This destruction gave a clear line of fire for the guns of *Helga* which was moored midstream opposite the Custom House firing at the General Post Office, the headquarters of the rebel forces.

The fires on the south side of the barricade spread into Wynn's Hotel through Hoyte's corner and then into the Hamilton Long chemist shop until an inferno raged from Eden Quay right through to Abbey Street and out to the south side of Sackville Street. As night fell over the troubled city the destructive flames extended to the Imperial Hotel, Cleary's warehouse and shop, the new bakery of Joseph Downes and on to the restaurant in North Earl Street. Every building from the south side of North Earl Street back to Eden Quay was on fire, and the fire brigade could not respond. The ambulance crews, however, made speedy forays into the area to remove seriously injured civilians in spite of objections from the military.

In the early hours of Friday morning Thomas Street fire station turned out to deal with a major blaze at the corner of Bridge Street and Ushers Quay and prevented the spread of the fire to other premises. Four houses and a barricade consisting of a city tram were destroyed in the flames. The brigade was now becoming active again, turning out to tackle the flames in the Eden Quay area some hours later and preventing the spread eastwards. Again working from Cathedral Place the firemen attacked the blaze in the Downes bakery and Hickey's restaurant. This dangerous work saved many premises on the north side of Earl Street, but by evening new fires had broken out on the south side of the street. By 6 pm the General Post Office, which had been continuously shelled for a number of hours, was now a mass of flames as was the newly opened Coliseum Theatre, and both buildings were eventually totally destroyed.

The fire brigade was officially informed by the army commander on Saturday at 3.40 pm that all active military operations had ceased, and that the firemen might now attempt to extinguish the fires that were still blazing in both the Sackville Street and Church Street areas. Every available fireman was immediately turned out. Within a short period of time excellent progress was being reported when without warning several shots were fired from the direction of

Dublin Fire Brigade at ruins of General Post Office, 1916.
Photograph, collection of Edward Chandler. © National Library of Ireland

Upper Abbey Street, the bullets hitting a wall close to where a fire crew was working. Immediately afterwards more firing came from the Aston Quay side of the river, six bullets penetrating the side of one of the brigade's aerial ladders.

The men were returned to their stations, their fire engines abandoned, and the fires were allowed to continue unchecked. But that evening, when Jervis Street hospital was threatened by fire, the whole brigade turned out, repossessed their appliances and, backed up by additional engines from the Guinness Brewery and Powers Distillery, they fought the flames. Crews worked throughout the night and continued in relays for the next week, cooling the smouldering ruins that a few days earlier had been the shopping and business centre of the city of Dublin.

One of the most spectacular fires was at Hampton Leedom in Henry Street on the Saturday some time after 10 pm. It was reported as an awe-inspiring sight as multicoloured flames from the oil stores shot skywards and oil barrels exploded. As if in some infernal region, the fire crews were surrounded by the gaunt blackened walls of the buildings and struggled in vain under a vast miasma of flames. It seemed a fitting end to a week of sombre glory.

Purcell's report on the week's events estimated £2,500,000 as the value of the buildings and stock destroyed by fires in the city. Over 200 buildings were destroyed, some of them among the finest landmark premises in the city. Included among them (in addition to those already mentioned) were the Royal Hibernian Academy, Hopkins and Hopkins, Nagle's and Sheridan's public

152

houses and Mooney's pubs at Eden Quay and North Earl Street, the Metropole Hotel, the *Freeman's Journal* office, Bewley's coffee house, the Oval Bar, Thom's printing works, Doherty's Hotel and Lawrence's Jewellers.

Of the many harrowing incidents dealt with by the Brigade the chief listed one on the Tuesday of Easter week. Fire crews were dealing with a blaze in Henry Street when a volley of shots rang out killing a man and a woman beside one of the fire engines. At about the same time a brother of one of the brigade turncocks was shot dead when he was conversing with an engine driver at the corner of Cathedral Place. On Thursday firemen at a blaze in a house in Harcourt Street discovered the body of a volunteer lying with his rifle and bag of ammunition. Chief Purcell had observed the spreading of the inferno in the centre of the city with his binoculars from the balcony of the headquarters tower, the brigade being at the time unable to turn out to that area at the height of the shooting and shelling. On Friday when the brigade attempted to stop the fires spreading into Marlborough Street in the face of continued gunfire, some of the firemen were threatened that they would be shot, and the crews had to retire. During operations in the following week firemen had several narrow escapes as ammunition and bombs exploded in burning buildings, and sniping continued for up to three days after the general surrender.

Following the Rising the fire brigade began to get back to some form of normality. A number of major businesses in the city, including Powers Distillers, donated monies to the fire brigades of Pembroke and Rathmines, but no such offerings were given to members of the Dublin Fire Brigade. The city coroner, Mr Louis Byrne, probably aware of these circumstances, wrote to the lord mayor on 8 June.

> *Now that the excitement of the Insurrection has passed, I deem it my duty to bring under notice the very great service rendered by the members of the City Fire Brigade, who worked the ambulances during this trying time. To my personal knowledge, these gallant fellows worked night and day, and several times under fire, bringing in wounded (Sinn Féiners, military and civilians) among the latter being women and children, and by their care and training saved many a valuable life. I consider some recognition of their services be made.*

The lord mayor replied in the following terms.

> *I quite agree with every word you mention with reference to the heroic work done by our Fire Brigade and Ambulance men during the trying time of Easter week .... under shell fire, and only for the great sacrifice they made (running grave risks to themselves) one of our largest Hospitals may have been destroyed .... I am sending your letter to the town clerk, and have asked him to bring it before the Council at its next meeting.*

The following memorial dated 10 June and signed by officers and men of the Dublin Fire Brigade was discussed by the waterworks committee.

> *During the period of the recent disturbances your memorialists were called upon to perform duties under conditions never imposed upon any body of public servants. During the first eight days of the trouble from 24 April to 1 May we were so hard pressed*

*that adequate rest or any change of clothing was impossible. Night and day our entire energies were solely devoted to saving lives and property of the citizens – removing hundreds of wounded to hospital, bringing a number of helpless people to places of safety, and attending conflagrations – all within the danger zone, and at imminent risk of our lives. Subsequently our work included keeping smouldering fires under control, demolishing dangerous remains in the devastated area, pumping work, and removing wounded people from houses who could not get out during the period of hostilities.*

The firemen went on to seek some reward and recognition for this work, and claimed that "neighbouring councils had accorded recognition to their fire staffs whose duties and risks were by no means comparable with those undertaken by us". In subsequent correspondence Chief Purcell maintained that he "had no knowledge of rewards granted by neighbouring councils, but Sir James Power informs me that his firm divided £100 between his fire staff as a reward for loyalty and good work on the occasion". He went on to inform the committee that letters of sympathy and congratulation had been received from many parts of

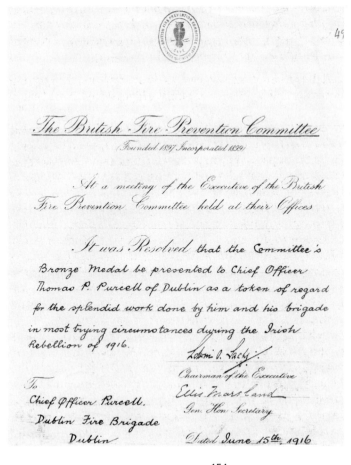

Commendation, British Fire Prevention Committee to Chief Officer Purcell, 1916. © Dublin City Council

the world, including the chief fire officers of London, Glasgow, New York, Boston, etc. He continued in the same vein.

> *Dublin treats its firemen well, better than any other city in the Kingdom. It expects and receives willing and efficient services at all times, but I submit that the month's work in question was so unusually arduous and dangerous as to deserve special recognition at the hands of the City.*

Purcell then suggested the grant of a special bonus of at least £100 to be distributed among station officers and men.

At the meeting of the waterworks committee on 8 August the chairman proposed that a chevron, carrying one shilling per week, be granted to the station officers and firemen on duty during the week of the Rebellion, and also that Turncock R. Lancaster be granted the same award. Also recommended was a bonus of £50 for the chief and for Lieutenant Myers. Dublin Trades Union Council sent a letter to the lord mayor

> *conveying their appreciation of the magnificent services rendered to the citizens of Dublin by the members of the Dublin Fire Brigade during the recent insurrection and expressing the hope that some permanent form of reward (both financial and decorative) may be afforded to the Brigade as a slight recognition of the ardour, courage and humanity displayed by them during that trying period.*

The council adopted the report of the waterworks committee, and the granting of this recognition put the Dublin firemen among the highest paid in the United Kingdom.

At this time an important research report by Alderman Thomas Kelly provided the corporation with a revealing account of the state of housing in the city. There were 25,822 families of 87,305 persons living in 5,322 tenements and 2,257 families of 9,812 persons living in 2,413 second-class or third-class small houses. The population in tenements was put at 128,000 and this was before the destruction to property during Easter week. These statistics showed that 229 persons out of every 1,000 of the city population resided in one room accommodation whereas Glasgow had 132, London 59, Liverpool 23, Manchester 7, Birmingham 4 and Belfast 3. It meant that 12,042 families with an average of 6.1 persons were occupying one-room accommodation. The statistics were a dreadful indictment of the authorities and property owners of the city. It was no wonder that the death rate from tuberculosis in the city for the years 1908-1912 (then the most up-to-date figures) was higher than in any other city or town in the United Kingdom.

Revenue in the city had dropped dramatically because revenue earning events such as the Horse Show had been cancelled. Against this drop, fuel and food prices had increased by up to 50% as a result of the war. Dublin was a city on the brink of disaster. The corporation again moved to try and get agreement with the wealthier townships of Pembroke and Rathmines to help alleviate some of the major problems in the city, seeking "the principle of amalgamation on such terms as might prove of mutual advantage to all concerned". It was a vain call, rejected by both of the boroughs.

Two further fires in 1916 remain to be mentioned. On 23 August the Dublin Fire Brigade received a call for assistance from Wicklow County. At about 2 pm a fire had broken out in the Marine Station Hotel in Bray. The fire was located in one of the staff rooms on the top floor of the three-storey building and the local brigade could not control it due to lack of water pressure. The flames soon spread into the roof void.

Miss Geraghty, the manageress, alarmed at the rapid spread of the fire insisted on calling the Dublin brigade. Purcell and a crew of twelve responded, taking less than thirty-five minutes to travel the twelve miles. All sixty guests and staff had been evacuated by the time the Dublin men arrived. There was some initial difficulty in getting an effective water supply. Fencing was dismantled and the motor fire engine was manhandled onto the promenade so that water could be drawn, luckily, from a full tide. The firemen then worked in a dangerous situation for almost three hours attacking a large fire above their heads in the roof void before eventually getting it under control. Most of the roof was burnt away but the two lower floors in the building were saved and not more than £20,000 worth of damage was caused. The efforts of the Dublin firemen received great praise.

On the bitter cold morning of 16 November at 2.19 am the fire crew of Tara Street station responded to a fire in Suffolk Street. The station being short-handed at that time –because of injuries and illness – a frightening situation arose for Chief Purcell, as later described by himself.

*The man who usually accompanied me was engaged driving the ambulance on that night, so that on coming down, finding the horse standing attached to my trap, I proceeded alone to the fire. On turning up the street, with the horse at a gallop, the reins came away in my hands, and I then discovered that not alone had they not been attached to the bridle, but even the bit was improperly put in at one side, fell away from the horse's mouth, hung down, swinging and hitting the horse under the jaw, irritating him. I had no possible control, and he galloped furiously through the streets, going up towards Grafton Street.*

*I shouted to two carmen at the corner of Nassau Street to stop him, but they failed to do so; then I made an attempt to reach his head by getting out over the dashboard and throwing myself forward on to his back, grabbed the tug-strap with my right hand, but in the darkness failed to catch with the left, and was slung off on one side. In the hope of a policeman or other person being in the street who might stop the runaway I clung on, hanging down beside the horse for the length of Grafton Street, being badly punished by the horse's hoofs all the time.*

*Seeing no hope of help, and before losing consciousness, I cast myself off backwards, and was flung across to the kerb on my head.*

He was taken to Vincent's Hospital suffering from injuries to the head, leg and body and was detained for some days. He was able to resume duty in early January 1917.

The chief laid a charge of "gross carelessness and culpable negligence for the accident" against Fireman Francis Redmond who was on night duty on the night in question. His duties involved making sure that the horse and trap were hitched for immediate response once a fire call was received in the station. The waterworks committee severely reprimanded and cautioned Redmond and added that any further neglect or disobedience would be dealt with severely.

Fireman Redmond had only joined the brigade in April 1916, and on 29 January 1917 he became seriously ill while on duty. He was removed to Vincent's Hospital where he was detained for three weeks and diagnosed as having "developed permanent heart disease, and is totally unfit for the fire service". The medical officer stated that "the incapacity was not occasioned in the discharge of his duty" and the law agent advised that the corporation had no power to superannuate him. However, under the National Health Insurance Act 1911 he was entitled to four weeks' pay, twenty-six weeks' half pay and twenty-two weeks' quarter pay, after which "he retired on medical grounds of five shillings per week until he reaches his 70th birthday". There the saga of the unfortunate incident rested for a time.

In December 1916 Patrick Reilly, James Walsh and Joseph Connolly were placed on the permanent staff of the brigade, having completed their probationary period. Joseph Connolly had been interned in Frongoch concentration camp for some months as a result of his participation in the Easter Rising, but he was reimbursed to the value of £22-10 lost wages for that time, under the same rule which compensated members of the British army who were away serving in the Great War.

Connolly was eventually to become chief of the fire brigade in 1930. This momentous year of 1916 closed with some further progress because the ambulance was fitted with pneumatic tyres at a cost of £28, the first vehicle to be so equipped.

In 1917 the corporation decided that the matter of granting licences to persons who were not competent to pass a rigorous test of management of motor vehicles should be examined. It also decided to grant a wage increase to turncocks of five shillings per week, and agreed that firemen whose wages were less than £250 per annum should receive the war bonus in line with all other corporation workers.

The number of street fire alarms had increased to fifty by this time, and the brigade was attending many hay fires. On 2 August more than £9,000 worth of hay belonging to the military was destroyed in a huge blaze on Gouldings Wharf at Alexander Quay, and more than sixty tons was destroyed in sheds in Raheny. These were to be among the last fires attended by Purcell as Chief of the Dublin Fire Brigade. He had again been on sick leave during the year and on 16 October he finally made application to the waterworks committee for permission to retire.

*Having occupied the position of Chief Superintendent of your Fire Department for over 25 years ... and now being in my sixty-sixth year ... I am anxious to retire from service. During my service I have devoted my energy to the improvement of the Department, which has been extended and completely reorganised and now compares in efficiency with that of any other city in the Kingdom.*

He went on to remind the the committee members of the serious accident which had befallen him in November 1916: "I still feel the effects of same". He agreed, however, to continue until a replacement was appointed. The committee decided to accept his request and recommended a pension equal to two-thirds of his salary and emoluments which stood at £702 per annum. This resulted in a pension of £468 per annum. The committee went on to place on record its

> *appreciation of the services of Captain Purcell at all times rendered to the Fire Brigade and the citizens ... and feel that the best thanks of the committee should be placed on record and conveyed through the Municipal Committee. They desired to place on record their appreciation of the successful manner in which various improvements initiated by him were carried out.*

Purcell's final annual report for the year 1917 showed no loss of life in fires. There were still two firemen on active army service since the beginning of the war, there were eight horses in service in the brigade, and ambulance cases remained at over 2,000.

The chief's retirement was agreed at the November meeting of the council. Thomas Purcell was surely the finest chief officer to serve in the Dublin Fire Brigade. It is sad to reflect that no memorial of any kind appears to have been presented to him from either the staff of the brigade or from Dublin Corporation. Such recognition had been given to lesser figures and for less service to the people of Dublin. He was to live for another thirty years and to see much change in his beloved fire service. It must have given him a great deal to reflect on. He lived out his life as an unsung hero who deserved greater recognition.

Purcell thanked the waterworks committee members for their assistance and courtesy.

> *I derived the greatest pleasure in initiating and completing the several improvements which ensured the successful working of the Brigade. I thank the staff for its loyalty and willingness on all occasions, and I am glad conditions generally for its members have improved. I leave the service much better than when I entered it.*

# 11 The terrible years

Following the council's agreement to Purcell's retirement John Myers wrote to the waterworks committee seeking promotion to the position of Chief of the Fire Brigade. The matter was discussed at the February 1918 meeting and it was decided that the vacant post should not be advertised and that Myers should be promoted as requested. The committee also recommended an immediate salary increase from £450 to £500 per annum for the new chief.

Lieutenant John Myers had been Purcell's deputy since 1897, and it was stated at the meeting that he had "proved his efficiency as a fire officer at more than one memorable conflagration". Nevertheless, the manner in which he was appointed to the position of chief seems a little strange and not in accordance with the forward progress of the brigade which had been evident throughout Purcell's leadership.

Three of the station officers applied for the post vacated by John Myers. Again without going outside the brigade Martin Jennings was given the position. Jennings had over twenty years' service; he was a tradesman, promoted foreman in 1911 and station officer in 1913. As an electrician he had carried out maintenance and repair work on the station alarms and lighting from time to time. He was a qualified instructor in both fire and ambulance training and had two medals for saving life. The wages on appointment in April 1918 were £200 rising by £10 per annum to £250, and there were allowances for uniforms, quarters, fuel and light.

In March of that year Firemen Joseph Lynch and Edward Doyle were presented with Royal Humane Society bronze medals and Carnegie Trust Award Certificates for endeavouring to rescue seamen from the *SS Cypress* who were overcome by gas in the ship's hold and unfortunately died. Both firemen were themselves overcome and had to be removed to hospital.

The political climate in the city had changed and the corporation was taking a full part in the changes. It warned the government against

> *The disastrous results of any attempt to force conscription on Ireland, and earnestly request them not to be driven by a hostile anti-Irish press campaign unto seriously considering any such insane proposal which, if put into operation, would be resisted violently in every town and village in the country. [It went on to instruct the lord mayor] to convene a meeting of John Dillon, Joe Devlin, E. de Valera, Arthur Griffith and representatives of the Irish Trade Union Congress to arrange a United Ireland opposition to conscription and to consider an all Ireland Covenant against it.*

While this massive political upheaval was taking place, the fire brigade was involved in more mundane business: insuring ambulances and motors for £59-18-10; paying £70 for a new horse to replace "Milton" who was put down. The firemen supported their union representative John Power to pursue their claim in conjunction with the other corporation unions for the application of the war bonus to all staff. Firemen were part of the workers' stoppage on 23 April in opposition to conscription, a decision approved by the city council. All those participating were granted leave with pay for that stoppage.

In July an industrial dispute arose between the firemen and Rathmines District Council. The Dublin firemen's union informed the corporation that their members would not turn out for duty in that township, but the Rathmines dispute was in fact settled before any call was made on the Dublin Fire Brigade. The Dublin service had established itself as a true lifesaving organisation even under the most dangerous of circumstances. But the return to a normal working environment for the firefighters during 1917 was not to last. The ending of the First World War in 1918 was the beginning of the armed struggle to achieve an Irish Republic, free from British imperial interference.

Technological advances were taking place at an increasing rate. Purcell had built up a network of contacts with fire services throughout much of Europe and in the USA where he was seen as an innovator who used his engineering skills to foster more modern techniques to combat fires. The use of sprinkler systems to protect major buildings was the kind of innovation that excited him. As far back as the eighteenth century a simple type of sprinkler, a pipe with holes bored at intervals was being suggested, and in 1806 a John Carey took out a patent for a system operated by pulling ropes, water being supplied from roof tanks which depended on rainfall. A French patent was registered soon after, and the first attempt to use a fusible link for a combined automatic/manual system soon followed. In London by the end of the century many buildings had installed even more efficient systems. The first true portable fire extinguisher was invented by George Manby in 1816. It consisted of a copper vessel containing three gallons of potassium carbonate with a quarter of the container holding compressed air. By opening a tap the air drove the contents out through a nozzle. Brass pump carbon tetrachloride extinguishers were manufactured by the Pyrene Company from 1912. They were still in use in the Dublin Fire Brigade up to the 1970s!

Celluloid had now become a major hazard in fire fighting. Following the deaths of nine people in a London factory where large consignments of celluloid were stored, an investigating committee recommended the introduction of very strict regulations concerning safer storage. No national legislation was enacted in Ireland until 1922 even though London passed a Celluloid Act in 1915. This slowing down of progress in the brigade was probably as much to do with the council as with the chief fire officer. At this time Irish political minds were focused on larger issues than fire fighting or fire prevention.

In his first annual report Chief John Myers called for complete motorisation of the brigade, a call he was to make throughout the remainder of his career with only limited success. At this time even spare parts for the motor vehicles were a problem. In September 1918 Myers tried to get hind wheels with artillery hubs for the Leyland ambulance but Leyland would not supply them. Neither would either of the two other companies in England who specialised in these products, so they had to be manufactured in Dublin. Although the waterworks committee that year would not sanction the purchase of a new motor fire engine it did give approval to the chief to furnish a specification for a motor car.

In March 1918, when other workers were getting wage increases, the firemen's union sought an award based on those paid under awards of the Committee on Production. The matter being referred to the law agent and rejected, it was agreed that the claim should go before the Interim Court of Arbitration (the first such move on a firemen's pay claim). As a result of that hearing in December 1918 an advance over pre-war rates was granted to officers and men of the brigade, and any war bonus already granted was to merge with it. This increase brought the maximum of the fireman's pay to thirty-seven shillings and sixpence per week. The last two firemen on military service, Patrick Bruton and John Murphy, who had both been in the army from October 1914, returned to the brigade and were paid £60-7-6 in lieu of lost wages. As there was no accommodation available for Bruton, a married man, he was paid ten shillings and sixpence per week from the time of his rejoining until accommodation was provided.

The 1919 chief officer's report noted that false alarms had risen dramatically to forty-two, an even greater figure than that for 1913. There were no deaths by fire, but the brigade had been busy with over 2,000 ambulance calls attended and a number of large fires, some of which were far outside Dublin. On 16 June at 6.10 am a motor pump with seven men responded to a call for assistance at Navan Flour Mills. This six-storey premises 1,609 ft x 30ft, was stocked with flour and contained a large amount of milling machinery, including an electric power station. The Dublin Fire Brigade crew pumped from the river Boyne but were unable to prevent the complete destruction of the premises with a loss calculated at £14,000. However, the firemen successfully prevented the fire from spreading to an adjoining building. That fire was one of eleven calls from outside the city boundary, and it must have been a hot summer because many of those calls were to gorse fires at Howth. The brigade declined to attend a fire in Wexford in November.

February saw the biggest fire of the year when Maple's Hotel in Kildare Street was burned to the ground. This four-storey structure formerly consisted of three residences, but was then being used as government offices. Within twenty minutes of the firemen's arrival all floors were alight and flames were leaping through part of the roof. It was more than twelve hours after their arrival on the scene that the last firemen returned to station. Another destructive fire took place at the *Irish Times* printing works in Fleet Street. Just after noon on 26 September a blaze broke out under one of the printing presses in the machine room and spread into a ventilation shaft right up to the roof of the three-storey building. Following this fire Sir John Arnott on behalf of the *Irish Times* sent a £50 cheque to be distributed among all the firemen who attended the fire.

Following the granting of a forty-seven hour week to all corporation workers in January 1919 the firemen also sought a reduction in their working week. Talks on the firemen's claim continued throughout the year and focused on the introduction of a two platoon shift arrangement. This was rejected because it would involve a substantial addition to staff, and additional expenditure which the committee could not afford to pay. A compromise was eventually agreed by the waterworks committee in January 1920 that the men should work two days on duty and have the third day off, with three weeks' annual leave for every man. This was a significant breakthrough by the firemen on the previous long hours system.

The chief had told the committee that such a new system would require the employment of six additional men and cost £1,052-18-9 per annum. It was an historic month for the brigade because at the same time as the waterworks committee was agreeing to the reduced working week, the lord mayor was receiving letters from Rathmines, Rathgar and Dun Laoghaire councils seeking an arrangement for the attendance of the city fire brigade in their districts when required.

But all was not well in the brigade The city council received notice of a dispute between the chief officer and the firemen signed by the union secretary. It claimed that

> In January the Municipal Council granted the members of our union one day's leave of 24 hours in three, but the Chief Fire Officer ordered all hands on leave to report for duty at 10 pm, this would mean a loss of 12 hours to each man on his days' leave. In the circumstances the Union decided not to comply with the Chief's mandate; but I am directed to assure you that our members will turn out to big fires if rquired.

Thus began the first major threat of industrial action by the firemen in the history of the Dublin Fire Brigade. The council referred the matter to the waterworks committee for their immediate attention, and the matter was resolved. Meanwhile, Martin Jennings, second officer, was retired on medical grounds in March 1920. When the waterworks committee attempted to fill the vacancy the recommendation was voted down, and a decision postponed for twelve months on the proposal of William Cosgrave.

The recent local elections with the extended electoral lists had totally altered the composition of the corporation, and even on the nationalist side divisions had arisen. It is significant to note that whereas all the Labour representatives on the council voted for filling the second officer post, all the Sinn Féin representatives voted against. However, they were all agreed that the section of the Dublin Corporation Act 1898, which covered the issue of funding the fire service from the water rate, needed amending. It was estimated by the waterworks committee that at least five shillings in the pound on the rate was required to fund the service. The council sought to have the limit on the funding abolished altogether, but it failed to get sanction for this change from the Local Government Board. This opposition may have influenced the reply to the townships of Dun Laoghaire, Pembroke and Rathmines regarding arrangements to have the services of Dublin Fire Brigade available in those townships:

> ...at present it was not possible to enter into an agreement with any of the outlying townships to attend fires in their districts, however the Chief Fire Officer would always be ready to render assistance having due regard to the safety of lives and the property of the city of Dublin.

The townships could obviously not depend on the city brigade as a matter of right.

The city and the country was by this time in turmoil as the military and political struggle for Irish independence intensified. Many republican councillors were missing from council meetings because they were in custody, and the pro-unionists were undermining the democratic work of the corporation by opposing the progress of the business. The divisions in society were obviously affecting the former unity of the firemen. Some officers had ceased to be members of the Dublin Firemen's Trade Union, having transferred their allegiance to the Local Government Officers (Ireland) Trade Union.

The first ever joint pay claim with the Irish Transport and General Workers Union, Building Labourers Union, and Irish Municipal Employees Trade Union in May 1920 probably signified that some of the firemen were now represented in one or other of these unions. This division of the firemen's loyalties into different unions was in the long run damaging to the pay and conditions in the Dublin Fire Brigade.

In July 1920 the waterworks committee again raised the matter of filling the vacant post of second officer, making the following proposal that

> ... *a competitive examination be set, with the Chief Officer of Belfast Fire Brigade, Professors of Engineering from both UCD and TCD and the City Architect acting as the examiners. The exam to allocate 40% of the marks for practical knowledge of the fire brigade and fire fighting, age not to exceed 35 years except in the case of a Corporation candidate. All candidates to be medically examined beforehand, and the position open to all Irishmen.*

At the council meeting the previous decision of not filling the post for twelve months was amended to six months. This proposal gave rise to the strange situation that the post of second officer was to be filled by competitive examination whereas the post of chief fire officer was filled without either examination or interview!

Meanwhile, the town clerk had written to ten of the other major fire brigades in the United Kingdom seeking detailed information on rates of pay, duty hours, annual leave, and clothing allowance for staff. This had arisen from a decision of the waterworks committee which "sought such recommendations as will make the Dublin Fire Brigade equal in point of economy to that of any other city, whilst no less efficient, and report on same to the Council".

On 14 December 1918 a general election took place and Sinn Féin contested every constituency on a policy of "Securing the establishment of a Republic".

- By withdrawing the Irish Representation from the British Parliament, and by denying the right and opposing the will of the British Government.
- By making use of any and every means available to render impotent the power of England to hold Ireland in subjection by military force or otherwise.

The result of the election was that Sinn Féin returned 73 of the 105 constituencies while the Redmondite Nationalists returned 6, leaving *The Times* to comment on "the overwhelming nature of the victory of Sinn Féin" and to concede that the general election "was treated by all parties as a plebiscite and admittedly Sinn Féin swept the country".

National politics would never be the same again – neither would local government. On 21 January 1919 Dail Éireann, representative of all republican parties, convened and issued a declaration which would cause conflict in Dublin Corporation: "We ordain that the elected representatives of the Irish people alone have power to make laws binding on the people of Ireland". Part of the strategy aimed at breaking down the British administration in Ireland was directed at local government. This included the setting up in June 1919 of arbitration courts. It was to such an interim court that the firemen's union referred their claim for payment of a pay increase under awards of the Committee on Production.

During the War of Independence the Dublin Fire Brigade was in the forefront of a humanitarian service to both sides. Dublin Corporation, as with all other local authorities, was directly responsible to the Dublin Castle administration through the Local Government Board. This body had power to veto all transactions and to withhold monies derived from public revenue for all public benefit carried out by the local authority. Thus was Dublin Corporation deprived at times of access to funding to carry out its statutory functions. Elections to local councils took place on 15 January 1920 under the proportional representation system, introduced to give the minority unionist population representation. But the result was an even greater defeat for British policy. Only one city in Ireland, Belfast, remained in Unionist hands. Dublin Corporation passed a resolution at its first meeting "that this council of the elected representatives of the city of Dublin hereby acknowledge the authority of Dáil Éireann as the duly elected government of the Irish people, and undertakes to give effect to all decrees duly promulgated by the said Dáil Éireann in so far as same affect this council". Funding for any new initiatives in the fire service was now almost non-existent.

By June 1920 Dublin was a scene of much intense military activity, heavily garrisoned, almost daily incidents of shootings, and military cordons thrown around areas of the city from time to time. It was in these conditions that the fire and ambulance crews of the brigade tried to cope and professionally carry out their regular duties. Soon inquests were abolished as republican attacks were followed by reprisals and the civilian death toll rose.

On 20 September, Balbriggan, a small village on the northern extremity of Dublin county, was extensively damaged when, as a reprisal following the shooting of a Black and Tan soldier in a pub in the village, soldiers from Gormanstown camp three miles away started a fire which was not extinguished until a small factory and twenty houses were destroyed. A request to the Dublin brigade for assistance was of little use because all the buildings were destroyed before the brigade could arrive. Herbert Asquith, addressing a meeting in Newcastle, cited the case of Balbriggan declaring that "wanton incendiarism there in one night reduced to smouldering ruins a factory employing up to 400 people". He said that he had seen the results of such things in Belgium and Northern France but never thought that "he would have lived to see carried out by officers and accredited agents of the law similar processes of wanton indiscriminate revenge and destruction against our own fellow subjects". As shooting and bombings continued in the city the military introduced a curfew from 10 pm to 5am. The empty streets were regularly filled with the sound of fire engines and ambulances responding to nightly incidents as acts of attrition continued to cause death and destruction throughout the city. Many times the ambulances took soldiers and police to hospitals and injured civilians to separate hospitals from the same incidents.

Probably the most harrowing day for the brigade during this whole ghastly period was Sunday 21 November. Urgent calls for ambulances flooded into headquarters early in the morning requesting attendance at various parts of the city where men had been wounded or assassinated. Before 10 am thirteen dead and six wounded had been removed to hospital. These harrowing scenes were exacerbated for the crews when soldiers at the scenes ordered the ambulance men at gun point to remove the victims to Bricin's Military Hospital and accompanied them with weapons drawn. In spite of this coercive pressure, all cases were dealt with professionally.

However the day's carnage was to continue when in the afternoon while 15,000 spectators attended a match in Croke Park, the military carried out a barbaric reprisal. Fire was directed at unsuspecting players and panic-stricken supporters, killing, wounding and seriously injuring many. There was further injury in the ensuing stampede through the exits to avoid the merciless shooting. Amid scenes of fear, panic and horror, the ambulance crews, when they arrived, dealt speedily with the trauma and bloodshed, driving packed ambulances to the nearest hospitals and returning again and again through the fear-filled streets until all those needing medical help had been seen to. In all the massive chaos and slaughter, the firemen – assisted by the Red Cross – had coped and saved many innocent lives as twelve victims died and sixty were wounded.

The following day, when the brigade had to tackle a massive fire, the city was in turmoil because all cars were banned from entering the centre. In addition, there was an ongoing railway strike caused by the dismissal of drivers for refusing to transport soldiers and munitions. The fire was at Paul and Vincent's chemical and manure works on Sir John Rogerson's Quay. It took over six hours to get it under control and two firemen were injured by falling masonry.

As 1920 was drawing to a close the city of Cork was placed under martial law. An ambush on a patrol of auxiliary police at Dillon's Cross was followed by an act of incendiarism by the military setting a number of houses in that district on fire. Not long afterwards auxiliaries and Black and Tans began arriving in Patrick Street and the civilian population made their way home well in advance of the curfew.

Gratton Street fire station was alerted to attend Dillon's Cross. As they travelled via the centre of the city they saw that Grant's drapery store, one of the city's largest shops, was on fire. They stopped to fight this blaze, but as they got to work they saw soldiers smashing the windows of other premises and entering the buildings armed with cans of petrol. Captain Hutson, Chief of Cork Fire Brigade, in his report stated that "all the fires were incendiary fires and a considerable amount of petrol or some such inflammable spirit was used in all of them". The fires raged throughout that night and next day the lord mayor requested assistance from Dublin and from Limerick.

Some time after 6 pm on 12 December a Leyland motor pump, with Chief Myers and a crew of seven Dublin firemen, was loaded on to a train to travel to Cork. After a lengthy delay at Mallow they finally arrived in Cork at about 2 am the next morning. By this time Chief Officer Hutson was seriously injured and in hospital, but the battle with the burning city buildings continued. The small force of firefighters showed no signs of despair even as the flames razed the city centre. The unequal struggle was described in the *Freeman's Journal* as follows.

> The heart of the business centre of Cork, including some of the most flourishing firms have been burned out. Damage estimated as up to two and a half million pounds has been inflicted on the inhabitants. The destruction caused by this deliberate act of incendiarism by the military mirrors the German burning of Louvain.

It was not until Wednesday 15 December that the gallant crew from the Dublin Fire Brigade returned home. Chief Myers praised the manner in which the exhausted Cork firemen had stood to their posts and continued: "The only way to bring home to the people of Dublin the extent of the destruction is to say that Cork is even worse than O'Connell Street and Abbey Street were after Easter week 1916".

The Dublin Fire Brigade had certainly played a significant role in stopping the Cork fires from spreading, and the town clerk wrote on behalf of the council "to return sincere thanks of the Corporation and citizens of Cork for the invaluable

**Cork and Dublin fire brigades working together during the burning of Cork, 1920.**
**Photograph, © National Museum of Ireland**

services rendered by your Brigade … please convey to the individual members of your Brigade our appreciation of their work".

A compromise eventually settled the railwaymen's dispute and trains began to get back to normal running. The Great Southern and Western Railway Company sent a bill for £110 to Dublin Corporation "being the cost of a special train for a detachment of the Fire Brigade to Cork". There was some opposition by the council to paying this bill, but after a heated debate a majority decided in favour of making the payment. Funding of the fire service had now become a major issue for the corporation, and the new political allegiances had aggravated the situation. With the council pledged to Dáil Éireann there was no chance of getting large funding from the Local Government Board, or any agreement from that body to alter the water rates receipts and so improve funding for the fire brigade. The councillors played for time, and controversial decisions on items which would increase expenditure were pushed from one committee to another without a definite decision being made. One such matter was the appointment to fill the vacancy for the position off second officer in the brigade. The age-old question of raising funding from the insurance companies again became a live issue as the Post Office rental charges for the street fire alarms was increased from £130 per annum to £314-10.

The Local Government Officers Trade Union sought an allowance for Station Officer Kelly for loss of leave and discharging portion of the duties of second officer. The Dublin Firemen's Union then wrote calling for the filling of the post in the interest of the safety of its members. The waterworks committee discussed these developments and then voted by a majority "to propose that Kelly be given the post because he has been assisting the Chief Fire Officer to perform duties which had hitherto been discharged by the Second Officer". But this decision was unanimously rejected by the council. Meanwhile Chief Myers having rejected the miserly offer of £50 which the council proposed giving him as compensation for loss of leave and carrying out some of the functions of second officer, agreed to have his case heard by arbitration. There was also an unsuccessful attempt by Councillor P.T. Daly, who acted as secretary and negotiator for the Dublin Firemen's Union, to have promotion confined to men with senior service. So while the corporation dithered on fire brigade matters, others issues were rapidly coming to a head.

Although Dublin was not under martial law, a virtual state of war existed with weekly shootings, bombings, death and destruction. Curfew was now imposed at 8 pm and the use of incendiaries as an implement of war was growing. The brigade's annual report for 1921 would record that at least fifteen cases of incendiarism against crown properties took place in Dublin, and there were almost as many other fires whose causes were unknown but which could be considered as attacks on British rule in the city. By far the most notable and destructive occurred on 25 May when the Custom House was destroyed.

The Better Government of Ireland Act came nominally into force on 3 May. On 19 May a new viceroy was appointed who immediately issued a proclamation summoning the new Northern and Southern parliaments to meet in June.

Elections took place for the twenty-six county parliament under a system of proportional representation. Neither Labour nor the old Nationalist party contested the election and all 124 seats were filled by Sinn Féin. The elected members together with with those returned in the northern election of 24 May refused to sit in a "Southern Ireland" assembly and instead constituted themselves as the Second Dáil Éireann. The following day saw their most effective act of defiance against British rule in Ireland. The Custom House was the most admired building in Dublin, and it was regarded as James Gandon's masterpiece of architectural beauty ever since its opening in 1791. In 1921 it housed the Local Government Board with its records and staff. It was also home to Inland Revenue, the Assay Office, the Income Tax and Estate Duty offices. Its destruction would strike a tremendous blow at British administration throughout the country. (See colour section, illustration number 10.)

Just before 1 pm on 25 May armed men entered the Custom House, held up the porters and other staff and removed all of them to the main hall. They then carried cans of paraffin, bundles of cotton waste and other inflammable materials into the building. Offices were ransacked with axes and papers were scattered on the floors and sprinkled with paraffin. Following the sounding of long blasts on a whistle the magnificent old building was set on fire.

Just after 1 pm a section of the IRA entered the central fire station, closed all the doors and set about removing vital parts from the fire engines. Simultaneously the three sub-stations were raided and similar action was taken. At Thomas Street station a party of IRA took the Leyland pump to Crumlin where it was held for over an hour. By 1.45 pm telephone calls were being received at brigade headquarters reporting a fire in the Custom House, but the brigade was prevented

**Burning of Custom House, 1921, with firemen tackling fire.**
**Photograph, © Department of the Environment**

**Burning of Custom House, 1921: dome has vanished.**
**Photograph, © Old Dublin Society**

from turning out. Five minutes later, just as a party of auxiliaries arrived to find out why the brigade was not responding, the ambulance drove out from the engine room with the IRA men concealed on board.

The fire engines were quickly made roadworthy again and turned out to the fast growing fire. Chief Myers, attending in the horse-drawn trap, stated that on arrival he found "the fire had made headway to a very great extent" and he ordered eight lines of hose to be laid down from the motor pumps drawing water direct from the Liffey. A gun battle was still taking place in and around the blazing building, grenades were being thrown, and a number of dead and wounded lay on the ground. The staff who had been inside also lay on the ground as the street battle continued. Despite all this the firemen set their jets to work. With all the floors and basement soaked with paraffin and a delay of at least an hour before the arrival of the brigade, it was apparent that the building was doomed.

During the fire fighting the chief officer was injured when struck by falling masonry and he was confined to bed for the following two weeks. However, crews from all four fire stations worked until 8.20 pm the following day to save the historic building. Unfortunately, according to one fireman's account, some firemen (with republican sympathies) on entering the building made sure that fires were started in areas that were not yet alight. Also, spare fire tunics were given to trapped IRA men allowing them to escape and be driven back to Tara Street fire station in the chief's trap.

Crowds of onlookers gathered to watch the attempts of the fire fighters to extinguish the vast inferno. The extent of the damage varied across the building, the centre suffering most as it was completely destroyed. Other sections did survive, smouldering for as long as two weeks. But Gandon's glorious masterpiece had gone, and British rule in the twenty-six counties was soon to follow. An area of 70,000 square feet had been involved in this awful and tragic spectacle. In the same year other huge fires were fought by the Dublin firemen, but none would figure in the history of the city with an impact such as that of the Custom House fire.

Following the burning of the Custom House several other buildings owned or occupied by the British government were set on fire. Chatham Row Vulcanizing Company had fifty tyres set alight, and the coastguard stations at Dollymount and at Pigeon House Road sustained serious damage. On 3 June the National Shell factory in Parkgate Street, housing motor cars and army lorries undergoing repair, was badly damaged by fire which extended to the Ordnance Depot before it was extinguished by the brigade using eight lines of hose.

On 9 June Dublin Corporation wrote to the Fire Offices Committee in London, referring to the Report of the Estimates 1921-22 pointing out "the benefits the existence of the Fire Brigade had for the Insurance companies" and recommending that they should pay towards this service and in the event of their not paying "that the service be reduced so as to bring its costs within a rate equitable for all citizens". On 11 August the Fire Offices Committee acknowledged in writing "the high standard of efficiency maintained by the Fire Brigade" and fully appreciated "the excellent work which the Brigade has done both in the City of Dublin and at times in the surrounding districts." But

> my committee hold very strong views ... and they maintain that they ought not be called upon to contribute, either annually or otherwise, towards such expenses .... The function of Fire Insurance Companies is to indemnify their policyholders against loss or damage by fire .... The protection of life and property from fire ... is as much a public duty of the governing body of a district as the protection of life and property from violence and robbery, the provision of proper sanitary arrangements, the maintenance of roads and other municipal matters, and as such it should be provided at the public expense.

The letter went on to maintain that the "Towns Improvement (Ireland) Act 1854 and all subsequent General Public Acts had authorised the levying of rates for the maintenance of Fire Brigades, and such Acts are fair and equitable". Special taxation on the lines suggested by the waterworks committee of the Dublin Corporation would, in the opinion of the insurers, be unfair and injurious to the public interest and opposed to all precedent.

> It has never been suggested that life insurance offices should contribute towards the cost of Fire Brigades or Sanitary improvements, or marine insurance offices towards the cost of harbours of refuge or lighthouses, or offices insuring against burglary toward the cost of the police force, but they are as much interested in such matters as fire insurance offices are supposed to be in arrangements for the extinction of fires.

The Fire Offices did not think it "unreasonable" for those outside the rated area who avail of the fire service to make some payment "but the Fire Insurance Companies are under no legal liability to recoup such expenses and whenever they do so the payments made by them must be regarded as ex gratia. For the above reasons, therefore, my Committee feel they are unable to comply with your request. This then was the definitive position of the Fire Offices to the Dublin Fire Brigade. The council in receipt of this letter sanctioned an increase "from 6d to 9d of the Public Water Rate ... to terminate on March 31st 1922, when the rate leviable will be sixpence in the £".

As a result of correspondence between the British prime minister Lloyd George and De Valera in June 1921, a truce was introduced on 11 July. All military action on both sides ceased, but the corporation continued its allegiance to Dáil Éireann and suffered the consequences of scarce funding. How this directly affected the fire brigade was clear when the post of second officer remained vacant and compensation claims by Myers and Kelly were put on hold even though the waterworks committee had recommended filling the post and compensating the two officers. The refusal of the Fire Offices to make any direct contribution towards the fire service had convinced the corporation "that the service be reduced so as to bring its cost within a rate equitable to all citizens".

Elsewhere changes were coming in the fire service. Since the ending of 1914-18 war, firemen in Britain had been pressing for changes in their pay and conditions. Trade union recognition was finally granted to London firemen in 1919.

The truce held and talks to settle the Irish claim to sovereignty were conducted in London. But the settlement terms divided the country and after confrontation between pro-treaty and anti-treaty supporters the nation was plunged into civil war when government troops laid siege to the republican-held Four Courts in June 1922. Again the Dublin Fire Brigade would have to work in a state at war, fires would increase by 142 on the previous year, malicious calls would jump to ninety. From 28 June to 5 July all firemen would be continuously on duty, often under gunfire.

One cannot doubt that the firemen were as divided on the question of the treaty as was the country as a whole, yet as a public body they put the job first and carried out all tasks asked of them. It is unlikely that any fire service in any capital city in the world had to begin each day in such terrible circumstances. The opening week of war began at the Fowler (Orange) Hall, Rutland Square (now Parnell Square), a building occupied by republicans which was attacked by Free State soldiers. Before evacuating the building the occupants set the ground floor on fire, but the quick response of the fire brigade prevented the fire from spreading to the rest of the house. Although the Four Courts was besieged, no attempt had been made to fortify the premises or resist attack on the building, yet it took three days and nights of heavy shelling before the surrender took place.

At 11.45 am on 30 June Chief Myers received a message from General O'Daly informing him that the Four Courts was on fire. He replied that the brigade would respond if combatants on both sides ceased firing, but the general said his attack must be maintained. Some ten minutes later a call was received via police telephone stating that the heavy firing had stopped, so all available men and appliances were turned out. The firemen found that the fire had started in the

Dublin Fire Brigade at Four Courts, c. 1922, fireman Bernard Matthews (centre) with
Chief Officer Myers (right). Photograph, © Paul Matthews

records office and was extending to the entire building with flames engulfing the
central hall.

At intervals small explosions occurred as bullets became heated. Seven lines of
hose were laid to tackle the growing blaze. The chief officer spoke to Rory
O'Connor, one of the republican officers, and was told that at least seven tons of
highly explosive material were strewn all over the building. At 3.15 pm a land
mine exploded injuring three firemen, necessitating their removal to hospital.
One of them, Bernard King, with sixteen years' service was on sick leave for 116
days before being retired from the service as a result of his injuries.

After this incident firemen continued to fight the fire from outside, where
some received minor injuries when coming in contact with networks of barbed
wire as the area immediately around the building became smoke logged. After 5
pm the great dome collapsed, and the gathering crowds watched the destruction
of another of Gandon's wonderful buildings. All brigade efforts were now
concentrated on preventing the destruction of surrounding buildings as masses
of sparks and burning embers were blown out of the roofless inferno on to the
adjoining premises. At about 8 pm the brigade began returning crews to their
stations, but relays of men continued to do duty extinguishing pockets of flame
until 4 July. Rarely had the dreams of a nation ended in such a disastrous episode.

The real inferno was left for Upper O'Connell Street, most of which had
survived the bombardment and fires of Easter Week 1916. Once the Four Courts
had been surrendered the Free State troops concentrated their attacks on the
republican stronghold in O'Connell Street. This began on 3 July with rifle and
machine gun fire, soon to be followed by shelling from armoured cars and

artillery. The fire brigade chief in his detailed report gives a vivid description of the occurrences as the firemen carried out their work.

*Brigade on arrival was directed to the Y.M.C.A. building, 43 Upper Sackville Street. I found the building strongly barricaded and, after failing to procure an entrance, resorted to playing two jets from hydrants through the upper windows from the street. In the meantime an aerial ladder from C Station was extended in Sackville Street as a water tower. The prolonged rifle and machine gun fire grew intense and bullets came dangerously near, a number striking the escape. We had to take cover, and our attempts had only the effect of temporarily allaying the outbreak. The extended ladder had to be abandoned, and next morning when firing had quietened down Station Officer Barry and the men of Buckingham Street recovered it.*

*On 5 July at 12.19 pm called to conflagration involving northern end of Sackville Street. All sections ordered out. Motor from central Station under Lieutenant Power and Station Officer Kelly stopped by armoured car at corner of Marlborough and Gloucester Streets and held for thirty minutes. Seizing the first opportunity two lines laid down into Thomas Lane were working right between cross-fire ... the greatest danger was from ricocheting bullets and at one time we had to abandon our appliances and take cover. The conflagration spread with amazing rapidity on the eastern side through breaches in the party walls of houses occupied by Republicans ... Thomas*

Purcell Escape used for demolition work, O'Connell Street during Civil War.
Photograph, © Edward Chandler

*Street section after doing excellent work were forced to retire. This outbreak left unchecked spread north to Gogan's shop and south to the Tramway offices and back into Marlborough Street. Next came Findlater's, the Brigade concentrated on this point and were successful in saving this building, under which was stored thousands of gallons of whiskey.*

*Eventually St Thomas's Church in Marlborough Street and the Concert Hall next door were destroyed in spite of heroic efforts by the fire crews. That long day ended for the Dublin Fire Brigade when they saved part of Stanley Street Workshops which had also been fired in the same savage conflict. A total of 80 buildings were completely or partially destroyed over those*

Funeral of Michael Collins passing Bank of Ireland, 1922. Photograph, © Michael Counihan

*eight days; sixty-five people were killed and twenty-eight wounded; hundreds were put out of work; hundreds homeless, and an estimated £3 to £4 million property lost.*

The whole block of buildings from Findlater's Place down Marlborough Street to Cathedral Street and around to Sackville Street, with the exception of the pro-cathedral, had been damaged in the flames. The pro-cathedral survived because a jet was constantly played on its roof, windows and doors throughout the conflict. The seeds of fratricidal strife were well and truly sown when Churchill wrote to Michael Collins: "If I refrain from congratulation it is only because I do not wish to embarrass you. The Archives of the Four Courts may be scattered but the title-deeds of Ireland are safe".

The year 1922 was an exceptionally difficult one for the firemen because apart from the punishing days and nights of that fateful week in June, many other fires and incidents arising from the civil war had to be dealt with. The chief officer's report lists twenty-five fires attended as being caused by armed men setting the premises alight. One of the most shocking was at the house of a TD (member of parliament) in Fairview, where the brigade arrived to find a mother and two young children severely injured as their home burned about them. All three were taken to the Mater Hospital where one of the children died.

The Rotunda Rink was in use as the temporary general post office because the

174

Funeral of
Michael Collins
approaching
O'Connell
Bridge, 1922.
Photograph,
© Michael
Counihan

O'Connell Street building was still in ruins since Easter 1916. It was held up by armed men who removed cash, postal orders and sacks of registered mail before torching the premises. Despite a gun battle taking place nearby as the military tried to apprehend the culprits, the firemen arrived promptly on the scene to tackle the flames.

Three sections of the brigade had to deal with a similar blaze at Cahill Printers of Great Strand Street where the roof of the two-storey building collapsed on to many tons of paper and machinery. Another fire caused extensive damage to the machine room and paper stores of the *Freeman's Journal* in Townsend Street. Railway signal boxes, Amiens Street post office, government property and houses of TDs were all targets as the vicious effects of the civil war continued throughout the city. At the height of this mayhem the Dublin Fire Brigade attended the funeral of Michael Collins, as it had attended that of Parnell some thirty years before.

175

On 29 August the brigade was requested by President Cosgrave, the new leader of the government, to respond to a serious fire caused by rioting republican prisoners in Maryborough (Portlaoise) prison. Lieutenant Power with a pump and crew made the fifty mile trip by rail. The prisoners were being held under armed guard in the recreation yards while the firemen entered the smoke-logged prison to extinguish separate fires in 150 cells on three floors. It was twenty hours later when the exhausted firemen returned to Dublin.

The huge workload on the fire brigade in that awful year revealed a problem relating to the ambulance service. Although the horse-drawn ambulances had been reduced to two-man crews, there was a shortage of trained ambulance personnel and the result was that some of the qualified ambulance men had to work continuously. To solve this, an important decision was made that in future all recruits would be trained in first-aid, as would the rest of the current staff. This led to an arrangement with the St John Ambulance Association to provide the necessary training courses. Thus began a fruitful agreement which was to last for over fifty years.

The long-running saga of compensation and promotion regarding Myers and Kelly continued. The Firemen's Union wrote to the lord mayor calling for the appointment of Kelly to the post of second officer because he had acted in place of the chief officer when he was absent as a result of injuries sustained at the Custom House fire. But the council insisted on the post being filled by written examination. This exam took place on 20 April with only seven firemen attending, which seems to indicate a boycott of the proceedings by Station Officer Kelly, the other station officers and the foremen. John Power, with fourteen years service, got 72% of the marks and was appointed second officer from 1 May 1922. This was on a six-month probationary period with a salary of £300 to rise by £10 per annum to a maximum of £400. The waterworks committee now tried to settle the outstanding claims of Myers and Kelly by recommending a £50 payment to the station officer and £150 to the chief. Kelly agreed to accept this offer, but Myers did not agree that the amount was adequate compensation. Eventually, almost a year later, the chief's claim was brought to arbitration by his union and settled.

Although the waterworks committee had previously agreed to introduce written exams for promotion to station officer (s/o), no such exam was proceeded with, and a vacancy had existed at that level for almost twelve months. Chief Myers and his newly appointed second officer decided to promote Martin Dowling to fill the s/o vacancy. This led to further unrest. At the council meeting the recommendation was carried by twenty-seven votes to twelve in spite of the opposition of the Firemen's Union and Labour representatives at the meeting.

The lord mayor had received a revealing letter signed by the three foremen in the brigade which stated that each of them had over nineteen years service.

*During the last three years (which has been the busiest period in the Brigade's history) each one of us, some time or other, had to take the place of an officer. We make bold to say that there is not a question regarding Fire Brigade work or apparatus that any of*

**14. A Dublin Firefighter. Portrait by James Conway.**
Collection Dublin Fire Brigade Museum

15. The Mansion House with Dublin Fire Brigade engines.
Transparency, © Dublin City Council

16. Fire at premises of The Irish Times in Fleet Street in 1999.
Photograph, collection Trevor Whitehead.

17. Water tender built by the Irish firm of Hughes on a German Magirus chassis in 1985.
Photograph, collection Trevor Whitehead

18. Fire drill in Dublin Fire Brigade Training Centre.
Photograph, © Dublin Fire Brigade

19. Hose drill with extension ladders in Dublin Fire Brigade Training Centre.
Photograph, © Dublin Fire Brigade

20. Dennis turntable ladder in Dublin Fire Brigade Training Centre.
Photograph, © Dublin Fire Brigade

21. The O'Brien Institute, Dublin Fire Brigade's Training Centre.
Photograph, © David Barriscale

22. The Dublin Fire Brigade Museum at the O'Brien Institute.
Photograph, © David Barriscale

23. Watch sitting on turntable ladder outside Tara Street Fire Station
in December 1996, a few weeks before the building was vacated
Photograph, © Alan Finn

24. One of Dublin Fire Brigade's fleet of Dennis/Magirus turntable ladders.
Photograph, © Trevor Whitehead

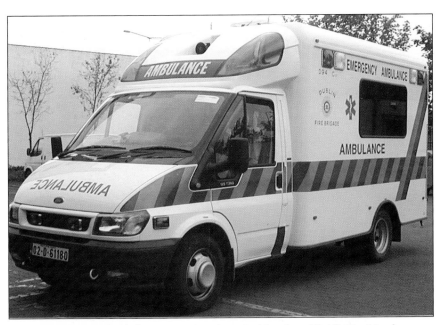

25. Provision of ambulance service is an important function of Dublin Fire Brigade.
Photograph, © Dublin Fire Brigade

26. Women firefighters have been an integral part of Dublin Fire Brigade since 1994.
Photograph, © Dublin Fire Brigade

**27. Author Tom Geraghty (right) with Cork firefighter Tom O'Brien
at Madison Square Garden for memorial held on 12 October 2002
to honour New York Fire Department colleagues who died
in 9/11 Twin Towers attack.
Photograph courtesy the Author**

*us would fail to answer .... Martin Dowling had been employed to look after the horses, a qualification which is now practically non-essential. Not one of us has ever seen him with a helmet on at a fire, or returning to his station with a wet uniform. He has no training at any time at fire or ambulance drill, an important qualification, you will admit, is a thorough knowledge of each. How can an officer teach men what he does not know himself? Lord Mayor and gentlemen we appeal to your sense of justice with a hope that you will see this appointment is made in a fair manner.*

A proposal to postpone the appointment until the chief and Mr Power came before the council to explain their decision was ruled out-of-order by the lord mayor!

It was reiterated by many at the meeting, however, that all future promotions should be filled on the basis of written exams. Martin Dowling was appointed as station officer and assigned to Buckingham Street station in November 1922. It is interesting to note that in 1924 Dowling and two other firemen in that station were found guilty of breaches of discipline by Lieutenant Power. Dowling was fined £10 and transferred to Thomas Street station.

At the July meeting of the council the establishment of a fund to be called "Dublin Fire Brigade Widows and Orphans Fund" was suggested. There was a proposal that the fund could be started by a donation from the corporation and then annually supported by voluntary contributions from the corporation, business people and the public. This, however, was voted down. Over the next few months two station officers and six firemen retired from the brigade on the grounds of injury or ill health. Purcell's stalwarts who had braved the guns and bombs during the real crisis years were taking their disciplined experience out of the service. Between June 1921 and June 1923 a total of fifteen long-serving staff left the brigade, an exodus exacerbated by the manner in which these faithful public servants saw themselves being treated by their employers and the government that they and their families had supported into existence. The firemen left but, in spite of the pleas of the chief, the horses remained.

The new year 1923 had started with some hopeful signs that an end to the destructive civil war would soon take place. But the first weekend of January was to be a trying time for the firemen. On 6 January armed men held up a goods train at Raheny and then set its engine careering along the track towards Howth. It collided with and de-railed a passenger train, injuring several passengers. The brigade attended the scene with ambulances and fire engines but luckily there were no serious injuries to be dealt with. Much worse was to come. On the next day some time after 8.30 pm a boy (imitating his elders) threw a squib into the greengrocer's shop on the ground floor of a four-storey tenement house on Summerhill. It immediately set fire to a paraffin oil tank in the shop, and within minutes the place was blazing. With cries of "fire" from bystanders many of the unfortunate tenants rushed from their homes. Others were not so fortunate as the flames engulfed the ground and first floors. Some children were dropped from second storey windows into an army great coat held by rescuers, while some adults on the first floor jumped.

The fire brigade arrived and immediately began rescuing trapped people from the top two floors. By means of an escape ladder four people were rescued. As other crews tackled the fire a further entry on to the top floor was carried out by Buckingham Street firemen from the escape and they removed the bodies of a woman and two children. The first and second floors collapsed before the blaze was extinguished. Five people died and six others were removed to hospital suffering from burns and shock.

It was another unnecessary tragedy in a city that had become hardened to loss of life in bad circumstances. The fire crews had done everything possible to save the victims, but the quickly spreading fire made it impossible for those on the top floors to get out or the firemen to reach all of them in the chaos of the situation. Chief Myers was recorded in the press as saying "the anxiety of the public to assist the Brigade hampered them in their work". But the quick thinking of some passers-by had been responsible for saving those trapped on the first floor. The *Freeman's Journal* in an editorial the following morning summed up the situation, showing what firemen were being confronted with on these all too regular occasions.

> The overcrowded tenement is not only a forcing house for disease, but a death trap in case of fire. Why should it be possible to store oil tanks and vend inflammable stuff on the ground floor of an overcrowded tenement?

The civil war officially ended in April 1923, and an uneasy peace gradually took its place. The government decided to resolve the nation's financial problems by reducing pensions and wages, and shedding jobs in the army and public service. The 1924 annual report of the Irish Trade Union Congress stated that the government "announced that high prices, high profits and high wages cannot be sustained, but no action has been taken in regard to high prices and high profits, but the government have given direct encouragement to drastic reductions in wages". The report claimed that the Minister for Local Government had succeeded in fixing low wages for workers employed repairing roads and was using these low wages as an argument for the reduction of the standard rate of wages paid to workers in permanent employment of local authorities. Pay and conditions in the Dublin Fire Brigade were soon to the fore in this government attack.

In February four incendiary attacks took place on government property. In all these incidents the speed of turn-out of the fire crews led to the fires being extinguished before any major structural damage was caused. This in spite of the fact that land mines were laid at two of the buildings.

Although there were government moves to reduce wages in local authorities, in March 1923 Dublin Corporation decided to grant a bonus to all Dublin Fire Brigade staff "who served during the period of hostilities in the city last summer because the work done by the officers and men, in circumstances of exceptional difficulty, deserved special recognition". The chief was given £30; Lieutenant Power £20; station officers £10; forty-four firemen and turncocks £5 each. The firemen and turncocks were each granted a chevron carrying pay of one shilling per week, and all ranks were granted one week's holidays in excess of ordinary leave during the year.

In September 1923 a comprehensive report on the future of the fire brigade was published by the waterworks committee. It pointed the brigade in the right direction for the coming years, outlining the urgent need to dispense with horses and to purchase a motor turntable ladder of greater capacity than the ladders currently in use. The report was divided into six sections, as follows: apparatus; rates of pay; numbers and conditions of service of staff; sick pay; sub-stations; public water rate and cost of fire brigade. Chief Officer Myers and Councillor T.J. Loughlin had visited cities in the UK to obtain relevant information.

It was at last recognised that "the retention of horse-drawn vehicles of any kind for modern fire brigade work can no longer be justified. Speed in arriving at a fire is of vital importance." The full details of the recommended new apparatus and alterations to existing apparatus are given below.

1 Leyland Motor Fire Pump. Six Cylinder, 1,000gpm, with first aid tank and hose reel, and 40ft "Ajax" ladder .......... £2,300

1 Morris-Magirus 85ft Turntable Ladder .......... £2,750

2 Leyland Chassis 30-32hp worn driven .......... £1,700
These two chassis will have to be specially constructed to carry the present (horse-drawn) aerial ladders)

1 20hp car (with special body) for Chief Officer .......... £550

First Aid Tank and Hose Reel, 40ft "Ajax" ladder and Lighting Set.
For the Tara Street Leyland RI 1090 .......... £170

First Aid Tank and Hose Reel, 40ft "Ajax" ladder.
For Thomas Street Leyland RI 2080 .......... £110

16 Two-Gallon "Fire Snow" Extinguishers at £3-10s each; 3 Carbon Tetrachloride Extinguishers at £4 .......... £68

1 "Proto" Oxygen Breathing Apparatus .......... £25

In addition to the above appliances and equipment, a 500 gallon underground petrol storage tank was to be installed at Tara Street headquarters.

The cost of the new items would be partly offset by the sale of the following old equipment: two steam fire engines, chief's trap, two hose tenders, one fire escape, two ambulances, and the chassis of two horse-drawn aerial ladders.

The Department of Local Government held an enquiry into the application to raise the necessary funds, and a government loan of £8,000, repayable over ten years, was sanctioned to cover all the expenses.

Under the heading rates of pay in the report the waterworks committee considered the special nature of the work which distinguished the fireman from all other classes of employee, and in view of the arbitrators award was satisfied that rates of pay should now be stabilised on a basis which would attract the very best type of men. The pay would now be: station officers £5 per week increasing by two shillings and sixpence per annum to a maximum of £5-7-6; foremen £4-17-6 and firemen £3-10 by annual increments of two shillings and sixpence to a maximum of £4-12-6. It was also recommended that staff should be increased by six to give a total complement as follows: Tara Street – chief, second officer, station officer and 30 firemen; the three sub-stations – one station officer and eight firemen each – giving an overall total of six officers and 54 men.

Under the heading sick pay it was stated that prior to 1922 firemen were paid during illness but since then each case was dealt with on its own merits. Because of the nature of the firemen's work, their hours and different duties the former practice of full pay during illness should be sanctioned for the future.

Under the heading fire stations, having considered the possibility of closing one or more stations, it was decided that the continuance of Thomas Street and Dorset Street was essential. Although Buckingham Street was close to headquarters it was felt that it was essential for access to the highly important North Wall area, and the station was also surrounded by densely populated tenements with the Clontarf/Fairview area now part of the city, and plans for extensive housing in that district. The advantage therefore was thought to lie in the retention of all three sub-stations.

Under the headings public water rate and fire brigade cost it was stated that the cost of the brigade was running at £23,445 and the limit on water rate at 6d in the pound was £22,380, leaving a deficit of £1,075 per annum. Adoption of the report would increase this to a total future deficit of £3,617-15. The report proposed an actual increase on the public water rate of 2d in the pound, which should cover all possible contingencies for a considerable time in the future. The report also recommended power to be sought for compulsory contributions for the fire brigade by the insurance companies.

This was a most interesting and comprehensive report but it faced into stormy waters even as it was being adopted by the corporation.

In October an exam to fill the vacant position of station officer was held. The two papers on "Building Construction" and "Fire Appliances" were set and corrected by ex-Chief Fire Officer T. Purcell and the city architect. Only six candidates presented themselves and Joseph Connolly, with seven years service, received the highest marks and was promoted to station officer on 4 December 1923. The chief, in his annual report for that year, mentioned that the brigade was still attending fires in both the city and the surrounding areas caused by incendiarism, some twenty-three being attended. False alarms were also a big problem, with sixty-seven being received, and they gave some interesting insight into how the judiciary maintained the class system.

A business man, Joseph Edelstein, already with two previous convictions was again fined on three separate occasions amounts of £2, £5-10 and £20 for sending bogus calls. Two tenement boys however with no previous convictions received a fine of £2 plus fourteen days in prison and one month in prison respectively for a similar offence.

Huge fires had occurred in the Mineral Water Factory on the South Circular Road and at the Davoy Cocoa Company in South William Street. Although there were some aspects of progress in the brigade in 1923 there were still seven horses in service. Early in 1924, when the first public telephone kiosk was being erected at the corner of Grafton Street and Stephen's Green, the government was rejecting the corporation's application for "extending the amount of Public Water Rate or raising the proportion for the Dublin Fire Brigade", but they did suggest that the cost of the ambulance service should be charged against the improvement fund instead of the fire brigade. The chief was being directed to inspect quarterly all fire appliances in the post office stores, and the system of the brigade reimbursing sick or injured firemen for medicine was being abolished. The corporation was allowing the erection of sixty-seven kerbside petrol pumps on the basis of an annual rent of £3-3 being paid, and the corporation was being indemnified from any liability for damage to life or property arising out of the erection of the pumps.

On 4 February, while the ambulance horses were being exercised in Aungier Street, the driving pole broke and the horses ran away, galloping down Georges Street and finally dashing into the window and glass door of Hyne's restaurant (a last act of defiance?). The unfortunate handler had to be taken to hospital and the horses sustained flesh wounds. The law agent had to make a settlement of £50 to cover the damage to Hyne's premises.

By March approval had been given by the council for the purchase of a new turntable ladder, and by the next month the city had a regular motor bus service. The army was demobilising and thousands of ex-soldiers were being put on the dole, and taxi meters were first introduced in the city.

The event with the most far-reaching effects for the city of Dublin was the completion of a government inquiry into the corporation. It showed that the corporation had a debt balance of £18,791 in revenue account and £31,229 in capital account – a total balance of £50,020 when the councillors were elected in 1920. But although there remained £87,019 on revenue account the capital account was overdrawn by £68,445.

Yet this was the period when the council supported Dáil Éireann, creating problems of funding and being isolated by the Local Government Board. Salaries and wages were not the major factor in increased expenditure in the corporation. The auditors account showed that, including overtime, the costs for 1921-22, 1922-23 and 1923-24 were respectively £146,001, £163,609, £159,172. Yet wages and employee numbers were listed for cuts.

The dramatic result of the inquiry was that Mr Cosgrave and his government dissolved the Dublin Corporation. Apparently what could not be politically controlled must be done away with, in the new order. Augustine Birrell in his memoirs *Things Past Redress* wrote:

> *Mr Cosgrave has abolished the Corporation of Dublin by the stroke of his pen. Any English Chief Secretary who had attempted to do the same piece of good work would have been compelled to resign by a combination of Unionists and Nationalists in the House of Commons.*

A new chapter in the administration of the city of Dublin was now about to begin. For the foreseeable future the Dublin Fire Brigade would be run by appointed commissioners.

# 12 A new administration and amalgamation

The government attack on wages in the Dublin Fire Brigade had begun even before Dublin Corporation agreed to grant the special bonus in March 1923. It decided to reduce the war bonus for all corporation workers by 50%. This would mean a reduction of thirteen shillings per week for the bulk of corporation employees and fourteen shillings per week for firemen. The trade unions opposed this savage reduction in earnings and threatened strike action if the cuts took place.

Following talks with a sub-committee of the corporation's financial and general purposes committee a deal was agreed to reduce the wage cut to nine shillings per week in three phases of three shillings each in April, August and December.

On 17 April 1923 when the firemen were notified of this proposal, the Dublin Firemen's Union wrote back stating that its men "would not agree to any cut in their present rate of pay" because they were not a party to any negotiations on this matter. But the corporation replied on 23 April as follows:

> the war wage reductions amounting to 9/- per week for the current year accepted by the other unions shall take effect in the case of the Fire Brigade men with the exception of the three Station Officers to whose rate of pay a 25% reduction will apply ... no small section of employees was allowed to refuse to accept reductions in rates of war wage agreed as reasonable by the majority of their colleagues, particularly when members of the Fire Brigade were saved portion of the increased cost of living by the fact that they enjoyed house accommodation, fuel, light and clothing allowances.

The Dublin Firemen's Trade Union then served strike notice on the corporation. A meeting was convened for 9 May between the firemen's representatives, the corporation and the ministry of industry and commerce. In the meantime the Minister for Home Affairs had written to the lord mayor

> regarding the fact that Dublin Fire Brigade had given notice of their intention to cease to work as a protest against the action of the Municipal Council in deciding to reduce the war wage of which the Fire Brigade members were in receipt in common with other municipal employees. The Ministry assumes the Corporation are unable in the circumstances to accept full responsibility for the protection of the property and lives of the citizens against possible outbreak by fire. Your letter having been considered in this light, I am directed by the Minister to state that the Government recognise that they have a primary responsibility for the protection of property and lives against fire when the ordinary protective machinery breaks down.

*I am to point out, however, that in discharge of this responsibility which the Municipal Corporation would ordinarily fulfil, the Government will of course, utilise when necessary the various appliances, the property of the Municipal Council.*

The live-in firemen were left with no option but to accept the pay cuts now imposed on them by a government that had just seen out a bloody civil war. In time the pay rates recommended by the water works sub-committee in its comprehensive report on the fire brigade would be ignored by the government commissioners who were soon to take over running the corporation's affairs.

Under the provision of the Pensions Act 1920 the government also ensured that pension adjustments were kept below cost-of-living increases, thus hitting at the retired firemen along with all other pensioned public servants.

In February 1924 the Minister for Local Government approved the £8,000 loan recommended for the purchase of new fire brigade appliances and equipment. This resulted in orders being placed with Leyland Motors for a 6 cylinder 1,000gpm motor pump, and with John Morris of Manchester for a Magirus 85ft motor turntable ladder. The latter appears to have replaced one of the three 66ft manually operated horse-drawn aerial ladders. The brigade still had seven horses in service.

During the year twenty firemen attended a St John's Ambulance training course in order to qualify as ambulance personnel. When nineteen successfully passed the exam all firemen, with the exception of drivers, were put on rotation for ambulance duties.

With the further move to motorisation the position of the brigade turncocks came under scrutiny and Lieutenant Power was directed by the commissioners to report on their workload. Prior to abolishing Winetavern Street Station in 1912, no turncocks were attached to headquarters and it was only when Thomas Street opened that the two turncocks were attached to the central station.

*Now four turncocks working six-hour shifts are on the station. Their working time is principally spent sitting in the kitchen, on very rare occasions they go out with the Brigade to regulate the valves. The new motors with first-aid tanks meant rapid reaching and dealing with fires, and the recent replacement of large number of small choked mains by larger ones rendered turncocks services unnecessary.*

As a result of this report the waterworks engineer was directed to provide plans of the arterial mains and valves with an up-to-date list of hydrant placings to the fire brigade, and the turncocks were returned to the waterworks section of the corporation.

Lieutenant Power made an interesting reference to "first-aid tanks" assisting the rapid dealing with fires. The two original Leyland motor pumps did not carry any water for fire fighting, but the report in September 1923 on future equipment for the brigade recommended that both these appliances be fitted with a first-aid tank of water (about forty gallons), a small pump and a 140ft reel of small bore hose. This apparatus soon proved very effective in the initial attack on small fires.

At this time all the senior fire brigade officers – Myers and Power of Dublin, Hutson of Pembroke and Whyte of Rathmines – were associate members of the

Institution of Fire Engineers (IFE), formed in 1918. It may perhaps have been that the IFE gave them the 'associate" designation in an attempt to expand the organisation's membership in the greater Dublin area. They certainly attended the annual conferences in the 1920s, following correspondence from Arthur Pordage, one of the Institution's founders.

The purchase of two motor ambulances was now approved, the chief officer and the acting city engineer proposing either the de Dion Bouton or the Arrol Johnston as being suitable. However, the commissioners rejected the proposal saying that "since the Public Health Section is to defray the cost of providing the new ambulances, the Dodge built by Callow and Sons of Kevin Street, Dublin, shall be procured as the price only amounted to the sum of £293".

In September 1925, following the arrival of the two converted motor fire escapes and one of the new ambulances, five of the remaining seven horses were sold off. Regarding the newly motorised aerial ladders, only one can be found in the motor taxation records (Y17587). It was originally proposed that two Leyland chassis would be purchased, but the one of which there is a taxation record was on a Peugeot chassis. It is recorded that two motor chassis were ordered from Peugeot Cars (Ireland) Ltd to which two aerial ladders were to be transferred at a cost of £1,334 with bodywork by John Howard of Sandymount. The original cost estimated for Leyland chassis was £1,700.

In March 1926 the remaining horses were sold when the second of the new ambulances was received. The glamorous spectacle that convinced small boys that they wanted to be firemen when they grew up was no more. The image of galloping horses pulling a thundering fire engine with smoke billowing, crewed by firemen with glittering brass helmets and shining high boots, was passing into history. The heroics of mounted men rushing to rescue people in distress would soon be confined to celluloid images of cowboys on silver screens.

The Dublin Fire Brigade changed towards the end of the 1920s with much of the old sense of corporate loyalty destroyed by division and neglect. By 1925 the Firemen's Union was no longer the sole representative body of the firemen and, while there were no pay claims made by the three unions who now acted on behalf of individual members, there were claims regarding station accommodation, with one union arguing a different approach to the next to gain advantage for its own members.

Such was the situation when in 1925 Fireman James Kennedy of Thomas Street Station, who had eight years' service, was taken ill and removed to Crooksling Hospital suffering from tuberculosis. He had a wife and four children, all under seven years of age. The medical officer in Crooksling informed the corporation that James Kennedy would not be fit to return to duty as a fireman and he was retired on fifteen shillings per week in spite of a request from the Firemen's Union to continue him on sick leave because of his family circumstances. His death occurred just five weeks after the decision to pension him off but the commissioners refused to extend his sick leave even for that period. That was in November 1925. Early in February 1926 the chief was directed "to order

Mrs Kennedy to give up her flat in the Fire Station". The union made representation to Captain Myers and he contacted the commissioners on her behalf but was given an order to have her family vacate by 18 February.

Mrs Kennedy wrote to the commissioners seeking employment as a cleaner/caretaker owing to the fact that she was destitute but the law agent on 1 March told Captain Myers to "demand possession and if possession was not given an ejectment summons" would be applied for. The commissioners informed Mrs Kennedy that "there was no employment". When she had not vacated the law agent brought a caretaker's summons against her for ejection but the magistrate, more humane than her husband's employer, "refused to make a decree" until they found her accommodation.

The Royal Hibernian Military School in the Phoenix Park housed some 200 soldiers of the Irish-speaking battalion of the army. On a morning in March 1925 smoke was observed on the third floor under the central clock tower of the three-storey building. The alarm was raised and troops proceeded to remove furnishings onto the parade ground whilst awaiting the arrival of the fire brigade. The top floor was well alight when the chief and two engines arrived and, soon after, the clock tower collapsed through the blazing roof. It was a bitterly cold morning. Witnesses reported that icicles formed on the stand pipe and leaking hose. Captain Myers sustained a blow to the helmet and serious foot and leg injuries when struck by an amount of falling masonry as the roof collapsed, and he was removed to hospital. By 4.00 am the Dublin Fire Brigade had the fire under control just as the army fire service arrived from the Curragh Camp. The top floor of the building was completely destroyed; there was lesser damage however to the lower floors and the fire had been prevented from spreading to the officers' quarters in the east wing. Sometime after 5.00 am the weary firemen rolled up their frozen hose and headed back to station.

On 31 March 1926 the Tara Street Central Control received a call at 8.57 am for assistance at a fire in Malahide. Two sections were despatched to the two-storey mansion "La Mancha" which stood in its own grounds surrounded by trees, gardens and fields. By the time the brigade arrived the roof was on fire (part of it had actually collapsed) and there were fires burning on the first floor. Drawing water from a large pond in the grounds the fire crews got to work quickly and were soon to be confronted with a grim sight. In the dining room of the house they discovered the body of a man which they removed pending the arrival of the ambulance. By then gardaí, who were also in attendance, had discovered a second body – this time in the grounds. Worse was to follow. When they forced entrance into the main front bedroom upstairs the firemen discovered the bodies of three women. A sixth body was discovered in the basement. At this stage the top floor was an inferno with all five bedrooms and the bathrooms fully alight. The remainder of the roof finally crashed down totally destroying everything on the upper floor. The ground floor was saved in spite of extensive damage in the hall and to the stairs.

On investigation it became obvious that separate fires had been started in different rooms and the doors and windows were closed so that the flames were fanned only when the rooms were opened up. It transpired that the fires had been smouldering since the night before but had made little progress because the oxygen had burned out in the tightly closed rooms. An employee was charged with the murders, found guilty and executed. To avoid early detection of his vicious crime he had arrested the progress of the fires he had started by closing all the doors and windows in the house. However the fire load and the flammability of the furnishings in some of the bedrooms had taken such a hold that the windows shattered and spread the fire through the ceilings into the highly inflammable attic contents in the old building.

The following year the brigade again experienced a situation involving murder. This time it was requested by the gardaí to drain a deep pond near Dundalk in County Louth. A foreman and two firemen spent six days pumping water before one of them pulled a sack from the almost empty pond and found that it contained the body of a missing housekeeper for whom the gardaí were searching. A youth of nineteen years was eventually found guilty of this brutal crime and executed.

There was a sequel to this incident two years later when Chief Fire Officer Connolly received a bill from the superintendent of the gardaí for Louth and Meath claiming payment for meals provided. "They [firemen] slept in the Barracks for a period of six nights and partook of their meals in the Mess. The Firemen left without paying their share of the Mess, viz. 18s-3d each". All the attending firemen's expenses had been paid to Dublin corporation by the garda authorities. However, the mess fund in the barracks was the responsibility of those stationed there and the "cute" Dublin firemen left without making their contribution. From the records it is clear that at this time the commissioners had relaxed the position of the Dublin brigade responding to calls outside the city, leaving the decision to the chief officer.

In January 1926 a new loan of £120,000 was approved by the government for improvement of the water supply to the north city and, under the new Police Forces Amalgamation Bill which finally abolished the Dublin Metropolitan Police, the police rate in Dublin would be reduced to seven pence in the pound from the financial year 1926-27.

The work of the fire brigade continued as normal in spite of staff cuts. Brigade numbers had been reduced by four through death and retirement but only two vacancies were filled and these by the transfer of "two redundant motor drivers from the Cleansing Department". A point of interest is that these two drivers had worked a 47-hour week for a wage of 76 shillings whereas they now had to work 96 hours for 75 shillings per week.

The brigade had notified the commissioners of the increase in "possible fire situations over which they had no jurisdiction, in motor cars, buses and charabancs transporting petrol without any fire extinguishing apparatus and wholesale unlicensed premises selling petrol and paraffin". Although 1,040 licences were granted in the city for storage of petrol and carbide, prosecutions were regular for non-compliance with the law, usually resulting in a fine and forfeiture of the product.

Chief Fire Officer John Myers died suddenly on 19 March 1927. He was the chief who had finally overseen full motorisation. He had led the brigade through some of its most trying years, and unfortunately both he and his staff had been victims of much neglect by their employers. There was no corporation to honour his untimely passing but the commissioners did express sympathy to his relatives. The Commissioner of the Garda Síochána, Eoin O'Duffy, expressed his condolences on 22 March 1922.

> *The news of the death of Captain Myers, Chief Fire Officer of the Dublin Fire Brigade, was a source of deep regret to the Garda Síochána, and I wish on behalf of the Force to tender sympathy to his relatives and to the members of the Brigade.*

Such were the times that his passing received scant attention in the national newspapers. Shortly before his death Myers had sold one of the steamers to the Main Drainage Department for £75.

John Power was interviewed and promoted to the position of chief fire officer from 1 May on a salary of £500 per annum, rising by £20 per annum to a maximum of £700. The post had not been advertised but his appointment was confirmed by the Department of Local Government and Pubic Health in September, the same month that the long-awaited Dodge motor car was finally provided for the chief fire officer.

The commissioners also gave permission at this time for the use of Dublin Fire Brigade ambulances for stretcher cases en route to Lourdes on pilgrimages organised by the Catholic Truth Society.

It was January 1928 when the "appointment of a fit and proper person to the position of Second Officer of the Fire Brigade at a salary of £200 per annum to £400 after 15 years" was approved. The promotion had been confined to the four station officers in the brigade and Joseph Connolly with "thirteen years service as a member of the Dublin Fire Brigade who for the past four years has occupied the position of Station Officer was appointed to Second Officer. A competent motor driver, he has good knowledge of motor transport, mechanical appliances and First Aid work."

The appointment was made following an interview and the commissioners reminded him on his appointment that he was

> *… responsible for maintaining the present standards of efficiency of the fire brigade, as well as maintaining and improving the service. In this connection it might be his duty to deal with cases of dereliction of duty by other members of the fire brigade, and, accordingly, he should disassociate himself from and hold himself completely independent of the men and their organisations.*

## The 1927 cinema fire

The fire call was received from a passerby and was made from a street fire alarm sometime after 11 pm on 25 October 1927. When the two brigade sections arrived, the Bohemian Cinema on Phibsborough Road was well alight with smoke billowing from the roof of the locked building. In trying to force an entry through a first floor window Captain Power was injured when the glass canopy on which he was standing partially collapsed. He received a serious wound to his thigh and an injury to his back that required his removal to the Mater Hospital for treatment.

The firemen eventually forced entry and noted that the screen and an organ were completely destroyed. The fire, which had reached into the roof void above the ceiling, was dangerous to deal with and it was very difficult to gain access to the location of the main blaze. The firemen eventually got to the roof area and, with a two-pronged attack from above and below, the blaze was extinguished after some some three hours effort. A huge volume of water had been used trying to prevent the flames from spreading and so the upper levels of the building became smoke-logged and dangerous for the men tackling the fire. The entire cinema from the balcony outwards sustained major damage. Much of the ceiling had collapsed and smoke and steam rose through the damaged roof into the night air. Captain Power did not return to full duties until ten days later.

In March 1928 examination papers were drawn up by T. P. Purcell and the city architect to fill the station officer post vacated by Joseph Connolly. Again confined to the brigade personnel, ten candidates took the papers on "Building Construction" and "Fire Appliances". James Howard received the highest mark and was promoted to the vacancy The same month saw the lord mayor's coach being stored in Thomas Street Station, the chief being allowed to "purchase a typewriter", two dozen foam extinguisher charges for the motors and a "Davy" automatic lifeline for rescues in burning buildings. The firemen's red shirts were listed at £1-12-11 and boots were £3-10 a pair in that year's estimate.

For twenty-five years, celluloid had been the cause of serious fires but fortunately for Dublin in the early years the amount in storage or circulation was small. However, with the rapid growth of cinemas, a native film-making company and photography, there was a need for stringent restrictions under enforceable legislation to protect the public. No such legislation had been enacted. This was in spite of the tragic Drumcollogher cinema fire on 5 September 1926 in which forty people lost their lives. In Britain there had been calls by the fire brigade for strict control of the manufacture and storage of cellulose and when a massive fire occurred in a film waste storage building in St Pancras, London, new regulations were introduced on 1 April 1928 governing the manufacture and stripping of cine-film. The only action taken in Dublin occurred when Dublin Fire Brigade was contracted by Metro Goldwyn Mayer and Warner Brothers to destroy film waste at "one guinea" per roll.

The chief fire officer successfully objected to the Dandy Boot Polish Company installing a 5,000-gallon petrol tank because the firm already had two 8,000 gallon tanks. In December the brigade was given permission to pay for the driving licences of their fifteen drivers at a cost of £7-10 in total. In comparison with previous years this was a quiet period for major fires in the city and Captain Power was given compensatory leave in lieu of leave lost between the death of

John Myers and the appointment of the new second officer. The year ended with the commissioners publishing a Dublin Corporation handbook edited by Bulmer Hobson in which was listed among their achievements "the provision of an efficient fire brigade".

A decade of rising prices, rising unemployment, falling wages and low public morale was continuing as the Dublin firemen faced into 1929. It would be another trying and eventful year that would present more questions than answers in spite of the versatility shown by the brigade in carrying out its work. Captain Power had pressed the commissioners to purchase a new modern fire engine to meet some of the problems being confronted by the brigade, particularly in regard to petrol and cellulose fires. Using the unexpended balance of £1,666 from the £8,000 loan of 1924 for fire brigade equipment, he was finally allowed in July to purchase a motor fire tender and foam generator from the world famous firm of Merryweather and Sons at a cost of £1,360. This appliance was built on an Albion chassis with a 275gpm pump and was the city's first fire engine fitted with pneumatic tyres.

John Power never lived to take delivery of the new fire engine. He died suddenly on 9 July. The pressure of the job in these times was obviously having a serious effect on the officers of the brigade. Unlike his predecessor he did not even warrant a considered obituary in any of the national papers and there is no letter of condolence in the archives from the Commissioner of the Garda Síochána, nor indeed from any organisation in the country. The status and recognition of the Dublin Fire Brigade was at a low ebb.

At a time when the brigade needed a new emphasis, better trained and educated firemen to cope with the increasing problems presented by the rapid advances in the sciences, the firemen were being pressed into handling modern engines of which they knew very little and to face fires and hazards in factories and workshops of which they had only scant knowledge. The service was showing all the signs of neglect. Joseph Connolly was appointed chief fire officer on 12 September and the post of second officer was filled the following month following a written examination confined to the four station officers. James Howard, who had been promoted to station officer only the previous year but who had joined the service in 1907, came first in the examination and was confirmed in the post.

Perhaps the legislators were too involved in other matters to worry about what was happening in the fire brigade. A Bill was being pushed through the Dáil by the government aimed at creating a new Dublin Corporation with much fewer councillors and a wider electorate, with boundaries extended to take in the townships of Pembroke, Rathmines and Rathgar, and some rural areas of the county.

This new body, representing all those in the extended area, would have twenty-one elected representatives only and would be controlled by a government-appointed city manager. The Bill was passed by the Dáil in December 1929 in spite of major opposition from representatives of the townships and a number of members of parliament. Elections to the new extended corporation took place the following September and another new era in local government had begun. The Lord Mayor of Cork, addressing a meeting in Dublin in opposition to this new type of council, referring to his own situation, said that "the Cork City Management Act was an insult to the people of Cork, a sneer on public representation, and a gross fraud on a good and decent people".

The firemen saw out the year with a dangerous and complicated rescue on the Liffey on the cold, stormy night of 5 December. Crew members of a Dublin University eight were returning from a training session on the flooded river when the fast flow drove them into the Islandbridge Weir, smashing the boat and throwing the crew into the freezing water. Two of the victims clung to the weir while the other six were washed over the ten-foot drop into the cauldron below. They managed to get onto the two small islands just below the weir and awaited rescue.

The brigade responded just after 6.00 pm and in the light from the chief officer's car headlights they succeeded in slinging drop lines to the two men on the weir and hauling them through the fast-flowing river to be removed by ambulance to hospital in a frozen and exhausted state. Then a cockle-shell boat was taken from a boathouse further up the river, carried on the roof of an ambulance to the scene, manhandled by the firemen over a wall and lowered into the river. The boat containing two rowers from the boat club and a fireman named Matthews playing out a drop line held by the fire crew on the wall made its way to the nearest island where another two of the victims were clinging to a tree as the rain battered down on them. With a second line secured to the tree the boat was then pulled back to the wall with the rescued two and the oarsmen on board. When the victims had been hauled up, the boat was pulled back to the island by the fireman.

In the meantime the fireman had slung a third drop line with a buoy to those on the other island. One by one they made their way across, gripping the fastened line. In this relay fashion all four got to the first island and, as with the earlier rescue, they were pulled to the river wall from where they were taken to hospital. Eventually fireman Matthews and the oarsmen were taken from the boat to the riverside. The rescue had taken over four hours during which all of the work was carried out in the headlights of the chief fire officer's car from the opposite bank of the river. Fireman Curran, one of the rescuers, had the following to say to a reporter: "I was at Scapa Flow and I think I know what currents are. The current around the Islandbridge islands was tremendous for a river the size of the Liffey". Newspapers the following day reported winds of up to 65 mph and heavy rain all along the east coast causing storm damage and the death of three Skerries fishermen on the night of the Islandbridge rescue.

A dangerous outbreak of fire confronted the Dublin Fire Brigade on the afternoon of 28 June 1929 at the auxiliary repair depot of the Irish Bus Company in Dominick Street, an organisation soon to sell off its business. Sometime after 3.00 pm, while thirty men were working in the premises, flame from a blow lamp set fire to a large sheet of tarpaulin beside the paint store. Before the staff could react a huge ball of fire burst upwards and the building had to be rapidly evacuated. The enclosed area was 200 feet by 90 feet containing six buses, spare parts, piles of tyres and stocks of paint, with a wooden roof covered by felt and tar.

Before the brigade arrived on the scene there were two loud explosions as petrol tanks on the buses exploded shooting flames towards the roof and throughout the rest of the building. The fire crews tackled this dangerous conflagration from Dominick Street and Mountjoy Street at the rear as gardaí arrived to take control of the huge crowds that were gathering. Soon the whole building was alight from end to end and, as the fire crews hosed water onto the flames, the roof collapsed in a shower of sparks and burning debris which threatened adjacent tenement houses. A gas pipe fractured in the workshop and spread further flames. Some two hours after their arrival the firemen had the fire contained and the gas cut off, and had prevented any real damage to other premises.

The company directors, who heard of the fire while attending a meeting at Aston's Quay, adjourned to attend at the scene. Their spokesman stated that "they appreciated very much the work performed by the Dublin Firemen". All six buses were destroyed in the flames as over £11,000 worth of damage was caused. The newspapers complained of water shortages and delays as fire crews first got to work.

This now regular press criticism of the fire brigade resurfaced when on 1 December it agreed to respond to a fire at Killua Castle, Clonmellon, Co Meath. The Leyland which was sent from Dorset Street broke down just beyond the village of Clonee some five miles from base. A second motor was dispatched from Tara Street but its engine gave up near Dunshaughlin some ten miles outside the city boundary and it had to be towed back to base. Fortunately the fire burned itself out without destroying the whole building. A major row broke out later about this incident when the corporation law agent sought payment to cover the Dublin Fire Brigade costs.

Training within the service was also the subject of criticism. When a fireman was injured while involved in a jump-sheet drill in Tara Street station the secretary of Dublin Trades Union Council complained in writing to the commissioners "about the antiquated method of drill which consisted of throwing a life size dummy 40 feet in a rescue sheet held by Firemen below". The writer demanded that this practice should cease, a call also supported by the Dublin Citizens Association.

The decade ended with another horrific cinema tragedy when on New Year's Eve a fire broke out during the penny matinee at the Glen Cinema in Paisley, Scotland. The premises were packed with children. Panic broke out when a fire was detected in the projection room. Although the fire was of limited intensity and was confined to the projection area, the children rushed for the exits and the confusion and panic that ensued led to a final death toll of seventy, most of the victims between the ages of three and seventeen. The incident brought the horror of Drumcollogher into the news again because of the comparisons that were made with what had now happened in Scotland. Paisley had a large Irish emigrant population, mainly from Donegal and the northern counties, and some of those who died were Irish or of Irish origin.

In the face of strong opposition the government finally passed the Greater Dublin Bill which was to give a new city manager control over the local authority. While this political debate raged on, the Dublin firemen were again being called upon to deal with an emergency in another county. A twenty-four-inch water main at Kilmacanogue burst, cutting off supply to Foxrock, Dundrum, Rathfarnham and many rural areas of the south county. Water from the burst main added to the torrential rain that had been falling for some considerable time, causing extensive flooding near the Wicklow village.

The new Merryweather with its full crew pumped water for more than twenty-four hours as a repair crew worked to effect repairs to the burst pipe. However, on its way back to station in wet conditions the following morning the fire engine skidded into a tram standard in Upper Baggot Street seriously injuring one of the crew and damaging the engine.

At that time the fire service consisted of a chief fire officer, second officer, four station officers and 52 firemen. Any long-term injuries or illnesses reduced this meagre staffing level. Two of the firemen were continuously on fire duty in government buildings and taking account of one man in each of the four stations doing phone duty, together with those on ambulance duty or on leave, the effective number for actual fire fighting was little more than twenty on any one day. Yet in spite of this small number, the commissioners allowed the city strength to be further reduced by responding to calls for assistance from outside the city. (See colour section, illustration number 11.)

Ambulance charges, which had again become an issue of discussion, were still calculated on the basis of £1 for "private cases" (non-accident) within the city boundary and two shillings per mile outside. Poor persons within the city were removed free of charge but this led to regular questioning of charges, refusal by some to pay and claims of impoverishment by many when billed. In all these cases a doctor's certificate had to be supplied to the ambulance crew stating that the patient was not suffering from any infectious/contagious disease. Maternity cases were treated as accidents and removed free of charge. Captain Connolly sought a change to this system because it tended to reflect badly on the brigade when bills were reduced following representation to the legal or finance sections of the corporation.

In November 1929 the chief wrote to the commissioners regarding serious staff shortages when he had "seven men on sick leave, three men on annual leave" and a vacancy at second officer level. He maintained that

> in 1924 the Chief Fire Officer had at his command 59 men including Officers. Today with two men on constant duty in Government Buildings and an average of 10 men on leave each day, and the ten above referred to, the remaining thirty men are divided between four stations, to carry out fire, ambulance, telephone or any emergency that arises.

He was supported in his claim by the Irish Local Government Trade Union, which demanded that the second officer position be filled and the overall staff levels increased. All he got however was the promotion of a station officer to second officer the following January.

In February the brigade had tackled a major fire in the Alliance Gas Company works in Pearse Street, drawing water from the canal to extinguish the blaze. The following month the Grand Canal Company claimed that the fire crews had used 700,500 gallons of water and demanded payment at the rate of a penny three farthings per 1,000 gallons used. This led to a long drawn-out dispute between the company and the corporation law agents before a settlement was reached.

Captain Connolly sought permission to attend the Institution of Fire Engineers Conference in Leicester and a further allowance owing to the fact that "he was using his holidays to attend and afterwards to visit some other cities including Berlin to inspect improvements there maybe in fire appliances". He informed the commissioners that he held a "diploma" from the IFE and that Captain Whyte of Rathmines was also attending the conference. He was given permission to attend with costs covered by the corporation. On his return he sought the purchase of another Proto Breathing Apparatus, for the reason that two smoke helmets and the one Proto set were insufficient for a fire service like Dublin with so much call on its service.

Early in the morning of 21 March a fire broke out in the Killeen Paper Mills in west county Dublin. Over an hour was allowed to elapse while some staff on the premises tried to deal with the fire before a call was sent to the Dublin brigade. When the chief and two engines arrived on the scene, the core of the extensive building was fully ablaze and lit up the area as palls of dense smoke rose into the morning sky. Problems arose for the fire crews when the suction hose became choked by partially burnt paper floating in the stream from which they were pumping water to tackle the fire.

Despite the problems the fire was confined to the "heating floors" of the mill where it had begun. Although the damage ran to thousands of pounds, the firemen prevented the blaze from spreading to the bag room and power plant. Some four hours after their arrival the crews were rolling up hose on the snow-covered ground as they prepared to send one engine home while the other damped down the remaining pockets of fire. Captain Connolly told reporters that less damage would have occurred if the caretaker had called the brigade earlier and that the premises should have had an automatic fire alarm system. His view was that if the brigade had received an immediate alarm it was likely that the fire would have been limited "to so small an area that the damage would have been almost negligible". It seems from this interview that Connolly was now fully accepting that the Dublin Fire Brigade would respond automatically to fire calls in the Dublin county area!

Elections to the new extended city council took place on 30 September and the following month the new council, lord mayor and city manager took over responsibility for the fire service. One of their first decisions was to sanction the payment of £40-7 to have a direct telephone line installed between Tara Street,

Rathmines and Pembroke fire stations. They also directed Captain Connolly to draw up a report on the future development of the new expanded city fire service. The suppressed townships each had their own traditions and of course they both had provided fire services and ambulance cover.

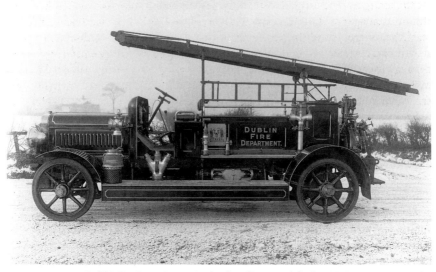

Dublin Fire Department, Leyland appliance supplied in 1912.
Photograph, © British Commercial Museum Trust Archives

Dublin Fire Department, Leyland appliance supplied in 1939.
Photograph, © Trevor Whitehead

194

# 13 The great tragedy

At the inaugural meeting of the newly elected corporation on 14 October 1930 the city manager outlined his powers and responsibilities. He was responsible for staff, their wages, conditions etc in this new hybrid form of local democracy; the mayor and the elected representatives had rights but these would be exercised through the manager. The chief fire officer would now be responsible to the city manager although fire brigade business would continue to be dealt with through the expanded water works, markets and public lighting committee. Gerald J. Sherlock, the city manager, made it clear to all present that his duties were the "efficient performance of official business and not the constitutional questions as to the City's government".

What that meant was to be made quite clear to the fire brigade within a very short period of time. On 9 December a fire broke out in a drapery store in Clones, Co Monaghan. The local volunteers assisted by the police were unable to control the spread of fire and at 7.52 pm the town clerk rang Tara Street fire station for assistance only to be informed by Lieutenant Howard that the brigade could not respond. At 8.35 pm a call was received from Alfie Byrne TD, the lord mayor, that transport was to be sent to collect him from the Mansion House and a fire engine was to be made ready to turn-out to the fire in Clones.

A station officer and six-man crew then accompanied the lord mayor on the fire engine which left headquarters for Clones. It was a foggy night but the motor and crew reached Dundalk some fifty-seven miles away in one-hour and thirty-five minutes, where they were stopped by the local gardaí and informed that Armagh Fire Brigade had crossed the Border and had the fire under control. Two shop premises were destroyed but the Northern Ireland crew had prevented the fire from causing any further damage. The lord mayor on his return to Dublin told gathered newspaper reporters that even though he had flown in an aeroplane and enjoyed that very much "travelling on the fire engine I have had one of the thrills of my life". What impressed him about the trip to Dundalk was "the quality of the driving".

Captain Connolly did not see the issue in the same light and wrote to the city manager: "I do not approve of sending an engine and men such a distance from the city, and would like the City Manager's ruling for the future". On 24 December he received a reply stating that

> only the City Manager has authority to send firemen outside the city boundary under the Local Government (Dublin) Act 1930 ... the Brigade cannot be permitted to attend outside the City except to fires in the neighbourhood of the City.

A similar letter was sent to the lord mayor.

This issue was eventually finalised after detailed correspondence from the corporation law agent when in November 1932 the city manager directed that a notice be sent to all other local authorities that the Dublin Fire Brigade was "legally debarred from sending the city Fire Brigade long distances outside the City, the responsibilities of such local authorities was for protection of life and property from fire in their respective areas". A bill had already been sent to Clones Town Council for £51-3, the amount claimed to cover the cost of the lord mayor's night ride to Dundalk on a fire engine.

Meanwhile Captain Connolly had set about consolidating the organisation of the expanded fire service. He recommended the retirement of Hutson and the suppression of his former post; also that four of the Pembroke men "be transferred to other Departments in the Corporation as they were unsuitable for the Dublin Fire Brigade". Both Hutson and Whyte, the former chiefs of Pembroke and Rathmines, agreed to take their pensions and were retired in February 1931. The corporation agreed to transfer one of the firemen listed as unsuitable but told Captain Connolly that the other men must remain in the fire service.

Following a fire in East James Street at which there were problems getting water for the fire engines, Captain Connolly wrote the following.

*The present system of having to send for a turncock from the scene of a fire is in my opinion an unsatisfactory one. Previous to May 1925 a turncock was permanently attached to the Brigade and always available.*

*I now beg to recommend that consideration be given to the revival of the original system, particularly in view of the extension of the city boundary.*

The water works engineer replied that he did not agree with the chief fire officer "as to the necessity for the provision of a turncock on continuous duty at the Fire Station". At the same time the chief sought permission to purchase a motor fire escape for £3,000 and leave to employ six extra men. The city manager rejected the request for an extra six men, but he did approve the purchase of a Merryweather turntable ladder costing £2,935. This new appliance, the second motor turntable ladder for the brigade, was demonstrated to the press at Tara Street HQ in September 1931. The *Irish Independent* reported that "although the members of Dublin Fire Brigade are quite new to it – the escape was only delivered yesterday – it was driven out of the station into the courtyard, and the ladder extended to its full length of 85 feet in less than a minute." The appliance was also fitted with a pump capable of delivering 275 gallons of water per minute.

In November the ambulance which had been presented to the fire brigade in 1915 by the Cinematograph Association was taken out of service. Its bodywork was then reconstructed on one of the 1925 Peugeot chassis from which the aerial ladder was now removed.

The Rathmines and Pembroke men received pay increases to bring them in line with those of their colleagues in Dublin but new problems arose for the chief officer. Some of the Rathmines men lived in council houses from which they responded to the station for fire calls while in Dublin there were married men but no married quarters to accommodate them. Connolly recommended the pulling down of some sheds at the rear of the former Rathmines town hall so that he could build married quarters for the firemen and he also sought to have three tenements in Townsend Street/Tennis Court Yard pulled down in order to extend Tara Street Station for the same purpose.

On 9 March 1932 when all the national interest was focused on the coming into power of the first Fianna Fáil government, with the support of the Labour Party and three Independents, the Dublin firemen were tackling a massive fire in Hawkins Street. A petrol tanker was delivering fuel to the National Service Garage when a conduit pipe burst and flaming petrol sprayed back into the tanker which immediately took fire. Soon a huge conflagration was taking place. The garage was alight in minutes and before the fire engines arrived the fire had spread to a tyre dealer's premises next door. Minor explosions followed one after another, throwing massive burning materials into the air and causing even more chaos. Water was plentiful from hydrants and the Liffey but the massive clouds of dense black smoke billowing from the burning premises caused major visibility problems for the fire crews.

As large crowds gathered the fire spread into the recently opened Catholic Central Library and threatened the Winter Gardens, Theatre Royal and other adjacent premises. Flames and large blazing embers spewed across the road breaking windows and forcing firemen to retreat to left and right. Soon the ambulance was removing two firemen burned by outrushing flames and a police constable injured when a roof collapsed. It was reminiscent of the infamous Theatre Royal fire except that this one was much more dangerous because of the petrol tanks beneath the forecourt of the garage.

Among the books being devoured in the flames when the library took fire was the Healy Collection (the books of the late governor general) which had recently been presented to the Catholic authorities. This was the same Tim Healy who had been responsible for procuring a pension for members of the Dublin Fire Brigade in 1902. In spite of the tremendous heat and suffocating smoke the firemen eventually surrounded the flames, saving part of the library and all other properties on the street from destruction. Night-time had fallen and still the fire crews worked to extinguish the flames. It would be the next morning before the final crew could be withdrawn from the dank eerie scene of statuesque high walls bare and dirty, held aloft by twisted steel girders surrounding smoking piles of rubble that had once been a prominent business premises.

On 22 March the owner of the Hawkins Street garage, who had been present throughout the major part of the firefighting, wrote to the lord mayor.

> *I wish to convey a very sincere expression of gratitude to Captain Connolly and the men of the Dublin Fire Brigade for the determined and courageous efforts to save my property and surrounding premises in the disastrous fire that destroyed my garage last week. The very nature of the fire, the stock and contents consisting of petrol, tyres, oil etc, made it almost impossible to subdue – yet the officers and men of the Dublin Fire Brigade went about their task with a methodical deliberation which gained the admiration of all who witnessed their efforts.*

When foreman John Markey died as a result of a job-related illness on the 5 July that year, leaving a young family to fend for itself, fires such as the Hawkins Street

incident were referred to as some humane councillors sought to bring ameliora-
tion to the families of firemen who died on duty. At the council meeting in
October the following motion was agreed.

> *In view of the dangerous nature of the occupation of the Fire Brigade, and in recogni-*
> *tion of their gallant services bravely and unstintingly given in the service of the*
> *citizens, a Fund to sustain widows and orphans of members of the Brigade who meet*
> *their death in the service, or whose death is occasioned because of something that*
> *occurred in the course of their employment, should be set up, to which insurance*
> *companies should be compelled to contribute.*

It was also agreed, with reference to the 1925 Pensions Act, that a two-thirds
pension would be awarded to members of the brigade on the completion of
twenty-five years service.

Throughout the fire brigade's history it has always contained men of
outstanding ability in spheres outside of the service. Many notable athletes had
lengthy careers as Dublin Firemen but in the 1930s the most noteworthy of these
talented individuals was station officer James Conway who throughout that
decade had paintings exhibited each year in the Royal Hibernian Academy of
Art. His work took its place at the annual exhibitions with those of Dermot
O'Brien, Leo Whelan, John Lavery, Sean Keating, William Orpen, Jack B. Yeats,
Sean O'Sullivan, etc, and all were painted in his family quarters in Dorset Street
fire station. His work had many subjects, from firemen to angels.

In August when the fire brigade refused to respond to a serious fire at St
Mary's College, Dundalk, the Irish Independent produced a series of articles criticising Dublin Corporation on the issue. The city manager in reply again asserted
the law agent's interpretation of the 1862 Act.

> *The D.F.B. can attend fires outside the city only in the neighbourhood of the city. The*
> *responsibility for the safety of the property and lives of the citizens of Dublin rests*
> *upon the Corporation of Dublin, while similar responsibilities devolves upon various*
> *local authorities in their respective districts. Many of these local authorities do not*
> *fulfil their obligations. I am unable to accept such a responsibility [sending a fire*
> *engine and a crew on a long-distance fire call] in the performance of an illegal action,*
> *in connection with a security service devolving upon other local authorities.*

He went on to call on the Department of Local Government "in the absence of a
National Fire Brigade service to draw attention to the responsibility of local
authorities to provide their own service".

The year 1933 was to be one of tragic fatalities involving the Dublin Fire
Brigade. On 13 April an ambulance proceeding to an accident in Church Street
struck and knocked down two young boys in Mary Street. One of the boys, aged
six, died later that day in the Richmond Hospital. He was the first civilian fatality
involving a Dublin Fire Brigade vehicle in the city.

Then on 16 April a fire engine returning to Rathmines station following a fire
call struck and killed a cyclist. The city manager was to inform the chief fire
officer in October that he could not fill two vacancies because the corporation

had been put to "very serious additional expenditure through liability for increased accident insurance premium owing to serious accidents in connection with the Fire Brigade". The old Pembroke ambulances were replaced, "being worn out and beyond repair, the makers now no longer in business".

In 1933 post office engineers connected a new system between Tara Street station and the four sub-stations "so as to enable each station to receive a message of fire call from central control simultaneously". Also the chief drew up a report on accommodation in the stations which showed that there were five firemen without housing accommodation. To resolve this growing problem the city architect proposed adopting the chief's earlier report of "conversion of buildings in Rathmines Town Hall yard to quarters for four firemen and a new two storey red bricked building, in place of the tenements, in Tara Street to contain married quarters".

The trade unions sought to have a change of clothing from red shirts, on the grounds of "safeguarding the health of the staff". Captain Connolly, just returned from the IFE Conference, concurred and recommended to the city manager clothing similar to that on issue in major British fire brigades. He ended the year by informing the manager that he had made provision in the estimates "for the employment of two further additional members for this Department. I would be glad to have sanction for the selecting of the two candidates as the staff is very much depleted." His estimate for the brigade in 1933-34 was £37,736, which included £3,500 for a new Merryweather Turntable ladder, but this appliance was not delivered until 1936.

With the expansion of the city Dublin Corporation sought a £500,000 loan for the purpose of building low cost corporation housing. Much of this would expand the suburbs by large-scale development in the former greenfield areas of Kimmage, Cabra and Irishtown, while flat complexes were planned for working class areas like the North Docks, Townsend Street and the Oliver Bond and Cook Street area. In his innocence the city manager thought that a massive shift of the working class into better built ghettoes would solve the housing problem and end the extensive vandalism the city was encountering at the time. His report dealing with "wanton destruction of civic property", which he laid before the council at the beginning of the year 1934, contained the following statement.

> *I must strongly complain of the complete absence of any sign of civic spirit in Dublin. Such wholesale destruction as that which I have instanced could not escape detection if citizens generally had such a spirit and were desirous of exercising it in their own interest .... destruction and pilfering of flower beds and parks, damage to public conveniences by dropping stones into closets, tombstones in parks being smashed, windows in churches (as many as 52 in St Mary's of Jervis Street alone), railings pulled down, daily breaking of windows in premises under construction, smashing of public seating and playground equipment, public lamps smashed and fire hydrants jammed with stones or bricks.*

He went on to point out the cost this vandalism placed on the corporation when so many pressing demands were being made on its limited resources.

The corporation did intend tackling the issue of safety in "places of public resort" by bringing in stringent byelaws and, under Section 16 of the Local Government Act 1892, "to make it an offence for any person to stand or kneel on a window sill above a specified height from the ground without adequate support or safety". Warning of the dangers of this practice had come from newspaper reports and details of brigade ambulances that were dealing with an increased number of fatalities and injuries from window cleaners' falls from window ledges.

The byelaws on protection from fire relating to theatres and other places of public resort within the city of Dublin were agreed in March 1934. These laws were comprehensive, stringent and detailed, as exemplified by the following.

> *[It shall be] unlawful to open or keep open any premises as defined under the byelaws containing an area for the accommodation of the public of not less than 500 square feet kept open for public performances of stage plays, public dancing, music or other public entertainment .... [until] the Corporation shall have issued a certificate to the effect that such premises are in accordance with these byelaws in so far as they are applicable to such premises. In addition to the requirements of these byelaws the provisions of the Cinematograph Act 1909 and the Regulations made thereunder must be complied with in connection with the use of any premises for the purpose of cinematograph exhibitions at which inflammable films are used.*

The new council was taking full account of the dangers, and the recent multiple tragedies had concentrated the members' minds sufficiently to press them into drawing up statutes with real impact. The byelaws covered forty-eight printed pages with 105 separate sections dealing with all aspects of construction, management, fire protection arrangements, heating and ventilation, gas lighting, electrical lighting etc. The fire protection section covered fire-mains or hydrants, chemical extinguishers, standard Dublin Fire Brigade fittings, bibcocks in theatres, music halls and cinemas, telephone communication with the central fire station, the training of staff employed in companies – including fire drills and the allocating of specific tasks in the event of fire, the importance of immediate communication with the fire service in the case of any outbreak however slight, the testing of fire hose at least once every year.

Premises where accommodation exceeded 500 persons were "to have sufficient frontage to at least two streets" and where numbers exceeded 3,000 persons "to abut on three streets" and "be isolated on at least one side other than the sides abutting on the streets by an open passage at least 12 feet wide". External walls were to be brick, concrete or stone "while partitions were to be of brick or stud" not less than three inches thick. "No timber partitions whether plastered or not shall be constructed." Gangways were to be not less than three feet six inches, floors and tiers were to be of fire-resisting materials and all doors for purposes of egress from the premises were to be clearly marked with the words "Way Out".

> *Two exits for each tier or floor, opening outwards, and where less than 300 persons are accommodated the width of each exit to be not less than 4 feet, but if more than 300 the*

*width of each exit to be 5 feet. Each exit to deliver the patrons into different thorough-*
*fares where practicable. Doors to open into a step or landing of at least 3 feet. All these*
*exit doors to be secured by automatic fastenings which will operate when the doors or*
*crossbar fastenings are pressed in the direction of exit.*

These were stringent precautions and showed that Dublin Corporation was taking very seriously the matter of the safety of its citizens when at leisure functions. If the regulations were properly enforced Dublin would be one of the safest cities in Europe for patrons of dancehalls, cinemas and theatres at that time.

In February the corporation agreed to the change of uniform proposed by Captain Connolly, leading to the changeover from the traditional "Garabaldi Red Shirts". At the same time new regulations were under consideration under a new road traffic act. The chief sought exemption for fire engines and ambulances on emergency turn-out under Sections 46 and 49 of Part IV of the Act dealing with speed limits, and sought an insertion "which would compel ordinary street vehicles, where possible, to clear the road for Fire Brigade or Ambulance on route to a call sounding gong or siren either by stopping or pulling into the side". He also sought "to have the parking of vehicles over street hydrants prohibited".

In June the practice of erecting hydrant indicator plates for the city as recommended by Captain Connolly was adopted by the engineering department. By then some of the ongoing problems of accommodation in the stations were coming to a head. An extreme case of how congested some of the married quarters had become was described as follows: "a married man with a wife and six children occupying in Buckingham Street station a bedroom and kitchen, the dimension of the rooms – bedroom 13 feet 2½ inches x 6 feet 9 inches and kitchen 14 feet 4 inches by 14 feet 11 inches". Conditions like these were a throw back to the days when Whitehorse Yard operated as a fire station on Winetavern Street. They needed immediate action.

Captain Connolly had examined systems in Britain where the firemen had moved out of the fire stations and were working a two-watch platoon system. He supported a shorter working week for the firemen but informed the city manager that a move to the above system "would require an extra 30 men in order to have at all times a reasonable fire-fighting unit available for the City". Staff at that time numbered seventy-five in total, distributed between all stations and Leinster House.

The brigade was at the time servicing an area of 18,670 acres with a population of 424,000. There was an average of over 400 fire calls and 3,400 ambulance calls per year. By the 1936 census the population had risen to 467,691. It was recommended that the firemen's working week should be reduced by one day (24 hours) off every third day instead of the current system of one day off every fourth day. This change could be introduced only if much more accommodation was provided on the stations for the expanded staff numbers.

On Sunday 10 March the brigade tackled a huge fire at the Old City Wheel Works in Pearse Street, now being used as the Electricity Supply Board stores, housing electrical fittings, cables, drums of wire, etc. In the high winds of that morning, burning embers blew about the street as the fire crews battled the blaze. Twenty horses had to be released from their stables in an adjoining property and taken to a place of safety as firemen hosed the inferno with thousands of gallons of water. Six lines of hose were operated for almost twelve hours before the fire was contained in the Hanover Street section and eventually extinguished.

The city was experiencing a water shortage as well as one of the longest and bitterest bus strikes in its history, a strike that was to last for over sixty days. On 12 April Dublin Corporation issued a notice that water would be turned off in parts of the city from 8 pm to 7 am. The River Dodder Scheme, which the city acquired with Rathmines and Rathgar townships, had made no great improvement to the water supply in areas of the city, so additional slow sand filter beds were being built at Roundwood to increase output. But even that measure when completed towards the end of 1934 did not provide sufficient extra supply to meet the daily growing demands on the piped water system.

On 25 April sometime around 7 am a fire was discovered in the "Plaza" building in Middle Abbey Street. This building, which extended into Princes Street, covered a concourse of many thousands of square feet and included a first-floor ballroom then nationally known because it was there that the regular Irish Hospitals Trust Sweepstake Draw took place. Thick black smoke was rising from the flames in the stricken building at the time the fire engines arrived. Within half an hour the structure was burning fiercely from end to end with the flames devouring everything trapped in their path, feeding a seeming ferocious hunger for destruction. Soon the roof collapsed, pulling one of the side walls down and exposing warped twisted girders and blazing materials where the day before hundreds of people were employed.

The Plaza was totally wrecked and had to be demolished in order to protect adjoining property and pedestrians. The city manager repudiated a claim of water shortage for firefighting purposes on the morning in question, stating that no water cuts were taking place in the old central city area. He further maintained that up to seven million gallons of water had been saved by cutting the water pressure at night. The fire crews fought the huge blaze for almost seven hours before it was safe to begin reducing numbers at the fire scene and send engines and crews back to their stations.

In November 1933 Dublin Corporation had endeavoured to secure an agreement with the ESB on a projected joint hydro-electric and water supply scheme on the Liffey at Poulaphouca, Blessington, Co Wicklow. The project was delayed by the ESB because of "economic deficiency" for some time to come in its electrical supply capacity. Since it did not have customers for the extra supply which would be generated it was seeking government subvention to cover any loss. Although meetings with government representatives did take place from time to time, progress was slow and it was not until 1937 that the scheme became fully operational and a dam was built on the Liffey. The project was not finally completed until 1948.

The variety of incidents regularly attended to by the fire brigade was highlighted when on 24 September 1934 Station Officer R. O'Gorman of Rathmines

Dublin firemen, army officers and Garda detectives at scene of Terenure plane crash, 1950s.
Photograph, © *The Irish Times*. Collection Old Dublin Society

gave evidence at a court of inquiry assembled at Portobello Barracks by order of
the commanding officer of the Dublin Military District. The inquiry arose following the crash of a Fairey IIIF aircraft at Terenure which caused the deaths of the
pilot and a crew member and seriously injured the other crewman. The inquiry
was held in order to establish the circumstances of the crash and whether the pilot
and crew were on duty at the time, whether they were qualified to fly the plane
and whether the flight was carried out in accordance with Air Corps regulations.

In a sworn statement Station Officer O'Gorman's sworn statement contained
the following:

> ... *we turned out at 3.05 pm in response to a call on the Rathgar Fire Alarm. When we
> arrived at Terenure Road East, we discovered burning wreckage in a front garden. The
> wrecked plane burned fiercely particularly around the pilot's seat and engine section.*

He saw the pilot's body and ordered his men to lay down "foam apparatus". While
this was being played on the fire they recovered the pilot's body and another
body lying behind the wreckage. Both bodies were removed to hospital as the
fire crew extinguished the fire. Under cross-examination he stated that he had

> ... *not seen a third man in or about the wreckage (he had been removed earlier), the
> aircraft was lying facing the house close to the boundary wall in the garden and the tail
> plane was about 10 feet from the gate. Apart from the burning fabric there appeared to
> be little structural damage to the wings. The engine was dug into the earth approxi-
> mately over the marks where it had struck and one propeller blade was standing
> perfectly straight up.*

A first for the fire brigade but dealt with in the usual professional manner.

On 8 December a plumber named Charles O'Leary was trapped part-way down in a forty foot well shaft at Kilquade near Greystones, Co Wicklow. Excavations dug in order to rescue the man showed that he was still alive so Station Officer Rogers and Firemen Bernard Matthews and John Gibney were despatched to the scene with oxygen equipment, to assist in a possible rescue. The victim was jammed up to his chest between an old ladder already in the shaft which had broken and trapped his legs and the ladder he had been using. Partial collapse had occurred and stones and soil had fallen in on him.

Soldiers from the engineering corps were called in to tunnel under the trapped man while the firemen, doctors, family and friends tried to keep him alive. The firemen lowered a helmet to him to protect his head and succeeded in getting a tube down to him for supply of air. At great danger to themselves the station officer and his two men attempted to get down to O'Leary and prise him from his entombment but to no avail. A rope and torch were lowered to him and he spoke to rescuers through a tube over a period of two days when there was another collapse of the walls of the shaft. Captain Connolly then withdrew his men as the military continued their tunnelling. O'Leary's body was eventually recovered on the morning of 12 December at about 2 am. The County Wicklow coroner, the garda superintendent and a Senator O'Hanlon commended the work carried out and the risks taken by the Dublin firemen in attempting the rescue but the chief fire officer would not recommend chevrons or other similar recognition for the three men involved because the "incident occurred well outside the city boundaries" and he did not want to set a precedent for such actions in the future.

In June 1935 approval was given to purchase the latest, all steel, 100 foot T/L Merryweather with mounted reciprocal pump for £4,722. However in March Captain Connolly was again writing to city hall regarding "the seriousness of not having a turncock with them going to a fire" and instanced a major fire recently attended at Bailey and Gibson of Dolphin's Barn, where it had been discovered that "the hydrant directly outside the factory gate had no water when the firemen made down to it" and on the same day "a considerable delay had occurred at a fire in the ESB premises at Hanover Quay".

> I had occasion some years ago to draw attention to the system of having to send for a turncock to attend at a fire call and condemned the system as being unsatisfactory but so far nothing has been done in the matter.

When the new form of local government was introduced in Dublin in 1930 it was agreed that the working of the arrangement would be reviewed five years later in 1935. All sections of the corporation and all aspects of its work were examined under this review. Captain Connolly used the review to again highlight the staffing levels and problems of accommodation in the fire service. He got agreement that all future recruits into the brigade should be under twenty-five years of age, five feet ten inches in height, pass a medical, preference to be given to single men, all recruits to be employed on twelve months probation and if medical and work were satisfactory then to be made permanent. Wages were fixed at seventy shillings per week rising to ninety-two shillings and sixpence; leave to be one day in four and, when staff numbers allowed, a reduction in the working week; annual leave of twenty-one days for all station officers and firemen; full pay when out sick and members to be supplied with free living accommodation, coal, light and clothing.

Relations between the fire brigade and city hall were not as cordial as they seemed. The ongoing issues of the provision of sufficient water for all firefighting incidents and assistance from the turncocks were causing friction. Following some of the major fires in the city and newspaper criticism of lack of water the city manager and town clerk undertook an inquiry into the matter on the direction of the council.

Two women lost their lives on the top storey of 16 Parnell Square on the night of 18 August 1935. The brigade arrived within five minutes of receiving the call and found the hall and staircase in a mass of flames with smoke billowing out of the upstairs windows. Two people were seen at a third-floor window and were rescued by the firemen. However there were two other women on the top floor. They died simply because the firemen and the gardaí present were informed that there were no other occupants in the building.

The advanced state of the conflagration at the time of the arrival of the brigade confirmed to the chief fire officer when he attended that in spite of the many people who gathered to watch the growing fire, no one thought to notify the fire brigade immediately and the two women who died were rendered unconscious by the fumes before they could get to a window.

The firemen had problems with water supply when they arrived, finding the first hydrant they approached was "capped" and took some two minutes to be opened. Furthermore in spite of a message to Tara Street "to turn out all sections" the turncock on duty did not receive a call to attend the fire. The brigade maintained that it received "no reply from Water Works", and sent an ambulance to Cornmarket to the residence of a turncock who lived there to bring him to the fireground. The water works engineers were unable to give any satisfactory explanation as to why the first hydrant was "capped", and there was a difference of opinion as to whether the turncock on duty was called or not. Turncocks on night duty in the water works yard responded on bicycles when they received a fire call.

Two fires in the same month were instanced where the system had been satisfactory. In the case of the ESB Stores fire, however, another major difference arose when the brigade maintained that it notified the turncock at 1.43 am and the turncock maintained that he was not notified until 2 am. At the "Plaza" fire more confusion took place. The chief officer agreed that he took some twenty minutes to survey the extent of the fire from both Princes Street and Abbey Street before requesting a civilian to go to a phone box to phone Tara Street to arrange for a turncock to be called because water pressure was not good enough to tackle and extinguish the fire. No record of any such call was found in Tara Street.

The city manager in a report agreed that "in the specific cases under review the existing system was not operated to the maximum degree of efficiency". He maintained however that the brigade was never better equipped and with the transfer of four of the Pembroke firemen, who were not considered suitable, and their replacement by four younger men, less sick-leave should be occurring and more men should be available. He stated that the chief wanted three extra men and as soon as he had a report justifying the request he would act on it. His report showed a number of major fire brigades in Britain that had no turncocks attached to them and furthermore the previous chief officer in Dublin did not consider them necessary. The result of his report was that the fire brigade was provided with a turncock on duty in Tara Street Station from 6 pm to 6 am each night.

Fire stations had been provided by Dublin County Council in Skerries and Balbriggan and because it was also intended to open stations in Malahide and Swords, the Department of Local Government approved S/O Rogers for "inspecting during his spare time firefighting appliances in the four towns" and "to submit regular reports to the Board of Public Health as to the work carried out by him". He was paid £10 per annum for this work.

Captain Connolly had attended the IFE Conference in Sheffield and reported to the city manager on "non-electric-shock Firemen's helmets and a new flare light acetylene type lamp". He recommended that the corporation purchase both of these items for the fire brigade. By the end of 1935 the solid tyres remaining on fire appliances and ambulances had been replaced by pneumatic ones. The new tyres were purchased from different suppliers depending on the cost, usually from the cheapest. During the year the chief had travelled to both Wexford and Drogheda advising both of those townships on the provision of firefighting equipment.

The brigade's third turntable ladder was commissioned in May 1936. This Merryweather was basically similar to the 1931 appliance, but the ladder was all steel and could reach a height of one hundred feet, and it was also equipped with a 275gpm pump. The cost was £4,722 but the corporation was also obliged to pay £958-4-6 import tax! This led the Dublin Council of Trade Unions to "protest emphatically against this imposition of tax on the importation of life-saving equipment for the Fire Brigade. We consider such equipment is essential to the safety of the citizens and should be admitted free of tax." It called on the minister to abolish the tax.

### The Pearse Street disaster and its aftermath

The year 1936 had begun as normal for the fire brigade – turning out to farms and factories on fire in the county area and the recipients of its assistance refusing to pay the amount agreed pleading "the poor mouth" and offering much lesser sums. Agreement was finalised to introduce a "reduction in hours to give leave off every three days instead of four" during the year.

Firemen McArdle and Nugent were both commended by the city coroner and the legal team at an inquest: Thomas Devlin died from shock and heart failure following scalds received on board the tug boat *Ben Eader* when a pipe burst in the boiler room. The coroner's verdict was "that sufficient care was not exercised on this occasion by the Officers of the boat. We wish to commend the two Firemen for their heroic efforts at rescuing deceased". Both received chevrons with one shilling per week.

On 16 July Fireman Leslie Crowe was involved in a dangerous rescue attempt when a car containing three people drove into the Liffey at the North Wall extension. In spite of the heavy swell on the river estuary, Crowe dived into the water on three occasions to try and locate the stricken car in the muddy fast-flowing river. Both gardaí and coroner commended "the conduct of L. Crowe" and he too received a chevron with pay.

In April 1936 the city manager announced his retirement on the grounds of ill-health, to take place on 1 October, and under the Local Government (Dublin) Act 1930 the council set about appointing his successor. There was opposition to this however and it transpired that the post would be filled by the minister under the Local Authorities (Officers and Employees) Act 1926. This led to a proposal "that as the Council will not have the appointment of its Manager, we recommend to the Minister the abolition of the office". The proposal was defeated by a single vote only, fifteen votes to fourteen. Eventually P. J. Hernon was appointed as the new city manager, with the council's nominee J. P. Keane as assistant manager.

At 8.45 pm on 25 August a major fire broke out in Rowntree's extensive factory in Kilmainham. Smoke was seen by people on the South Circular Road issuing from the windows as staff continued working in the factory. The workers were soon rapidly evacuating the building and massive flames were shooting into the evening sky as rolling clouds of dense black smoke enveloped the upstairs of the building. Fire engines attended quickly, laid down hose and directed deluges of water into the rumbling building. It was to be 4 am the following morning before the exhausted fire crews could reduce the number in attendance to one engine while the rest returned to stations.

The location of the factory on over an acre of ground on a hill, with open space in front, produced an amazing spectacle when in the throes of the flames. The illuminated sky could be seen from all over the north side of the city. So intense was the convected heat generated that families had to be evacuated from their homes in Albert Place as windows cracked from the heat. The fire had got such a hold when Captain Connolly arrived that he decided to concentrate on containing it and stopping it spreading to oil tanks and ammonia in the factory grounds. This was successful.

Up to 600 people worked in the factory, founded some twenty-five years earlier as the Savoy Cocoa Company, purchased by Rowntree in 1926 and expanded. It was a landmark building because of its size and location and the fire attracted thousands of sightseers. At the height of this fire an engine and crew had to turn out to deal with a fire confined to a room in Capel Street, and another engine later had to respond to a hay barn fire in Lucan which took four hours to extinguish. A really trying night for the fire crews.

The premises of 163 and 164 Pearse Street were three storeys over basement, brick buildings adjoining each other some fifty yards from the corner of Westland Row.

Pearse Street Fire, 1936, showing Leslie Crowe on ladder (second from top) who was injured. Photograph, © the Carroll Family (William Carroll, district officer, Dublin Fire Brigade, deceased)

Number 163 housed a barber's shop at ground floor level and a private hotel occupying the upper floors. Number 164 had a retail shop belonging to Exide Batteries Ltd on the front ground floor, vacant offices on the first floor and a family of seven living on the top floor. The basements, although not connected, were the location of a factory in which Exide batteries were assembled, charged and filled with battery acid. Fully charged batteries were stored at the rear of the ground-floor retail outlet.

The Exide business was an extensive one and to give more space in the basement area the rear walls had been removed up to ground level and a timber-framed annex, with a felt roof and three glazed skylights, had been extended outwards. This construction had opened the rear of both premises to possible fire spread. On the night of 5 October 1936 employees worked late in the factory, the last of them leaving at 8.30 pm. On the premises that night were five gas cylinders, two were 40 cubic foot coal-gas cylinders and the other three were 150 cubic foot oxygen cylinders. Housekeeping generally in the factory was poor and there were boxes, cartons and their inflammable contents thrown around the factory floor.

At 10.50 pm a Mr Kelly who occupied the top floor in No. 164 noticed fire in the rear annex and immediately evacuated his family. The fire brigade received the call at 10.54 pm and responded with the second officer and eleven men, a motor pump and a turntable ladder. The turncock turned out with them. They were informed that everyone was out of the premises but a search of the upstairs of both buildings was carried out and the ladder was pitched to the front of the buildings. A hose reel was made down into 164 Pearse Street and hose was laid down into the burning building. However, almost immediately there was a problem of water supply. None of the three hydrants tested provided an adequate supply of water to fight the growing fire in spite of the presence of the turncock. He then left the actual fire scene to check valves in Merrion Square and Lieutenant Howard went around into Westland Row to check the Grosvenor Hotel which was reported to be threatened by the spreading fire.

At 10.58 an explosion occurred and three of the firemen, Tom Nugent, Robert Malone and Peter McArdle, were never seen alive again. Two others had been able to leave No.163 and escape into the roadway outside, while Tom Potts, the youngest of those in the premises survived the blast by leaping onto the flat roof of No. 165 from where he was rescued and taken to hospital. Lieutenant Howard on his return to the fireground called out two more fire engines but by then a second explosion had taken place and the buildings were a blazing inferno. The explosions were so loud that they were heard by a garda inspector on duty in Castle Street and he immediately alerted gardaí to attend at the incident.

Soon after this Captain Connolly arrived on the fireground and summoned another fire engine from Thomas Street. Water pressure was dreadful. Later evidence at the inquiry was to show that although the three men who died at the incident all arrived as part of the first turn-out, all three had come on different appliances and no water came at any time from the hose they had brought into the building. The time of the second explosion was put at from three to five minutes after the first one.

In the pandemonium taking place it seems that no-one in authority realised that the three men were missing inside the buildings because a lengthy delay took place before the chief officer first notified the garda inspector at 12.30 am that "two men were missing". It was after 1 am when he told him there was "no trace of the men and a third man was also missing". The fire had developed rapidly and the upper floors in both buildings were enveloped in massive flames. Large numbers of people had gathered and they watched the puny efforts of the firemen as the roofs collapsed, crashing down through the top floor and knocking part of the back wall onto the annex. Flames rose from the dusty debris and more collapses occurred until mounds of burning rubble beneath treacherous walls was all that was left of the two buildings.

The fire burned on until sometime after 2.30 am. It was said later that it burned itself out, the brigade had failed to extinguish it. Once it had cooled sufficiently, frantic efforts were made by the members of the brigade to locate their missing comrades. Hampered by smouldering debris and falling masonry, firemen entered the dangerous conditions even climbing through a narrow aperture to get into the basement. During this hazardous work Fireman Leslie Crowe was badly injured by falling masonry while two other men received lesser injuries; the three were removed to hospital.

With grim determination the firemen, heedless of danger, continued and tons of rubble were shifted in the futile search in the light of their electric lamps. Sometime after 4 am in the sombre darkness just before the dawn, the smoke-blackened men discovered part of the hose which their missing colleagues had taken into the building.

Renewing their desperate work they soon came upon a charred and battered body, damaged beyond recognition. As the morning dawn threw some natural light into this macabre scene, a second corpse was unearthed. Both bodies were removed by ambulance to Sir Patrick Dun's hospital in the presence of the lord mayor and the other civic dignitaries who were now on the scene.

Still the sad work continued and about thirty tons of rubble and smouldering timbers were removed before the third victim was located just after 10 am. Because both buildings had collapsed into the basement the enormous quantity of rubble had to be shovelled outside as the search proceeded. Off-duty firemen had reported to the dreadful scene and worked tirelessly, assisted by members of the public, until the grim work was completed.

Messages of sympathy poured into the brigade from all quarters once the extent of the tragedy became known but simultaneously the newspapers began an outcry demanding answers to allay public fears. Typical was the editorial in the *Irish Times* of 7 October.

> *More courage is demanded in the ordinary work of a Fireman than in almost any other type of work … The public knows that accidents will happen but while it honours their sacrifice, it will ask whether the sacrifice cannot be avoided. The fire in Pearse Street has elements which call for most careful inquiry … The City Manager denies that there was any failure in the water pressure. Eyewitnesses said there was. The Firemen risk their lives whatever happens but they must not be asked to risk them with empty hoses in their hands.*

Amid the growing outcry the lord mayor called a special meeting of the corporation on 8 October. At this meeting it was agreed that the bodies of the three dead

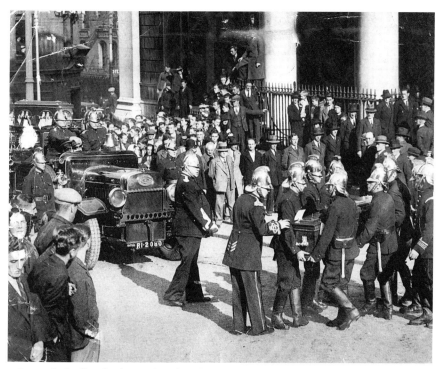

Removal of coffin of Robert Malone from lying-in-state at City Hall. Pearse Street Fire, 1936.
Photograph, © the Dublin Fire Brigade Historical Society

firemen would lie-in-state in City Hall, the funerals would take place on Saturday 10 October and work in the corporation would cease at noon to allow all workers to attend the funerals. The acting city manager, Mr Keane, assured the councillors that he would place before them at the earliest possible time the results of inquiries then being conducted regarding the fire by members of his senior staff.

Dublin was brought to a standstill as the civic funerals made their way through the centre of the city, attended by the surviving firemen, public dignitaries, church leaders, business representatives and officers and men from fire services throughout Ireland and Britain. Behind all this public display of sympathy was a void. Firemen McArdle and Malone had left widows with nine children between them, Nugent was single and due to be married. Representatives of the firemen sought a pension for the dependants but this was turned down by the corporation. Instead a fund was launched by the lord mayor which he claimed would become a permanent fund. After adequate provisions had been made for the dependants of the deceased men the fund would be built up for the relief of other Dublin Fire Brigade families who suffered through the death or injury of their loved ones in the future. The fund was launched in a blaze of publicity on 12 October. (See colour section, illustration number 12 & 13.)

It failed in all its objectives. The principals of state, trade and commerce had salved their consciences by attending the funerals. With the massive poverty and unemployment in the city at the time the workers could afford little so, in spite of daily reports on the fund in the national press, by the end of the month only £1,622 had been donated and the fund was soon closed. In spite of a plea from the Dublin Council of Trade Unions for special pensions to be granted to the widows neither the corporation nor the government would act on the problem. A leading trade unionist summed up the situation:

> ... if Firemen were soldiers bent on destruction they would get a pension. The depen-
> dants of Firemen should not be dependent on charity. They should be compensated as
> a right.

Fine sentiments, but the matter was allowed to fade away as other issues took the headlines and firemen were not granted pensions for widows and orphans until 1977.

A wider investigation than that being pursued by the city manager was demanded. Letters appeared daily in the press and articles were published which fuelled the debate. The public suspicion that water pressure in the city was inadequate was heightened when Dublin Corporation cancelled a scheme for slum clearance because it could not supply water to the builders of the planned replacement housing and the Dáil pushed through the second reading of the Liffey Reservoir Bill which up to then had been stalled in the usual political debate.

Finally on 25 November the Minister for Local Government announced the setting up of a tribunal of inquiry to report and make recommendations on the following.

- The cause of the fire at 164 Pearse Street on 5 October.
- Whether the fire was rendered more destructive by negligence on the part of the user of the property.
- Whether there was sufficient water pressure to deal with the fire.
- The steps taken to extinguish the fire and the efficiency of those steps.
- The causes of the loss of life.

The tribunal consisted of a senior counsel and two senior civil servants. It sat from 20 January to 19 March, thirty public sittings in all, and it heard seventy-seven witnesses, including fifteen from the brigade and eleven members of the gardaí. The usual array of legal talent was mustered by the state, the corporation, the property owners and Exide Batteries Ltd. What was not buried by the terms of reference would be wrangled away in legal jargon. The legal representative for the McArdle family asked at the outset that nothing be done by the tribunal "which might in any way prejudice or influence a court dealing with claims of the next of kin against Dublin Corporation or Exide Batteries" because "it was necessary to show that the next of kin had been precluded by the action of the Corporation from being represented at the present proceedings". The point was made because of the initial refusal of the corporation to fund legal representation for the victims' families.

The inquiry exposed serious deficiencies in the fire service and water works organisation and it highlighted the lack of legislation to control manufacturing processes in domestic and residential dwellings. Major areas of contention concerned the water supply and the persistent corporation denial of any negligence regarding this, the conflict of evidence between the turncocks and the fireman in the central control on the night of the fire, and the corporation refusal to accept responsibility for "any breach of statutory duty in the control of the business of Exide Batteries". Officership in the fire brigade was also found to be defective with senior officers seemingly not fully in control of their men on the fireground.

The tribunal findings were as follows.

- The most likely cause of the fire was a cigarette end or ends left behind by one or more employees of Exide Batteries.
- The fire was rendered most destructive because large quantities of inflammable material were left lying around on the premises and gas cylinders were not stored in a place of safety. However, Exide Batteries Ltd was not in breach of statutory duty.
- There was not at any time on the night of 5 October sufficient water in any of the hydrants in Pearse Street, or in the neighbourhood of the premises involved in the fire, at an adequate pressure to extinguish the fire.
- The steps taken by the brigade to deal with the fire were inefficient. Although the officers and firemen acted throughout the fire with great gallantry and personal bravery, there was a lack of supervision, direction, and control on the part of the two senior officers of the brigade over their own men.
- The circumstances in which Robert Malone, Peter McArdle and Thomas Nugent lost their lives were "sudden and violent explosions … increasing immediately the violence of the fire and creating a trap from which these men were unable to escape".

The recommendations were the following.

- The use for industry or factory purposes of premises occupied as dwelling houses ought to be prohibited or regulated.
- The use of basements for industrial or factory purposes should be prohibited or regulated.
- The storage, use and transport of compressed gas cylinders should be improved.
- The water supply in the city for firefighting should be improved.
- The arrangements on the water mains at night should be improved.
- Staff in the water works department should be improved for the purpose of assisting the fire service at fires.
- The brigade and its administration should be reorganised.
- A regular system of government inspection should be introduced to ensure maintenance of proper standards of efficiency.

Control of "dangerous buildings", though not strictly within the scope of the inquiry, should be dealt with on the lines of a list of recommendations outlined to the tribunal on 15 March by the city manger, the main features of which were that "all persons about to erect a new building, or convert or use an existing building or part thereof for making, storing or selling any substances to which the Explosives Act or Petroleum Acts apply, should be required to give notice to the Corporation. Supplemental to these provisions the analogous provisions of Sections 99 and 143 of the London Building Act 1930 should be enacted."

Captain Connolly in his own defence stated that he had requested more men and had recommended the purchase of a new pump and electric shock helmets to replace the brass ones. It was a feeble enough defence by one who knew his days as chief were now numbered. The acting city manager's contention that there was no "failure of water pressure" was found to be less than truthful but he would carry on in his post.

As the brigade tried to get back to some form of normality following this most tragic chapter in its history, the chief updated the rules for private ambulance cases on receipt of a call in control room.

- Can you pay a fee of £1 (in the City) and 2/- per mile each way outside the City boundary?
- Have you a medical certificate to the effect that the case is not infectious?
- Are arrangements made in the hospital for the patient to be accepted?
- If the ambulance can be spared the officer-in-charge gives sanction to its attendance. If the person says they are poor and could not afford the fee, the vehicle is sent free but the ambulance attendant ascertains the business of the patient when attending.

The chief had a sample helmet from Merryweather tested and, being satisfied, the city manager gave approval on 2 November of expenditure of £109-13-9 to purchase eighty-one of the non-electric shock "Super Safety Helmets". Later that month a fire engine was in collision with a bus on Phibsboro Road, injuring and hospitalising four of the crew and six bus passengers, further adding to the brigade's problems.

On 30 March 1937 Captain Connolly wrote to City Hall regarding the widows and children of Firemen McArdle and Malone "still in occupation of quarters in Tara Street Station since the Pearse Street fire and have been in receipt of coal free as allowed to married members living in houses in the Station". The following day an order from City Hall arrived stating that "no coal can be guaranteed subsequent to 31st March". The chief wrote also to the city manager suggesting

> arrangements to encourage members of the Brigade to qualify for the Diploma of the Institute of Fire Engineers and I would suggest that an allowance of 1/- or 2/- per week be granted to any members who pass examinations. I have arranged with the co-operation of the Secretary of the Technical School Department that examinations be held in Dublin.

This proposal was not acted on but was left over for decision by Captain Connolly's successor. In due course the new chief rejected the idea, stating that "these exam qualifications would be considered for those seeking promotion".

# 14 Re-organisation and the "Emergency"

In July the first pay increase for fifteen years was conceded to all personnel other than the chief fire officer and the second officer. The two shillings and six pence per week increase was backdated to 31 March and it was agreed that the issue of the introduction of the "two-platoon system of working" would be examined. A new Merryweather fire engine was purchased in August and agreement was reached to purchase two new ambulances from John O'Neill of Delgany, Co Wicklow. As Captain Connolly attended the IFE conference that September at Cheltenham he could be happy that improvements in the service were definitely taking place.

On his return he reported to the city manager, having visited Manchester, Liverpool and Port Sunlight fire brigades to observe drills, equipment, training and leave arrangements, and he sought sanction for the purchase of "six self-contained breathing apparatus sets and four resuscitating machines".

Joseph Connolly was retired from the Dublin Fire Brigade on 1 April 1938. He was forty-six years of age. He had been a victim of his own drive for personal advancement and of the system under which the Dublin Fire Brigade was forced to operate, particularly under the commissioners. He was an honourable and honest man working at a level beyond his competence, without adequate support or back-up from his employers. His post was abolished and in its place the position of Chief Superintendent of the Dublin Fire Brigade at a salary of £650 per annum, rising by increments of £25 to £750, was advertised for appointment under the Local Government Appointments Commission. In the meantime, the reduction of hours from 112 to 84 from 1 January had not taken place as agreed with the unions representing the firemen so the corporation agreed payment of "a half-day's pay in respect of each week since that date until the hours of the members of the staff are reduced as a result of the introduction of the platoon system". This matter was to be a priority for the new "boss".

On 24 April Major J.J. Comerford BE, BSc, MICEMI, was appointed

> to take charge of the Dublin Fire Brigade and to carry out all the necessary schemes or reorganisation in the Brigade at a salary of £750 p.a. subject to adjustments contained in Clauses 1 and 2 of the letter of March 1938 from the Secretary, Department of Defence. It is agreed that he will live in his own private house, not in the quarters occupied by Captain Connolly. Appointment to be temporary and for a period of not more than one year.

A new era had arrived and soon the new regime would be in place for a massive re-organisation of the service.

Major Comerford travelled to England in May to examine the organisation of brigades in Liverpool, Birmingham and London. He also visited the Home Office

Fire Department. Comerford was a commissioned officer in the Irish army, chief instructor in engineering subjects at the Curragh Military College from 1931 to 1934 and chief assistant director of military engineering since 1934. He would stamp an imprint on the brigade in ways not seen since the innovative period of his fellow Kilkenny man, Captain Purcell, and he would be the first of the long line of ex-army officers from the corps of engineers to head the Dublin Fire Brigade.

With the appointment of Major Comerford began the re-organisation of the brigade as envisaged by the Pearse Street inquiry and within eighteen months the service had been transformed beyond recognition. Soon after the agreed retirement of Captain Connolly the two senior station officers, Kelly and Dowling, also retired and it was agreed that Lieutenant Howard would retire when the planned changes were completed; he did so, citing age and ill-health, on 18 September 1939.

The immediate issues tackled by Comerford were staff shortages and the introduction of the two-platoon system by abolishing married quarters in the fire stations. Six recruits were drafted in but unlike their predecessors these men were to undergo three-weeks intensive training before they went on shift duty – and recruitment was set in train to take in another twelve firemen within six months. None of these entrants would be provided with residence by the fire brigade. Before the end of the year, twenty-five married men had moved from their accommodation in the stations to corporation houses near where they worked. They were allowed twelve shillings per week to cover rent, fuel and light in their new homes at a time when the rent on an average corporation house was seven or eight shillings per week.

Lieutenant-Colonel Symonds was invited over from England to inspect breathing apparatus and to train staff in its proper use. Symonds reported that "the Proto sets had been cared for in a rather haphazard fashion ... personnel knew practically nothing about the proper use of the sets and the dangers of allowing insufficiently trained personnel to use them". Symonds also acted as advisor on re-organisation to Major Comerford.

A new post of third officer was advertised and the brigade was to be divided into four districts, each with a district officer. Examinations were set for these posts and they were thrown open to all firemen with over seven years experience. R. O'Gorman became the first third officer in Dublin Fire Brigade. All these officers including the station officers would live in the fire stations, working 116 hours per week, while the firemen's hours were reduced to eighty-four per week.

The widowed mother of Thomas Nugent was given £100 under the Workmen's Compensation Act. Her solicitor was paid £426-10-9 by the corporation. Robert Malone's widow and young son received £502-1-8d between them from the same source and a letter from the city manager that she "surrender her quarters within a fortnight's time or legal proceedings should be taken against her". In the case of Mrs McArdle, she was to receive £600 for herself and young family and at "the earliest possible date, arrangement should be made with the Housing Department offering a tenancy of 43 Pembroke Cottages". The three deaths in Pearse Street had ceased to be an issue for Dublin Corporation.

The changes were coming fast. Before the month of October 1939 all married firemen had vacated the fire stations. A Miss O'Leary, short-hand typist, the first clerical assistant ever to be employed by Dublin Fire Brigade, was transferred from the housing department to join the staff. New fire fighting gear – consisting of the "Liverpool" waterproof coat and rubber boots – was issued to all personnel. Major Comerford commented that up to then, "all firemen were drenched through at a fire of any magnitude. This is both unnecessary and avoidable". With the introduction of the two-platoon twenty-four hour system, another twenty-two recruits began their fire brigade careers.

In all, forty new recruits were employed in the first eighteen months since the re-organisation had commenced and on 3 May 1939 the Department of Defence had written to approve of a further one-year secondment under the original terms to allow Major Comerford to continue as acting chief superintendent of the brigade.

A plot of land in Dolphin's Barn was acquired by compulsory purchase on the junction of Parnell Road and Crumlin road for a "new four bay station with accommodation for Firemen, hose drying, drill tower and two residences, for a District Officer and Station Officer, with small plots adjoining and a drill yard". The design was to be similar to that seen by the major recently constructed for the fire service at Speke Aerodrome, Liverpool, at cost of £24,000.

Thomas Street station was closed on 8 November 1938. Major Comerford had written to the city manager in August

> *with the establishment of the District system of fire brigade organisation in connection with the re-organisation scheme (and the new Dolphin's Barn Station) the necessity for Thomas Street as a permanent Fire Station no longer exists. S/O Walsh of Buckingham Street is still in occupation but due to move to Number 4 and I am of the mind to use Thomas Street for training A.F.S. and A.R.P. but this would only arise in the case of an Emergency.*

The brigade had suffered another fatality just as these large-scale changes were beginning. Fireman John Darmon, a motor mechanic who helped to service the brigade engines, died in the engine room pit in Tara Street. He was electrocuted and died on 23 August 1938, when repairing an ambulance engine. He was thirty-nine years of age and left a widow and four young children. The unfortunate accident was witnessed by the son of one of the firemen who also lived in the station and it was he who raised the alarm. The widow was paid £600 under workman's compensation. Major Comerford clearly intended to stamp his code of discipline on the brigade when he terminated the twenty-five years employment of a fireman found guilty of being drunk while on duty in Leinster House. He used the incident to convince the authorities to do away with the situation of having two firemen attached to Government Buildings, proposing that "two civilians be trained to do the work at a much lower wage".

Comerford modelled his re-organisation plans on Liverpool and Bristol: "Bristol has a 400,000 population, is not highly industrialised and its general problems are something similar to ours". While in Bristol he had met the chief engineer at Bristol Port, the chief fire officer, and the chief fire advisor to the Home Office. He intended a clean sweep in Dublin, writing on 24 November 1938 to the city manager.

> *I have discovered that there are large quantities of old record books which date from the latter end of the last century. These record books are now of course useless, value-less and are simply waste paper. I would be obliged for authority to have all records of more than 7 years destroyed.*

His main proposals for the new Dublin Fire Brigade were

1   The employment of thirty-two new firemen
2   A third officer and three district officers to be appointed
3   A second officer to be appointed by the Local Government Appointments Commission at a salary of £400 per annum rising by increments of £20 to £600, and third officers to be appointed by Dublin Corporation at £350 rising by £10 to £450, plus quarters, fuel, light and uniform
4   Rathmines, Dorset Street and Buckingham Street be designated as District Stations 2, 3 and 4 with district officers in residence
5   Standard uniforms for all officers and firemen, to be worn at all times when on duty
6   Second officer to reside in former chief's house and third officer in former second officer's house
7   Entry age to be reduced to twenty-three years and height to be five feet eight inches
8   In future the chief superintendent not to respond to any out of area fires and
9   New standard charges for officer or sub-officer: £1 first hour plus five shillings per hour after that; firemen three shillings per first hour plus two shillings after that; fire engine £6-10 for first hour plus £2 after that, with a 50% extra charge for all night calls
10   Two-watch system to come into operation on 9 November 1938.

He also abolished any further presentation of chevrons or monetary payments.

In January 1939 the Minister for Local Government fixed the salary for chief superintendent of the Dublin brigade at £750 rising by increments of £50 to £900, plus £60 car allowance with free telephone and uniforms. On 16 August Major Comerford was appointed permanent chief of the brigade and on 1 September Valentine Walsh BE was appointed second officer. Walsh had worked in local authorities in Ireland and on underground railway construction in London and for the British Office of Works on Munitions.

The new brigade was taking shape. The major declined an invitation to attend the IFE conference and no other officer was recommended to attend, but in company with the engineer of the water works he again visited fire brigades in major cities in England, returning to report

> *there is little use in building a new modern fire station if the present obsolete and unre-liable equipment in use throughout the City generally has to be used therein .... [in reference to the fire alarms] hardly a day goes by but one of them is out-of-order. A new*

*system cannot however be installed until Central Station has been reconstructed and a suitable type of modern control is provided.*

He proposed expenditure of £19,500 for modernising headquarters, £8,000 for provision of a new fire alarm system and £8,500 for reconstruction in Dorset Street and Buckingham Street stations. He sought and got two new Dual Purpose (D/P) appliances, with fifty-foot wheeled escapes, a new five-seater Dodge saloon car and a rescue van to replace the 1909 Leyland.

The Dolphin's Barn station was not built and opened until February 1964 and the new control room in Tara Street was not provided until the late 1960s. The street fire alarms remained as they were until they were completely dispensed with in the 1940s. But the delays, it could be argued, were because of our "Emergency", which represented a world war elsewhere and presented new priorities to the government and local authorities.

The major's first annual report listed 532 fires attended in the county borough and six attended outside; there were 3,576 ambulance calls, and staff consisted of six officers, eleven station officers and sub-officers, and ninety-eight men. There were six motor pumps and three motor fire escapes, four ambulances, one station car plus one motor tender. Two men were on permanent duty in Government Buildings. The report warned that "many of the above vehicles are obsolete".

## The "Emergency"

On 3 September 1939 the British prime minister announced that the "long strug-gle to win peace" had come to an end and Britain was now at war. The Irish government declared the country to be in a state of emergency and a new defining role was necessary for the Dublin Fire Brigade. In August the auxiliary fire service (AFS) had been initiated by the Department of Defence and Major Comerford was appointed as officer-in-charge. He immediately set about organ-ising this section of the civil defence organisation. He appointed Third Officer Gorman as training officer and promoted s/o Leech as acting district officer to work on AFS training with two instructors from the army, while four firemen were selected and promoted to leading firemen and seconded as instructors.

Although the re-organisation scheme for the upgrading of the fire stations was put in abeyance because of the new national situation another D/P appliance was ordered from Leyland and another batch of recruits was prepared for entry to the brigade. There was also now a disciplinary code in place and all personnel had to sign a document accepting its contents.

The attempt by the Dublin Firemen's Union in the early 1930s to institute a pension scheme which covered widows and orphans had been rejected by the Minister for Local Government who "was opposed to amendments to purely local codes" but he had suggested that Dublin Corporation would get the views "of the other Borough Councils with fire fighting services ... If agreement resulted it would be open to the Councils concerned to join together for the purpose of promoting a Bill to effect the objects outlined."

The corporations of Cork, Dun Laoghaire, Limerick and Waterford were written to. In each case the reply was the same, "not interested in the promotion of a Bill on the lines suggested". There was huge resentment among the Dublin firemen at this rejection of their legitimate claim and, seeing no way forward on their own, some of the shorter service members decided to join the growing numbers in the IMETU.

Under the 1939 Air Raid Precautions (ARP) Act, Dublin Corporation were given responsibility for the "organisation of an emergency fire fighting service, including provisions for auxiliary fire stations, fire patrols and fire posts". As already outlined the recruitment and training of this new force was handed to the Dublin brigade. At the same time discussion was taking place on the enactment of a Fire Brigade Bill to deal with issues such as provision of appliances, equipment and organisation etc on a national basis. This Bill, which was based closely on the 1938 British Fire Brigades Act, had no provisions regarding pensions for the firemen. Under these circumstances the majority of firemen agreed to be part of a pay claim served on 22 January 1940 by the IMETU for "an increase of 8/- per week for all employees in the service and employment of Dublin Corporation". This was a cost of living pay claim based on figures going back to 1937. When Dublin Corporation on 24 February confirmed that its final offer was two shillings per week and that it would not apply to the firemen the offer was rejected and the union serviced strike notice to commence on 29 February at 6 pm.

In spite of reservations by many of the firemen and an unsolicited attempt by Jim Larkin to get a compromise settlement the Union executive decided to carry out their members' mandate and proceed with the strike. They also insisted that the firemen should be directly involved in the dispute.

The city manager in reporting on the strike to the minister said that "the withdrawal of the services of the Fire Brigade personnel (except Officers) is a perturbing state of affairs when one considers the possibility of a serious conflagration in the City". He went on to outline how the two-platoon system had been introduced at a very large cost to the citizens: "I regard the lack of an adequate fire fighting force as most disquieting and I feel constrained to so report". The minister in his reply was "gravely concerned with the position which had arisen owing to the withdrawal of members of the Fire Brigade staff" and he sought "details on what steps you are taking to maintain a fire fighting unit which will be adequate for the protection of life and property".

Major Comerford had already sought assistance "from the two military fire brigades at the Curragh and Athlone, in addition to at least six gardaí on continuous telephone duty, etc". The brigade operated on the basis of eight senior officers and four station officers providing one motor during the day and two motors at night along with the military crews. The minister then proposed the following: "you should take the requisite steps to recruit and train a new staff if the men on strike do not recognise their obligations to the citizens and return forthwith". On foot of this letter Major Comerford notified the striking firemen "that unless they returned to duty not later than 10.00 a.m. on Monday 11th, arrangements would be made for an augmented Emergency Fire Service pending further decision".

On 16 March a Catholic bishop intervened and the general workers returned to work on 19 March on the basis of "restoration of pension rights". No increase on the two shillings per week offered prior to the strike was attained. Major Comerford refused to allow the firemen to return on duty at the same time as their striking colleagues in the corporation returned to work and when he finally allowed them resume in the fire stations, it was only after he had given them a dressing down and warnings in regard to their future conduct. This isolation of the firemen, who received no pay increase as a result of the strike, demoralised and divided them, the Dublin Firemen's Union soon ceased to exist, and trade union organisation in the fire service was weakened for the next twenty years.

A drill book was drawn up by senior staff and all drills were conducted in accordance with the new orders outlined in the drill book. Standards for turn-outs for fire calls were set in place and all operational orders were now in written form in a brigade order book.

Under the 1940 Fire Brigade Act, new arrangements were put in place for attendance at fire calls outside the borough boundary and the provisions of the 1862 Dublin Fire Brigade Act were set aside. Although the Act did not immediately lead to the setting up of fire brigades throughout the country it did initiate moves which would eventually see that happening. A sum of £40,000 was approved by the minister for improvements and extra equipment for the Dublin Fire Brigade.

### The Fire Brigades Act 1940

At long last national legislation providing for fire brigades was enacted into law. The 1940 Act dated 18 April was titled:

> *An Act to require sanitary authorities to make provision for the extinguishing of fires occurring in their sanitary districts and for the protection and rescue of persons and property from injury by such fires etc.*

The Act put paid to the 1862 Dublin Fire Brigade Act and resolved the vexed issue of attendance at out of area fires. It left a great many loopholes however – it failed to prescribe proper training and professional standards for fire services, standards of fire cover or procedures for dealing with potentially dangerous buildings, and it made no provision for fire inspectors.

Section 2 (1) contained the following statement.

> *Every sanitary authority shall make reasonable provision for the prompt and efficient extinguishing of fires occurring in buildings and other places of all kinds in their sanitary district and for the protection and rescue of persons and property from injury by any such fire.*

The main section on provision of fire cover set no standards for the type of service required, nor did it set standards in regard to training or efficiency. Other sections of the Act continuously used the word "may" instead of "shall" in order to allow local authorities latitude to opt out of provisions, while fines of "not exceeding fifty pounds and in the case of a continuing offence a further fine not exceeding five pounds for every day during which offence is continued" left unscrupulous owners of places of public resort who contravened "fire notices" to ignore the legislation over succeeding years.

Section 4 (1) helped resolve the controversial issue of out-of-area responses.

*A Brigade authority may authorise the Officer having command of their Fire Brigade, to send such Fire Brigade to fires occurring outside the sanitary district of such Fire Brigade's authority and may, in such authorisation, specify the circumstances and the conditions (including conditions as to distance under which such Fire Brigade is so to be sent outside such district).*

That and Section 16 (1) represented the main advances of the Dublin Fire Brigade.

*The expenses incurred by a sanitary authority under the Act shall be raised and defrayed in like manner as expenses incurred by such authority under the Public Health Acts, 1878 to 1931.*

Gone were the days when the financing of the Dublin Fire Brigade was confined to a small proportion of a water rate.

The Auxiliary Fire Service was approved as a separate and distinct organisation, except that it was to be an adjunct to the regular Dublin Fire Brigade in emergency conditions, especially in regard to the preparation of air-raid

**Auxiliary Fire Service under instruction from Dublin Fire Brigade c. 1940.**
**Photograph, © Dublin Fire Brigade Historical Society**

precautions. Also the Dublin Fire Brigade had the responsibility of organising the training and equipping of the AFS and the officers of the fire brigade had the same powers and authority over members of the AFS when in training as they had over members of their own brigade. Personnel of the Dublin Fire Brigade on rotary leave were expected to volunteer their services in the case of emergencies, ceremonial parades or funerals.

Conditions in the service were no longer uniform. Annual leave was a case in point: all those with service prior to 1 October 1938 had twenty-one days consecutive leave per annum; those employed after that date had only fourteen days annual leave during the first ten years of service and probationary firemen (first year) were not entitled to any period of annual leave. The chief superintendent had also taken on the power "at his discretion, in case of emergency to suspend or change all or portion of any period of annual leave". These were some of the new wartime conditions under which the firemen were expected to work.

Although initially the brigade had been divided into four districts as part of Major Comerford's 1938 re-organisation, further major logistical changes occurred as the war years continued. New permanent and temporary fire stations were opened at Thomas Street, Fitzwilliam Place, Stannaway House (Crumlin), Annamoe Garage on Old Cabra Road. This led to a wholetime establishment in 1941 of 1 chief superintendent, 1 second officer, 1 third officer, 1 divisional officer, 5 district officers, 11 station officers, 13 sub/officers, 9 leading firemen and 101 firemen, a total of 143 personnel.

### 1941– a most eventful year

The brigade had its first involvement with aerial bombing on the nights of 2 and 3 January 1941. At 6.15 am on 2 January Dublin witnessed its first aerial attack when a bomb was dropped at Rathdown Park in Terenure which destroyed two houses and damaged many others. Two sections of the brigade and two ambulances responded on a cold morning with a thin layer of snow on the ground. On arrival at the scene all but one of the victims had been removed from their damaged homes by the area wardens of the ARP. Within twenty minutes the fire crews had released the last victim and although seven people were removed to hospital none was seriously injured. The offending German plane seems to have circled over the area for some time because firemen on arrival decided to extinguish some of the local street lighting in order to frustrate another bombing. There was no further action however.

At 3.56 am on the following morning a heavy calibre high explosive bomb struck the pavement outside 91 and 93 South Circular Road completely demolishing the houses and causing serious damage to the synagogue and at least fifty other houses. A second bomb fell some thirty yards away. Lesser damage was caused to the nearby boxing stadium, Boyne Linen Company, the Presbyterian church, the White Swan laundry and Wills tobacco factory. With thirteen people still trapped on the arrival of the brigade extra motors were called in together with a third ambulance. The area was quickly sealed off by gardaí and the ARP as the fire crews and the AFS extricated all those trapped in the wreckage of their homes and had them removed to hospital. During this rescue work one fireman was overcome by escaping gas and was removed unconscious to hospital where he was detained. It is strange to relate that no major fire occurred at either of the bombings in spite of the damage to property and the ruptured gas pipes.

The *Irish Times* editorial of 4 January praised the heroism and application of both the fire brigade and the ARP but questioned the failure of the city air-raid warning system while suggesting that the time had come when "the matter of a 'black out' in Eire ought to be considered". Steps were taken to reduce lighting in Dublin with a further extension of the lighting restriction order but the city manager maintained that sounding alarm systems when the country was not at war would be unwise in that it might upset "people over a wide area because of what is perhaps an isolated incident".

On 24 January at 1.30 am a serious fire broke out on the floor above the State Apartments in Dublin Castle. Seven sections of the brigade attended and many ladders were pitched along both sides of the buildings as the flames advanced upstairs sweeping through the upper windows and engulfing a large section of the roof. Soon the night sky was illuminated by masses of sparks and burning debris as large sections of the roof of the ancient structures collapsed.

The whole of the cross section between the upper and lower Castle yards seemed to be on fire as crews, enveloped in volumes of dense smoke, tackled the blaze. Men were sent to the narrow streets outside the walls to check on the dwelling houses where most of the flying sparks were falling. Attempts were made to rescue valuable paintings and furniture as the firemen surrounded the burning area. It took over two hours of constant endeavour before the fire was brought under control and another four hours before the conflagration was reduced to burning pockets.

The brigade successfully prevented the dangerous blaze from extending into Saint Patrick's Hall and the banqueting room in the building but over a hundred feet of roof had collapsed and two upper storeys were destroyed along with extensive records from the Revenue Assessors of the Public Departments, irreplaceable paintings and furniture. Although the Italian Room was extensively damaged two extremely rare Bossi mantelpieces were saved. The newly reorganised brigade had worked efficiently and effectively under the eyes of many prominent politicians and civic dignitaries who were present at the scene.

The brigade's annual report for 1939-40 states contains the following revelation.

> *Notwithstanding the withdrawal of labour by the personnel of the Fire Brigade on the evening of 29th February, which continued for over 10 days, during which time the emergency service was run by the 12 Officers and Sub/Officers remaining on duty, with military assistance, the morale and discipline of the Brigade improved, but the fact that the personnel belong to several different trade unions, whose interests at times are apparently of a conflicting nature, means that the complete cohesion, so essential in an organisation of this type, is, in effect, practically impossible of achievement.*

The chief went on to thank "all the personnel for the willingness they showed to adapt themselves and co-operate in the greatly changed conditions which have been introduced".

### The Belfast Blitz 1941

Belfast faced its first night bombing raid on 7-8 April 1941 when a small force of German bombers was responsible for the deaths of thirteen people, including two AFS men, and serious damage to a portion of the Harland and Wolff ship-yard. The raid exposed the lack of defences for the city and led the fire service to complain of industrial and commercial businesses whose premises were closed, empty and ablaze with no fire watching precautions in place and insufficient water supplies to speedily attack the flames.

Dublin enjoyed its biggest ever military parade on Easter Monday, 14 April, as Belfast's citizens enjoyed the greyhound racing in Celtic Park and shows in the Opera House, Empire Theatre and the many cinemas throughout the city. On the following night just after 10.40 pm air-raid sirens sounded as some 180 German aircraft began to rain death and destruction on an inconceivable scale on the defenceless civilian population of Belfast. Actions that were considered crimes against humanity in the first world war were now accepted as part and parcel of "total war". Aerial bombing on built-up areas which included domestic dwellings, hospitals, schools, cinemas, churches, convalescent homes and hotels were now as much an accepted target of war as the sinking of civilian passenger vessels became in 1914-18.

Whole streets were shattered and houses blown to dust as hundreds of tons of high explosives landed on congested back-to-back terraces of working class houses. The ground vibrated from fearsome explosions as tons of incendiary bombs and parachute mines wrought havoc and terror on the stricken area of terrified inhabitants. The blazing holocaust lighted the heavens, turning the quiet night into a ferocious beacon which exposed the devastated area to further death and destruction from the ever-incoming bombers.

For financial rather than security reasons, necessary orders for extra fire-fighting equipment had been cancelled and manpower in the full-time fire service at 230 men left Belfast with fewer resources to deal with a major bombing raid than any other city in the United Kingdom. With roads impassable and one extensive fire extending into the next, buildings collapsing, water mains smashed and gas escaping from fractured pipes, help was summoned from each of the civil defence regions in Northern Ireland.

As the night blazed on, and the city faced total destruction, ever more fire pumps and crews were called for. At 4.25 am as the bombs still rained down, an urgent telegram was sent to Westminster for assistance for the stricken city. Eventually forty-two fire pumps with 400 firemen would be despatched by destroyer from Liverpool and by Admiralty ferry from Glasgow. Sometime after 4 am a momentous political decision was made. Brian Barton in his book *The Blitz:Belfast in the War Years* maintains that the original urgent message was sent by the Belfast Commissioner of the Royal Ulster Constabulary to the War Room at Stormont to seek assistance from Dublin.

John MacDermott, the Minister of Public Security, decided to support this urgent request for assistance and telephoned Sir Basil Brooke at 4.15 am asking "for authority to order fire engines from Eire". Brooke agreed, noting in his diary "I gave him authority as it is obviously a question of expediency". At 4.35 am a telegrammed request was sent to P. J. Hernon, the Dublin city manager. The request was sent by telegraph because the telephone lines between Belfast and Dublin had been severed by the bombing. A memo in P.J. Hernon's hand lists the sequence of events that followed.

- 5.10 a.m. Ministry of Public Security request urgent assistance from Dublin Fire Brigade in fighting fires in Belfast, Supervisor Public Telephones Belfast. Phone confirmed and I got in touch with Major O'Sullivan. Phoned An Taoiseach (Eamon De Valera) at the same time and informed him of this communication. He said it was a serious matter and would get experts to advise him.
- 5.50 Phone message from An Taoiseach – said "go ahead with Brigade and give any assistance possible" – phoned Comerford to this effect. An Taoiseach rang again said "to make sure it was of sufficient importance". Rang Belfast again, numerous fires from 10 o'clock enemy action, unable to go ahead and to check who definitely sent message.
- 6.10 Major Comerford said he would send 3 crews (2 F.B. and 1 A.F.S.), men to volunteer.

    Rang Belfast regards the message and was informed by Supervisor Telephones who confirmed the message before sending to get into communication with Town Clerk Dublin (City Manager).
- 7.30 Informed An Taoiseach to this effect and informed him that 3 pumps with crews were now gone – 2 more getting ready, 1 from Dun Laoghaire going and 1 from Dundalk gone.

A further entry in his memo lists fire brigade assistance sent to Belfast.

- 3 Regulars and 3 Auxiliaries (from D.F.B.) 1 Dun Laoghaire, 1 Dundalk, 1 Drogheda, re request, humanitarian ground, could not refuse, felt certain sent with your approval, Volunteers. In an accident quite certain of necessary legislation passed to compensate in case of accident. I feel certain that if it were necessary the Government would introduce legislation enabling compensation to be paid ...

From his entry it is clear that he felt that the state would if there was an accident face up to its responsibilities in the matter but obviously all was not straightforward on this issue of cover for those who travelled to Belfast because another note states "Minister for Finance wants it to be understood no commitment to pay all cost".

Some small gear had been removed from the pumps to allow for stowage of extra hose. Two portable pumps loaded with hose had been hitched to two of the motors for the journey to Belfast.

Within half-an-hour of receiving the message Major Comerford was addressing the Dublin firemen gathered from all stations at a meeting in Tara Street station.

He told the assembled firemen that Belfast had been heavily bombed, causing death and injury to large numbers of civilians, and that De Valera had instructed him to give as much assistance as they possibly could to their unfortunate fellow countrymen. Some of those present remembered the name of Cardinal MacRory being mentioned in the request. The major went on to state that because "they were being asked to perform tasks outside their agreed terms of duty, he was asking for volunteers but he hoped the men would not disgrace the great traditions of the Dublin Fire Service".

Practically everyone on duty responded to the chief officer's request for volunteers but then the vexed issue of pension cover was raised, especially in the case of married men who might be killed or seriously injured in Belfast. The men were given an assurance on this matter by Major Comerford who obviously contacted the city manager on the problem. Major Comerford then personally rang Dun Laoghaire, Drogheda and Dundalk, requesting them "to prepare to send fire fighting appliances to Belfast to assist at fighting fires the result of an air raid of incendiary and H.E. bombs" and telling the officers in these stations that "An Taoiseach has given permission for any available appliance to be sent to Belfast".

The firemen clung to their open vehicles thundering at 60 mph through the chilly morning on a journey of the brave into the unknown. The drivers kept their hands warm by sitting on one hand while driving with the other and changing over when the driving hand became frozen, while the crew alternated stuffing a hand inside their jackets while holding on with the other. Apart from the fact that none of the men could comprehend the extent of the destructive mayhem they faced, hardly any of them had ever been to Belfast before. Very little talk took place on this epic journey as each man kept his own counsel, apprehensive as to what he would encounter. From Killeen Border Post the pumps were escorted by military motor cycles right through to Belfast.

Once beyond Dunmurray the crews got a foretaste of the chaos and terror they were approaching as more and more army trucks joined them and thousands of evacuees, mainly women and children were to be seen resting by the roadside or in surrounding fields. The bombing raid had lasted from 10.45 pm until sometime around 5 am and now the city was a scene of great sweeping flames and voluminous palls of filthy grey smoke. The famous Ulster actor and playwright, Joseph Tomelty, said of that night "I don't think any of us noticed the dawn, for in a sense it had never been night".

The Dublin firemen arrived in Belfast just before 10 am but there was a long delay before they were eventually taken to Chichester Street fire station. Even there it was hard to find a senior fire officer to give them instructions on where they would be deployed. Eventually Dublin and other Eire fire crews were assigned to work in various parts of the stricken city. The men were later to describe the conditions they faced on that terrible day as a widespread pall of dense grey smoke locked out the sunshine and they toiled to the sounds of ambulance bells, random explosions of delayed action bombs and the voracious roar of flames devouring all in their path. They listened to the heavy thuds as massive walls collapsed and everywhere there was the stench of smoke and leaking gas and the frantic calls of rescuers as they tried to extricate injured or dead from smouldering mounds of masonry, timber and slates. It was like a city struck by some awful man-made earthquake.

In Flax Street where some of the crews worked, great gaps were visible in the lines of buildings where all that was left were heaps of rubble and charred skeletal interiors which were filled with rising smoke that caused painful breathing to those exerting themselves in their task. When one area was damped down the men were moved to another, no less unfamiliar than the place they had recently left. Among those who toiled that day grew the story of the unfortunate woman dug from the ruins of her house who, hearing the strange accents, inquired if she had been captured by the Germans. No, she was told, they are the Dublin Fire Brigade, to which she cried "Oh God, I've been blown down to Dublin".

No proper arrangements were put in place to provide food or rest periods for the Dublin crews. Begrimed, hungry and thirsty, they kept their pumps charged with water relays from rivers as town pressure often dropped to zero. In some areas provisions were provided by grateful civilians but in the chaos of the day many simply did without.

As exhausted Northern Ireland firemen snatched a few hours' sleep in Chichester Street station, still wearing their clothes, Dublin firemen assisted the AFS and soldiers, damping down fires while rescue work went on. The enormous effort of the Northern Ireland fire crews could not unaided either contain or extinguish the overwhelming number of large fires rapidly destroying major parts of the city. Sometime after 6 pm the fire crews from Eire began making up their hose lines and ladders to head for home. As each crew was ready it left the stricken city. By then most of the major fires were under control and the British firemen were arriving. The Irish government was concerned that if German bombers arrived again over Belfast, Eire's relationship with Germany might become extremely complicated, particularly if Irish firemen were to be killed by German bombs. At least once during that fatal day German reconnaissance planes flew over the stricken city, causing further panic to an exhausted population.

The return journey to the Border from what had been hell was itself something of a nightmare. Apart from the cold then affecting the exhausted bodies of the famished men, the southern fire engines were not equipped with black-out covers for their headlamps and after sunset they had to drive in the dark following the tail lights of a motor cycle despatch rider. They needed to remain awake, so there was much talk and singing as they headed through the night for Dublin. Two of the crews received refreshments in Banbridge before they sped on to the Border, others were entertained in the Ancient Order of Hibernians hall in Newry. It had been a long day, one that none of them ever forgot.

The *Irish Times* editorial writer was inspired to produce the following on 17 April.

*Humanity knows no borders, no politics, no differences of religious belief. Yesterday for once the people of Ireland were united under the shadow of a national blow. Has it taken bursting bombs to remind the people of this little country that they have a*

*common tradition, a common genius and a common home? Yesterday the hand of
good-fellowship was reached across the Border. Men from the South worked with men
from the North in the universal cause of the relief of suffering.*

Seventy-one firemen with their fire engines from Dublin, Dun Laoghaire,
Drogheda and Dundalk had travelled into the unknown in spite of the risks, to
help their colleagues and the distressed of Belfast. Val Walsh, second officer in
the Dublin Fire Brigade, followed his crews to Belfast later in the day to check on
their welfare.

John MacDermott, the Minister of Public Security in Northern Ireland,
thanked the Irish government for the prompt response of its fire brigades to
Belfast's call: "Their services were very valuable and we much appreciated the
quick response. They rendered invaluable service at a critical period." Belfast's
Lord Mayor McCullagh spoke of the amount of fire brigade equipment
destroyed or damaged and piped water supplies wrecked as he thanked "the visiting fire brigades for their prompt professional assistance and humanitarian
work". There is no complete list of the casualties of this blitz because many of
those killed in the devastating blasts could never be identified, but official figures
released some years afterwards estimated at least 900 dead and over 600 seriously
injured. John Blake, the historian of Northern Ireland's part in the second world
war, wrote the following: "No other city in the United Kingdom, except London,
had lost so many of her citizens in a single night's raid".

The *Northern Whig* on 17 April reported on the role of the fire fighters from
the south: "it was confirmed in Dublin this morning that units of fire fighting and
ambulance services from some towns in Eire assisted to put out fires resulting
from Northern Ireland's blitz". In Castlebar on 19 April De Valera expressed
public sympathy with the victims of the Belfast air attack.

*This is the first time I have spoken in public since the disaster in Belfast and I know
you will wish me to express on your behalf and on behalf of the Government our
sympathy with the people who are suffering there … they are all our people, they are
one and the same people, and their sorrows in the present instance are also our
sorrows. I want to say that any help we can give them in the present time we will give
to them wholeheartedly believing that were the circumstances reversed they would
also give us their help wholeheartedly.*

Two days later the Irish government minister Frank Aiken justified his government's position to journalists in Boston and asserted that the people of Belfast
"are Irish people too". On Wednesday 23 April an emergency powers order came
into effect in the twenty-six counties banning all illuminated sky signs, fascia or
advertisements during the hours of darkness and all lights on premises or hoard-
ings for the purpose of advertisement or display. It also decreed that during the
hours of darkness there should be no lights outside cinemas, theatres or other
places of public resort and lights could only be switched on in shops while they
were open to the public.

On the question of Irish neutrality Robert Fisk, in his book *In Time of War,*
states that Eduard Hempel the German ambassador to the Irish Free State made

the following claim in relation to the fire brigade response to Belfast: "we could have protested but ... nobody from Germany protested".

On 4 May a second major air raid took place on Belfast, this time causing greater destruction to business and centre-city properties but much less loss of life. The city was better prepared for this raid but assistance was again requested by the Belfast fire chief and was despatched from Dublin. Some fire engines and crews on duty were sent as soon as the request was received and details of the raid became known. These attending crews were organised to work alongside local firemen and were immediately on arrival given food before being directed with escorts to various areas in the city.

The sequence of events as minuted by the Dublin city manager P. J. Hernon on 5 May 1941 were as follows.

- 12.30 a.m. Taoiseach phoned and stated that Belfast had again been attacked. Told me to be prepared to send assistance if called upon to do so.
- 12.40 a.m. Phoned Major Comerford and told him to be on the alert.
- 12.45 a.m. Taoiseach rang again and stated that the Brigade should travel in daylight, if possible, but if an urgent message was received, we would have to take the risk and go. No men to be sent who had not volunteered. I told him that some members of the A.F.S. would have to go. In the case of accident to any of these men legislation would have to be introduced to deal with the matter.
- 2.25 a.m. Taoiseach rang and stated all available fire assistance which could be spared was to be sent to Belfast. Stated that the Police and Military were to be informed so that the way would be clear for the Brigade. Also stated that we should confine our activities to rescue from private houses rather than military objectives.
- 2.35 a.m. Rang Major Comerford to this effect. He stated he would send an ambulance with 4 men and Superintendent (Third Officer) Gorman and an engine with 8 men, making a total of 13 men, in 15 minutes after receipt of my call.
- 2.45 a.m. Call from Belfast. Line very bad. Informed Supervisor, Telephones, Dublin, to convey message re Military and Police. This had to be done via Portadown.
- 3.15 a.m. Rang Major Comerford and told him to instruct his officers regarding rescue from private houses rather than military objectives, as stated by the Taoiseach. Major Comerford stated it would be impossible to do this, but the utmost care would be taken to do it diplomatically if possible, and he was instructing the officers to this effect. Rang Mr. O'Mahoney (Town Clerk) Dun Laoghaire, and asked him to be ready to assist.
- 3.30 a.m. Rang Mr. O'Mahoney and told him to get ready to send an ambulance and to see that his men volunteered for this duty.
- 3.40 a.m. Rang Major Comerford again. He stated that 13 men had gone on 2 vehicles and 11 were ready to go on 2 more, that 1 crew each from Dun Laoghaire, Drogheda and Dundalk were proceeding to Belfast, together with 2 AFS crews and 2 pumps from Headquarters, Tara Street.

Hours of Departure:

| | |
|---|---|
| Dual Appliance and Ambulance | 3.21 a.m. |
| 4 Units | 4.40 a.m. |
| Dun Laoghaire | 4.40 a.m. |
| Lorry and Medium Unit | 9.55 a.m. |
| Staff Car (Major Comerford) | 9.55 a.m. |

25 Regular and 29 A.F.S. Volunteers.

Appliances Sent

9 pumps, consisting of

6 pumps and ambulance from Dublin with 53 men (29 AFS and 24 RFB)

1 pump from Dun Laoghaire

1 pump from Drogheda

1 pump from Dundalk

Message received from Superintendent Gorman from Belfast "Fires out of hand. Send 17 men and 1 pump". The men and appliance were dispatched and are included in total set out overleaf.

The following ambulances were also sent:

6 a.m. 2 Ambulances from St. John Ambulance Brigade containing 10 men

8 a.m. 3 Ambulances from Red Cross.

In some streets where the firemen worked there were obstacles lying around everywhere which had to be surmounted and just to walk about demanded great exertion. The stifling conditions in the strange surrounds caused anxiety to some of the men – but they got their work done. They had been cheered by civilians as they sped into the stricken city and they worked willingly alongside soldiers removing debris in the search for bodies or in the effort to keep streets passable. A problem arose when one of the Dublin firemen was injured by exploding cellulose while dealing with a cinema fire and required hospitalisation. When the fire crews headed home before night-fall frantic efforts were made to get the injured Fireman Byrne transferred from Belfast to Dublin. The consequences of the death of a Dublin fireman while working in the Belfast Blitz were more than the neutral Irish government wanted to contemplate. Eventually the injured hero was returned home for further medical treatment. He had sustained serious burns.

On 6 May the Belfast Presbytery thanked Dublin "for its invaluable assistance and generous help in the emergency just passed". The men who braved that first terrible night were later given five shillings each, said to have come from the Belfast authorities to compensate them for the cost of a meal while on duty in that city. They were told by Major Comerford that sometime in the future they would receive a suitable recognition from the Northern Ireland authorities for the great assistance they provided, but whatever he was referring to never came. However much they might have been subsequently overlooked or ignored by the Belfast establishment the famous journeys led to social contact between the firemen of both cities and as a result an annual football match was played between the two brigades up to the early 1970s.

Belatedly in 1995 on the fiftieth anniversary of the ending of the second world war an invitation was received by the Dublin Fire Brigade addressed to any survivors of those two historic days to attend a function at Hillsborough Castle and meet Prince Charles. Only four of those who were there were still known to be alive at that time and only one, Tom Coleman, travelled north to receive some recognition for his colleagues' solidarity at such a critical time.

The arrival of German planes over Belfast had finally brought the war close to home and it was an experience that no one wanted repeated. Dublin became home to thousands of Belfast evacuees and refugees fleeing from the destruction of their homes. The neutrality of the Irish Free State would mean peaceful circumstances even if impoverished ones. But the war followed them south and Dublin became a victim on 31 May. Even the support for the anti-conscription campaign by the hierarchy and the Nationalists in Northern Ireland failed to save the people of Dublin from German bombs. Cardinal McRory's anti-conscription pastoral of the time fell on deaf German ears.

*... an ancient land made one by God was partitioned by a foreign power against the vehement protest of its people and with conscription it would now seek to compel those who still writhe under this grievous wrong to fight on the side of its perpetrators.*

### North Strand bombing

Between 1.30 and 2.10 am on Saturday 31 May 1941 four bombs were dropped on Dublin. The first demolished two houses in Summerhill, damaging many more. The second, falling almost opposite O'Connell Schools on the North Circular Road, destroyed a shop and dwelling houses, trapping many inhabitants inside. The third fell near the Dog Pond in the Phoenix Park destroying one cottage and breaking the windows in both Áras an Uachtaráin and the American Legation. The fourth and most destructive bomb fell in North Strand Road just south of Newcomen Bridge, wrecking houses on both sides of the road and causing fierce fires in the piles of debris of more than 200 houses.

In all thirty-four people died, ninety were injured, requiring hospital treatment, and up to 300 dwellings were destroyed or seriously damaged. The tragic scenes of Belfast were recalled as distraught relatives called at the city morgue and city hospitals seeking news of missing family members. The ARP were quickly on the scene and wardens were at work as the fire engines and ambulances arrived. The fire crews were assigned to tackle and extinguish all pockets of fire while the gas company crews set about cutting off supplies to the area and assisted in rescuing those crying out or wandering dazed in the rubble-strewn lanes off the main road.

Gardaí and military sealed off the area as the rescue work proceeded through the morning dusk. All the emergency services stuck grimly to their toil while the extent of the numbers of dead, injured and homeless grew. This was an area of the city reminiscent of the back-to-back terraces in North Belfast and the explosions had been equally devastating. Piles of dusty rubble, heaps of bricks and old timbers filled gaps between roofless cottages and shattered homes where many of the poorest of the city had eked out an existence only hours before.

Soon all of the large fires were extinguished and rescue work was able to proceed unhindered. Injured and maimed were removed from beside the dead as the stillness of the summer morning was shattered by the clanging of ambulance bells, the roars of fire engines and pumps, the thudding sounds of collapsing buildings and, strangely above all, the plaintive cries of the injured and the shouted orders of the rescuers. The oncoming watch in the fire stations joined those already at the scene and those on leave were called in from their homes to cover the fire stations. Throughout the long day people gathered in the area to watch the ambulances make their way to the hospitals or the city morgue. It was early afternoon before the firemen were finally relieved from duty and allowed to make their weary journeys to their own homes.

Many opinions have been expressed as to why this outrage was perpetrated on neutral Dublin, some claiming it was a response to Eire's provision of rapid humanitarian assistance to blitzed Belfast, others that it was an attempt to force Eire into the war. There was even a view expressed that it was an unforeseen result of British interference with radio beams used by the Luftwaffe. Whatever the reason, on 19 June the German government expressed regret and promised to pay compensation. It was the last bombing to occur in Eire for the duration of the war.

The following morning just after 10.00 am two tenement houses in Bride Street suddenly collapsed without any warning killing three of the unfortunate tenants and injuring fifteen others. The city did not have to depend on German bombers to kill, maim or make homeless its struggling poor. When the fire

Richmond Cottages, nos. 24, 25, 26, and 27 following North Strand bombing.
Photograph, © Dublin City Council

233

Clarence Street North, nos. 33 and 34 (tenements) following North Strand bombing.
Photograph, © Dublin City Council

brigade arrived on the sorry scene the two houses were smouldering rubble while a third was in imminent danger of collapse and nine of the victims were still trapped in the debris. As other rescuers arrived the firemen rescued four and were searching desperately for the others. Panic spread and before the gardaí arrived the fire crews were being hampered as terrified neighbours removed their meagre belongings onto the narrow roadway for fear of further collapse. As Sunday mass-goers gathered in their hundreds to comfort each other and survey the rescue work and the damage more victims were taken from the ruins. If the firemen were traumatised by the events on the North Strand they had no time to dwell on it. They worked under the gaunt, threatening walls, free-standing sentinels above the destruction surrounding them.

The Bride Street housing collapse was described as "an outrage on the civic sense of the City". Had the North Strand bombs dropped on an area of decayed multi-storey tenements rather than on an area mainly of ancient cottages, how much more death and destruction, homeless and maimed would the city have endured? Some twenty-three families from the area joined the hundreds from the North Strand on the emergency housing list as corporation officials belatedly surveyed the adjacent tenements for structural faults.

During the year the city boundary was again extended following an agreement with Dublin County Council and in May the council agreed to the erection of the new fire station at Dolphin's Barn. Second-hand trucks were purchased as towing vehicles for portable pumps because Dublin could not procure any new fire engines from England and waterproof gear for the firemen was now made in Rathmines. At the same time Dublin Corporation had to replace over forty hydrants on water mains formerly in the county area because of different size outlets being used in the county.

On 2 June as the last body was finally recovered from the North Strand bombings the fire brigade was again dealing with a major fire. This time the premises of Kennedy's Bakers in Stephen street went on fire. This extensive building in one of the most congested areas of the city was to test the fire brigade organisation to the full. Horses had to be released from the stables as the frightened

animals stomped and neighed in the swirling dust and smoke. The flames devoured a major part of the bakery premises. Six sections of the brigade working over three hours were required before the area could be declared safe. The fire was extinguished without damage to any of the closely surrounding tenements but again some of the inhabitants had removed their few belongings on to the streets for fear of losing all they had.

With Germany now mainly engaged in war on the eastern front and no more bombing on Britain, Dublin began to return to normal. Large numbers left the AFS for various reasons, not least because of failure to provide them with uniforms, and the Air Raid Precautions Committee was disbanded because of non-attendance of the councillors. Howth urban

**Fire at the Stephen Street Bakery, early 1920s.**
**Photograph, © National Library of Ireland**

district became part of Dublin and the retained station and crew there came directly under the Dublin Fire Brigade.

In 1943 the city manager conceded a seven shillings per week pay increase to the firemen. Shortly before that award the brigade fought a serious fire that destroyed a large section of the Daisy Clothes Market. In November 1943 the corporation agreed to purchase an extra ambulance for use in the fire service.

In March 1944 the city manager got approval to spend £15,000 on reconstruction work at Tara Street and £30,000 for the new station at Dolphin's Barn where part of the site was to be allocated to the provision of a library. The war was slowly coming to an end. The lord mayor reported that the street fire alarm system had deteriorated to "such a condition that the maintenance of all the alarms is no longer a practical proposition". He pointed to large numbers of malicious calls and false alerts due to mechanical faults in a system costing £600 per annum. The brigade, on the other hand, "received 750 fire calls from other sources". It was agreed that from 1 January 1945 the number of street alarms were to be reduced from forty-eight to ten.

The period of the the "Emergency" ended with the payment of a "gratuity to those who served in the AFS in a wholetime capacity". As their services were terminated the brigade began returning to normal day-to-day routine. At the height of the period, immediately after the North Strand bombing, the AFS had grown to over 700 officers and men, equipped with three self-propelled pumps and seventy-seven trailer pumps, operating from fifteen auxiliary fire stations. On 31 December 1945 the Auxiliary Fire Service was formally disbanded and one of its Tangye heavy pumping units was transferred to the Dublin Fire Brigade; it was to see service for about ten more years from Buckingham Street Station. The brigade was reduced to its peacetime numbers and those in acting officer positions reverted to their pre-war status.

### Cavan orphanage fire

On 23 February 1943 a fire at St Joseph's Orphanage for girls, Main Street, Cavan, caused the death of thirty-five of the children and an elderly cook. The government set up a tribunal consisting of Justice J. A. McCarthy (chairman), Major J. J. Comerford and a Mrs Mary Hackett, to inquire into the unspeakable tragedy. It transpired that in spite of the 1940 Fire Brigades Act, Cavan had no local authority fire service and the AFS could not be called out for peacetime fires. Major Comerford, using Howth retained station as the example, wrote to Justice McCarthy explaining how a retained system was organised, trained, staffed and costed. As the only fire officer on the tribunal he also drafted a lengthy memorandum for the other two members detailing "an approach to the problem" of fire service cover nationally. The significance of this far-seeing document is that Major Comerford outlined what he considered as the best way forward: "The future of the Fire Service generally lies between a Regional Fire Service, with a certain amount of centralised direction and technical advice, and a full National Fire Service".

His memorandum favoured a regional system for a peacetime brigade:

*... improvements need to be made generally in the fire fighting organisation through-out the country .... such facilities cannot be satisfactorily provided by the numerous sanitary authorities as at present statutorily established. Any such facilities must of necessity be built around the provision of 'cadres' of wholetime permanent fire service personnel, organised throughout the country in the larger centres of population. All other personnel could be part time and paid small retaining fees on an annual basis.*

This proposal if adopted would require major amendments to the 1940 Fire Brigades Act. The tribunal report stated the following

*the Fire Brigades Act 1940 as at present in force, is hardly the most suitable enactment for the purpose .... The operation of Section 7 of the Fire Brigades Act, as designed, also appears to us to present some considerable practical difficulties. More specific standards for dangerous buildings should be set up ....*

No amendments were made to the 1940 Act and Major Comerford's regional fire service was ignored.

### The end of the "Emergency"

On 13 June 1945 Major Comerford outlined the proposed new establishment for the Dublin Fire Brigade. Over the three previous months the "temporary" fire stations including Thomas Street were closed and the wholetime auxiliary fire service was soon to be disbanded. The new lean fire service was to consist of a chief superintendent, a second officer, a third officer, one divisional officer at brigade headquarters, 5 district officers, 7 station officers, 8 sub-officers, 8 leading firemen and 90 firemen, a total of 122 persons to provide fire, ambulance and emergency cover for the expanded, growing city of Dublin and its surrounds.

By an order on 14 December Major Comerford issued the following decree.

*The City of Dublin A.F.S. will be disbanded as from December 31st 1945. Members of the service may retain badges, uniforms and caps. All other items of service equipment must be handed into Area Stores and AFS District Officers will arrange accordingly. Official expressions of appreciation and thanks from the Minister for Defence and from the City Manager to the Volunteer members of the Service for their work are to be formally read on parade by District Officers.*

*"On behalf of the members of the Regular Fire Service I would like to add my expression of sincere gratitude to all ranks of the Auxiliary Fire Service for their work during the past few years, for their very generous co-operation, for the amount of valu-able time they gave to their Fire Service duties, for their enthusiasm and for the high standard of efficiency which they attained which not alone reflected credit on them, but on the members of the Regular Fire Brigade who trained them.*

Earlier in the year Second Officer Val Walsh had resigned from the D brigade to take up the post of chief officer in Cork. His position in Dublin was filled by the appointment of Captain P. J. Diskin from the army corps of engineers.

Under the nationalised fire service in Britain the Home Office had established a section to manage and oversee the running of the fire brigades. When the 1947 Fire Services Act handed the fire brigades back to the councils and county boroughs the Home Office held on to the right to make regulations concerning pay, hours of duty, discipline, establishment numbers for staff and machinery and also the standards of training. To carry out the latter a new training college was soon established.

To oversee all these new arrangements a team of Home Office fire inspectors was appointed and the government agreed to a grant of 25% towards the cost of the new fire services. It was following the publication and adoption of these new regulations in Britain that the Irish Department of Local Government decided to appoint a fire advisor. However no inspectorate was formed and neither were training standards or establishment numbers to be introduced in this country.

### The new peace-time brigade

In spite of the major reduction of fire cover in the city and county the Dublin Fire Brigade was still despatched out of its area. On 3 March 1947 a huge fire gutted the entire west wing of Moore Abbey in Monasterevan. The abbey had been founded as a great seat of learning during the sixth century. Pillaged and suppressed at different times it had survived and was the home from 1927 to 1937 of Count John McCormack, Ireland's greatest tenor. It had then passed into the hands of the Sisters of Charity.

Two sections from the Dublin brigade responded to the fire call and made the forty-mile journey to assist brigades from the Curragh and Portlaoise. Drawing water from the nearby Barrow river the experienced firemen from Dublin set about their task of containing the blaze and eventually extinguishing it. The fire engines, together with a trailer pump, had taken just over an hour to cover the journey from the city. In spite of the trojan effort, the west wing with all contents and its beautiful oak-panelled walls was gutted and damage was sustained by both the adjoining chapel and central portion of the abbey.

On 2 September of the same year the brigade was confronted with a massive fire at the premises of Noyek and Sons in King's Inn Street. Four sections under Major Comerford battled for hours to contain this dangerous fire. The extensive building was stocked with huge amounts of timber, plywood and wall board. As huge flames roared into the early morning sky, residents of the surrounding densely populated tenements had to be evacuated and firemen drove lorries out of the smoke filled yard of the threatened premises. The twelve-foot high piles of timber took over nine hours to extinguish and the corrugated roof of the building collapsed, pulling down part of the supporting walls. Only the continuous deluge of water and a standing brick wall prevented the flames from engulfing an adjoining storage area also packed with timber.

It was a night that stretched the brigade to its limits because just half-an-hour after the response to Noyek's building two sections were sent in response to urgent calls from Naas and Curragh fire brigades for assistance at a major fire in a licensed premises in Main Street, Naas, which had got out of control and threatened to burn down much of the main street. Assisted by the other two brigades this fire was extinguished with the destruction of the public house and damage to both of the adjoining premises.

Since the end of the war, major fires in the city of Dublin had increased, testing to the full the strength and organisation of the reduced fire cover. As 1947 came to a close another massive fire destroyed machines, tons of flour and over 2,000 dozen loaves of bread when a major part of Kennedy's Bakery in Ringsend was consumed in flames.

The Chief's Report for the period March 1946 to March 1947 listed 2,135 fire calls received in that year, an increase of 848 over the all time record of the previous twelve months. Ambulance calls were down by 217 to 5,173 said to be as a result of the increased number of cars on the road, the drivers of which frequently brought injured persons to hospital. Special service calls were up to 214 and the cost of running the service was put at £57,500. In his report Major Comerford reported the loss of two lives in domestic dwellings.

> The number of fires in tenement property still continued to show an upward increase, and it was fortunate indeed to have to record that actually no lives were lost in these fires, although a number of rescues were effected ... it has been found in many cases where tenement basements were condemned as unfit for dwelling purposes, under the Public Health Acts. They were then let by the principal landlords as small factories or workshops – an ever more unsatisfactory and dangerous use ... A further difficulty is the case of premises occupied mainly as a series of tenement factories with tenants on the upper floors ... [which] could not be redressed with structural separation of the residential and business portion because of the acute housing problem in the City.

In March 1948 Major Comerford resigned from the brigade to take up the new appointment of fire advisor to the Minister for Local Government. One of his last brigade orders highlighted the shortage of staff in the Dublin brigade: "Number 2 District Ambulance (Rathmines) will be put out of commission while the first turn-out (fire engines) is engaged on a fire call". He had introduced a system of having a leading fireman on each ambulance and a bucket of chimney rods and a stirrup-pump also on board to deal with chimney fires.

His impact on the brigade, particularly during his early years, had been monumental, and much of the needed re-organisation that he carried through was still in evidence up to forty years later. However progress was not maintained in his later years in the service. His proposals for a new control room, updating of fire stations, introduction of radio controlled vehicles and the complete replacement of street fire alarms remained unfulfilled, while his plan to construct a new fire station at Dolphin's Barn did not take place until 1964. He was remembered by many of those who worked under him as a "dictator" because, as a strict disciplinarian, he imposed military discipline on the firemen. He was the man for that time, the most qualified fire officer in the country, with a tremendous organising ability, a fire engineer of international repute. He died suddenly in 1950 at fifty-one years of age.

# 15 A modern fleet

Captain Diskin was acting chief superintendent from 1 April 1948 until the appointment of Val Walsh to the post on 1 May 1949. Diskin resigned in February 1950 to take over the vacant position of chief fire officer in Cork. With the untimely death of Major Comerford further moves saw Walsh replacing the major as fire adviser and Captain Diskin returning to Dublin in October 1952 as chief superintendent with a new second officer, Brian Larkin, also a former officer in the army corps of engineers.

During Walsh's stewardship the corporation agreed to employ six extra firemen. The city boundaries had been extended to take in another 7,000 acres. All of these lands – Donnycarney, Walkinstown, North Crumlin and Ballyfermot – were soon the scene of massive house building programmes, adding greatly to the built-up suburban area covered by Dublin Fire Brigade. The brigade needed modernisation but the only progress made in 1951 was the first installation of radio-telephone equipment. It was however a year of some infamous fires, tackled and extinguished with pre-war equipment.

On Saturday 26 May, five sections of the brigade tackled a spectacularly dangerous blaze in the paint stores of Guinness Brewery at Cooks Lane. Before the arrival of the firemen the flames had spread to the newly erected escalator which brought coal from the quays up to the new £2.5 million power station. Adjacent to the paint shop were storage tanks of thousands of gallons of oil. It took much hard work and ingenuity to extinguish the fire on the high escalator while protecting the oil tanks and confining the paint shop blaze. Soldiers from Collins Barracks were brought to the site to assist the fire crews as large numbers of gardaí cordoned off the area from the thousands of spectators that gathered, crowding in from many of the surrounding pubs. It took the brigade over four hours to bring the fire under control. Although the power station and oil tanks were saved the escalator and paint shop were completely destroyed.

On the night of Wednesday 18 July the Abbey Theatre was destroyed by fire. It was a night remembered by most Dubliners of that time but particularly remembered by the overstretched men of the brigade as their busiest night of the year. The fire call was received just after midnight and the first sections to arrive were confronted by a locked building with smoke issuing from the roof. When the main door was forced firemen were met by a growing inferno engulfing the dressing rooms and stage. Valuable paintings were removed from the vestibule and office as hoses were laid into the premises. By then the rafters burning fiercely spat and groaned before their weakened beams sagged and crashed down under the weight of the roof, exposing the starlit sky to the smoke-begrimed figures playing out a real life drama inside.

In spite of the efforts of the nine fire crews who eventually fought the fire, most of the historic building was destroyed by flames and water. What had been

the former Mechanic's Institute and City Morgue was just a gaunt dangerous skeleton festooned with *The Plough and the Stars* posters, and the ghost of Yeats was left to haunt an eerie smoke-filled chamber. Some costumes, wet and bedraggled, were recovered on the following morning, repaired by willing if shaken staff so that the show could go on in the smaller one hundred seat Peacock.

Lennox Robinson – author, playwright and then director of the Abbey – summed up the night.

> *We people who had known and loved the Abbey Theatre for fifty years knew that it had to be taken down brick by brick and rebuilt as a new noble National Theatre and we should hate that. But the Abbey Theatre took the matter out of our hands. It decided that it would go up in smoke and flames … expire in glory.*

It was the year of a general election and the outgoing coalition government had been discussing a scheme which entailed spending £200,000 on a new national theatre.

From 9.40 pm to midnight on the night the Abbey burned down, six fire calls were attended to in various parts of the city by the fire brigade. Two sections were dealing with a gorse fire on the railway embankment at Raheny while another dealt with a similar blaze near Binn's Bridge. Two sections were dealing with a fire in a warehouse off Merchant's Quay and another crew was at a small fire in Westmoreland Street. All sections with the exception of a turntable ladder were out of stations when the call to the fire in the Abbey Theatre was received at 12.25 am, but all soon responded.

A third landmark fire in that year destroyed much of the machinery, records and files of the *Irish Times* in a spectacular blaze just after midday on 17 September. This fire started in the ground floor machinery and despatch rooms,

Fire at premises of *The Irish Times* in Fleet Street in 1951. The appliance on the left is a rescue tender.
Photograph, © Trevor Whitehead.
(See also colour section illustration number 16 showing fire at the same location in 1999.)

spreading with alarming speed once the highly inflammable inks, oils and paper became involved. A massive torrent of flames rushed up through the Fleet Street building as staff continued to work in the Westmoreland Street section. In spite of the speedy arrival of crews from Tara Street fire station the roof collapsed, filling the narrow street with burning embers and thick black smoke.

Onlookers, as usual, gathered in their hundreds and had to be moved because of the danger and because firemen were impeded from getting fire engines and ladders into place. In spite of the energy expended and risks taken, almost half of the four-storey building was destroyed along with historic records and tons of newsprint. Luckily there were no casualties. Much of the premises had been rebuilt only two years before, linking the triangle of Westmoreland Street, Fleet Street and D'Olier Street. As with the nearby theatre the show went on and in spite of the spectacle of the grim blackened walls surrounding piles of smouldering debris on the following morning and still attended by weary firemen, the *Irish Times* for 18 September was on sale in the shops, thanks to the co-operation of the *Evening Mail*.

In January 1953 the corporation, being fully aware of the destructive effects of major fires in the city, received sanction to borrow £12,000 towards the cost of purchasing modern fire engines. Dublin's first post-war appliance was delivered from Dennis Bros of Guildford in that same year. The Dennis F12 was rapidly becoming the most popular front-line appliance for all the large fire brigades throughout the UK and abroad. It was capable of carrying either a thirty-five foot extension ladder or a wheeled escape. The Dublin machine, with its Rolls Royce petrol engine and one hundred gallon water tank, was the first of a modern fleet which, over the next three years, replaced fire engines dating from 1924 to 1940. As the speed of fire engines had increased, the risk of being seriously injured by falling off the fast-moving machines had also increased, so the designers turned their attention to crew safety. The new appliances were of the enclosed limousine type inside which the firemen sat in comparative comfort and safety.

In 1955 a Dennis F8 pump and a Dennis/Metz one hundred foot turntable ladder went into service in the Dublin Fire Brigade, followed in 1956 by two F8s and two F12s. The F8 was a smaller bodied machine than the F12 and had a 200 gallon water tank. It did not carry an escape. The new turntable ladder replaced the 1931 Merryweather. The ambulance fleet was brought up to date with new Fargo and Dodge ambulances. By 1956 every ambulance and almost every fire appliance was equipped with radio. This resulted in greatly increased efficiency in mobilising and rapid communication between mobile appliances and brigade control.

Staff in the brigade in the year 1954 consisted of one chief fire officer, one second officer, one third officer, 6 district officers, 7 stations officers, 4 sub-officers, 6 leading firemen and 92 firefighters; a total of 118 personnel divided into two watches at an annual cost of £86,561.

The year 1953 is remembered because of the considerable number of fires in the city which involved fatalities (ten during the year), serious injuries, and major rescues by the Dublin firemen. Two young sisters lost their lives in a tragic fire in a four-storey over basement tenement at 26 Harcourt Street just after 11 pm on the night of 9 March. Four sections from the brigade attended this dangerous fire, which started in an unoccupied room on the first floor. It spread rapidly, smoke-logging the stairwell and the upper floors. Firemen rescued three people from the top floor by turntable ladder while five others were removed unconscious from within the lower floors. All were removed to hospital but the two young sisters were dead on arrival.

On 7 December another two young children died and a third was unconscious when removed from the upper floor of a domestic dwelling on Collins' Avenue. On 22 December six women survived by jumping to safety from first-floor and second-floor windows when the three-storey premises that housed a skirt factory, home furnishing firm and four flats in Dorset Street went on fire. The building was extensively damaged in the serious fire and the women only survived because of a quick thinking bus driver who, seeing the blazing shop on the ground floor, drove his bus onto the pathway thereby enabling the terrified women to leap onto the roof and roll down into blankets held by neighbours on the street. The brigade extinguished the fire in an hour. The women were taken to hospital for treatment.

That December was a specially busy month for the brigade because a week later a huge fire destroyed Portmarnock House in the most spectacular blaze in the county area for years. This four-storey mansion was a fine example of Queen Anne architecture, being over 300 years old and particularly noted for its ornate stained-glass windows. For centuries it had been the home of the Plunkett family, whose ancestors included Oliver Plunkett, but some five years earlier it had been purchased by Portmarnock Estates and was in use as a store for a company called Emerald Food Products. With the nearest water supply at least half a mile away there was no chance of the fire crews saving the building after their nine-mile dash from the city. The roof eventually collapsed and the ruined structure would in time be demolished in clearance for redevelopment.

The most potentially dangerous fire tackled by the brigade that year was on board the 10,700 ton Caltex tanker *Colombo* on 3 December. This tanker was docked within a few yards of tanks containing 5,000 tons of petrol. Just before the cargo was due for discharge a major explosion shook the engine room, causing flames to shoot up onto the deck. As fire engines sped to the dangerous scene the crew of sixty-seven left the vessel but the captain and his officers stayed on board to assist the arriving firemen. The docks were cleared and one crew member, injured in the explosion, was removed to hospital. The main generator in the engine room had exploded and set fire to that area of the ship, turning the welded steel plates into red hot metal. Water was immediately poured onto the deck of the vessel to cool the plates and firemen in breathing apparatus entered the ship's passageways in an effort to contain and extinguish the fire. Ropes had to be slackened as the ship listed from the volume of water being poured into her engine room. After an hour the flames were extinguished and repair work could commence to restart the pumps and steady the vessel. Large quantities of foam were used on this occasion, much of it provided by the Port and Docks Board which afterwards paid tribute to the skilful work of the firemen in preventing a catastrophe in the port. It was situations like this, including dock warehouse fires, which led in 1963 to the Dublin Port and Docks Board installing firefighting monitors and, later, foam equipment on its tug boats and making them available to assist the fire brigade.

In September 1953 the working week of the Dublin firemen was finally reduced from eighty-four to seventy-two hours, in line with Belfast and other major brigades in Britain. This reduction in hours was at an estimated cost of £9,000 per annum and would entail employment of eighteen additional men.

While still awaiting the delivery of the two new Dennis fire appliances in 1954 the brigade was confronted by two of the largest fires in the city since the end of the Emergency. On the bitter cold evening of 5 January just after 6 pm a fire developed in the four-storey Central Hotel building at the corner of Dame Court and Exchequer Street. Within fifteen minutes of the discovery of the fire a huge inferno was rushing through the building, which housed four shops at ground level, some thirty bedrooms upstairs and storage areas with stocks of drapery, fancy goods, footwear, toys and general merchandise. Hundreds of spectators on their way home from their workplaces gathered to watch the spreading flames. This confluence of people hampered the fire crews in getting along the street to the doomed building.

As fire crews tackled the massive fire the roof collapsed through the top floor, creating a huge inferno inside the upper walls. Glass exploded at ground level breathing fire into the lower floors. Staff were evacuated from the adjacent central telephone exchange, cutting off all trunk calls to the country from 6.45 to 9 pm. At the height of the fire seventeen jets were in operation attempting to surround the blaze and protect the phone exchange. Had the fire spread the entire national phone system could have been out of operation for days. As it was the one hundred operators were able to re-enter their workplace just before 9 pm while the fire crews continued their relentless struggle to extinguish the fire. Most of the hotel building was destroyed and a fire crew was still in attendance extinguishing pockets of flame on the morning of 7 January.

An even more extensive conflagration was to destroy three premises at the North Wall on 24 June, those of the B&I Stores, the carton making department of the Temple Press and the warehouse of E. Browett and Sons. This horrific fire was fanned by an exceptionally strong wind which spread blazing embers over a large area of the north docks. causing minor outbreaks of fire in the timber yards of T&C Martin and Brooks Thomas.

All fire crews in the brigade attended, supplied with thousands of gallons of water from the Liffey by Dublin Port and Docks Board tug boats in the struggle to contain the massive fire. Onlookers on the south wall of the Liffey watched the explosions of carboys of sulphuric acid bursting through the burning roofs and showering the firemen with sparks. In the absence of walkie talkie radios the fire crews each took the responsibility of tackling and surrounding a section of the fire while working towards the central core of the blaze, backed up by two sections kept mobile to deal with any extension of the furious inferno to premises outside the ring of jets. Some four hours after the brigade arrived the fire was

Dennis F8 fire engine, 1957: first limousine motor purchased by Dublin Fire Brigade
(firemen rode inside). Film still, © Gael Linn

contained and begrimed firemen dragged hose and branch pipes towards the
centre of the blaze. The devastated area was still being patrolled by busy fire
crews on the following day. Without the ready supply of water from the Liffey
this fire could have caused even greater havoc all along the extended yards and
warehouses of the north city docks.

As if reluctant to let the year out without further destruction to the impoverished of the city, radio and newspapers recorded the worst night in living
memory on 8 December as winds and rain wrought havoc on many families.
The Belfast line was closed as the Great Northern Railway bridge over the Tolka
river was carried away by floods and pounding seas which caused huge destruction along the coast. The Wexford train was also cancelled because of flooding
and coastal erosion. Floods swept through the streets, forcing families to seek fire
brigade assistance as their homes were flooded and old and young had to be
carried to places of safety. Power in much of the north city failed as lines and telephone poles came down, adding to the chaos. The worst hit areas of the city
were along the banks of the Tolka from Glasnevin Bridge to Fairview where the
flood waters raged above the arches and banks, swamping roads and nearby
properties. Trapped cars had to be pushed off the flooded roads by fire crews
responding to assistance calls.

There were long hours of pumping and bags were filled with sand or clay as
the firemen, knee deep in water, tried in vain to answer as many emergency calls
as possible. Often in chaotic situations fire appliances had to be pushed to shal-
lower places before they could be driven off to help somewhere else. The Dodder,
Poddle and Camac all caused flooding on the south side, requiring fire engines to

pump out flooded areas in Dr Steeven's and St James hospitals and to ferry the infirm and the elderly from their destroyed cottages along the Camac. Areas of the city recorded floods of up to four feet which frequently destroyed the humble homes of many of the poorest of the city. Seven fishing boats were sunk at Ringsend as the brigade attended over a hundred calls for assistance, hampered by the rising waters and sixty-five mph winds. The recently arrived Dennis fire appliances showed their worth in support of the greatly extended emergency crews working incessantly to try and bring succour to the hundreds of victims.

In 1955 there were three incidents involving the Dublin Fire Brigade which, for different reasons, merit recording. The first occurred on 12 April when two women and two children all from one family tragically lost their lives in a tenement house fire at 41 Cuffe Street. This old four-storey tenement premises housed fifteen persons, members of four families, as well as a clothing factory and shop. Sometime after 1.30 pm fire was noticed in the back parlour on the ground floor. The alarm was raised by a worker who ran upstairs to the top of the building to warn the owner who was tarring the roof. He claimed he alerted the family on the third floor, the occupants of the top floor luckily being away at the time of the outbreak. The other tenants all vacated the building on seeing smoke or hearing the cries of "fire". The factory girls were on their lunch hour. For some reason the unfortunate Long family did not rush out of the burning building.

Large numbers of passers-by had gathered outside as brigade sections arrived from Rathmines and Tara Street. An extension ladder was pitched to the second floor window where a woman and child were calling for help. Fireman Tom Coleman, a veteran of the Belfast blitz, mounted the ladder and was handed a young boy by the woman, who could not herself get out of the window because of an iron bar which divided the window space. The boy was handed down the

Cuffe Street Fire, 1955. Photograph, © *The Irish Press*

ladder and removed to hospital wrapped in the shawl of a neighbour. The three-year-old boy died the following day.

Fireman Coleman had to chop the bar and window frame before he could pull the hapless Mrs Long onto the ladder from where she was lowered to the ground. Removed to Vincent's Hospital she died some time later. The turntable ladder was then raised when neighbours shouted out that there were two more people still in the room above. When firemen with hose entered the smoke-filled premises they discovered the bodies of a young woman and her five-year-old daughter. Their bodies were removed from the building on the turntable ladder. While this drama was enacted in view of hundreds of spectators the fourteen year old daughter of the younger woman was being rescued by neighbours from a second-floor window at the rere. Much criticism was levelled at the fire brigade for the way this fire was dealt with and the attempted rescue of the family.

Evidence at the inquest on the victims was questioned by legal representatives for the family. It seems that when the turntable ladder was being extended a fireman proceeded to ascend it before the pawls were housed and had a leg trapped between the rounds. This led to the ladder being lowered while the injured fireman was being removed, causing some delay. The final evidence showed that this accident occurred after the boy and his grandmother had been removed from the window. Attempted rescues at times like these are not always successful in spite of the best endeavours of fire crews.

A more successful episode took place at a multi-factory premises in Exchequer Street on 29 September when seven people were rescued by turntable ladder from the third floor as a much fiercer fire than that encountered at Cuffe Street destroyed a lower floor of these premises. The predictions of the late chief officer, Major Comerford, regarding the terrible danger of fire in multi-storey tenements that also housed factories had been ignored with fatal consequences in Cuffe Street. It was to be repeated on several other fatal occasions before the massive clearance of tenements was undertaken by the corporation and the use of these treacherous buildings as factories was outlawed.

The second notable incident, in August, involved an extensive fire at Clonsast, County Laois. Six fire brigades, including the Dublin brigade, fought a fire which consumed two-mile-long turf ricks, threatened a forest and damaged acres of Bord na Mona bog. Dozens of tractors and trailers, hundreds of bog workers, FCA men, gardaí and civilians were all at one time or another involved in trying to save thousands of acres of stacked turf and bog and several thousand trees in nearby forests as heavy winds and lack of water allowed an out-of-control fire to spread destruction on the bog. The fire had been burning for five days when it suddenly got out of control and all hands were called upon to prevent a catastrophe. It had been an exceptionally dry summer and the country had witnessed an unusual number of bog fires.

The third incident concerned the prosecution of the owners of the Leinster Cinemas Ltd, Dolphin's Barn, for overcrowding and failing to keep the means of escape exits clear at the Tivoli Cinema, Francis Street. District Officer A. McDonald told the judge hearing the case that when he visited the premises he found overcrowding in the gangways to the stalls and a row of people standing in the intersecting gangway. The cinema manager was questioned by the corporation law agent:

*You have heard of Drumcollogher?*
*– I have indeed.*
*You know that the burning of people there took place because there was overcrowding?*
*– Yes.*

Imposing penalties the judge gave as his opinion: "the maximum penalty of £5 provided under this Act [1940 Fire Brigades Act] is completely out of date". The owner was fined £10 for two breaches of the then outdated Act. In spite of the judge's relevant comments the Act remained in force unamended until replaced by the Fire Services Act 1981.

The passing of the 1955 Factories Act put much extra work on an already overstretched fire service. Within a year the two senior officers of the brigade were being instructed to inspect plans of over 250 factories by the city architect, deal with over 200 factory certificates as well as keeping up-to-date with ongoing inspections of places of public resort, gaming and lottery premises, buildings to be inspected under petroleum regulations, inspections of corporation offices and other institutions etc. All this, while running an operational fire service, was bound to lead to the growth of serious problems somewhere.

As if to exacerbate the growing problems the government decided to re-establish the auxiliary fire service (AFS) in 1957 with responsibility for training this new AFS laid on the chief fire officers throughout the country. As a start, training

Central control room, Tara Street fire station, 1957. Film still © Gael Linn

courses for AFS instructors from all over the country were to be conducted at Tara Street fire station under the direct supervision of Captain Diskin. This emphasis on training for the AFS led to a serious reduction in brigade training at a time when the number of recruits annually began to be substantial.

The equipment upgrading continued over the next few years with two more Bedford ambulances, a Morrison trailer pump and a service van, and by the end of the decade radio telephones had been installed on twelve vehicles. The 1950s saw the abolition of the position of leading fireman and the end of the payment of fuel and light allowance to those who had still been in receipt of it since their removal from quarters in the fire stations.

Pay and conditions in the brigade at this time led to many within the ranks seeking and gaining transfers to better-paid employment in other departments of the corporation while the granting of a small wage increase to firemen and sub-officers in December 1957 led to some station officers seeking to revert to the sub-officer rank and very few of that rank being willing to seek promotion. It was not until April 1960 that finally the district officers hours of duty were reduced to seventy-two hours per week in line with junior staff and both district officer (D/O) and station officer (S/O) ranks received a wage increase to restore much of their differential on pay. As part of this agreement the D/Os and S/Os finally moved out of their fire station homes. The new duty hours for district officers led to an immediate reduction in the number of inspections of places of public resort during performance, a duty which up to then was mainly carried out during the period of the longer working week of that rank. The reduction in hours had not led to any increase in the staff and there had not been any change made in the operational duties of the district officers.

At the latter end of 1960 Captain Diskin went on sick leave, and he retired on medical grounds in January 1961. He was a native of Galway who graduated in civil engineering at University College Galway and joined the army at the commencement of "The Emergency". He was promoted to captain and served as O/C of the Curragh Camp fire brigade. For a brief period in 1945 he served as chief fire officer in Limerick before taking up the post of second officer in Dublin. He was responsible for equipping the Dublin Fire Brigade with the most up-to-date post-war appliances, a humane man who laid a lot less emphasis on military discipline than his predecessor but had lost control of the brigade before his early retirement.

The 1960 annual report lists 2,814 fire calls attended, 298 "special service" calls and 9,149 ambulance removals in a brigade costing £101,728 and manned by a chief officer, second officer, third officer, 7 district officers, 7 station officers, 10 sub-officers and 116 firemen (wholetime) with 15 retained personnel at Howth.

For a period after Captain Diskin's sudden retirement Third Officer Gorman was acting chief officer but former Second Officer B. Larkin who had left in 1960 to take up the post of chief fire officer in Louth returned as second officer and acting chief.

During this period many questions were asked in the corporation regarding the "adequacy of the fire brigade" and whether the disposition of units was the best to deal with the major fires in the widely dispersed areas of the city. A decision was made to replace Dorset Street station and to finally go ahead with the construction of the long overdue fire station at Dolphin's Barn.

When nineteen lives were lost in a night club fire in Bolton, more questions were put to the city manager on whether measures in place in Dublin were adequate for "guarding against fire hazards and their consequences in clubs, institutions, theatres or other buildings to which the public have a right of entry", to which be replied

> ... *places of public resort are subject to byelaws made by the Corporation under Section 55 of the Dublin Corporation Act 1890 ... There are powers under the Fire Brigade Act 1940 ... These powers are considered adequate and the fire department carry out regular inspections of such places.*

Acting Chief Officer Larkin then compiled a comprehensive report on the current state of the brigade and recommended major changes regarding the forward development of the service. His report proposed the closure of the old stations and the building of new ones, a new control room, the purchase of new replacement fire engines to a time scale and a substantial increase in the establishment. All of these matters had been highlighted by both of the previous chief officers but the most innovative section of his report dealt with the long-neglected area of fire prevention. He was the first to propose a special section of the brigade to deal with necessary progress in this area.

Dublin Fire Brigade officers with retained firemen from Irish county brigades on training course, 1958. (Front row, left to right: Paddy Gordon; Tom Kavanagh; Michael Rogers; Jarlath Diskin; Val Walsh; Bernard Larkin; Richard Gorman; Nicholas Bohan; Larry Carroll; Mick Delaney).

# 16 Fires, bombs, death and progress

In April 1961 the Local Appointments Commission interviewed candidates for the post of chief fire officer, but failed to appoint anyone. The salary for the job was increased from £1,700 to £2,030 in an attempt to attract more suitable candidates and further interviews took place in September. As a result, the position was offered to Thomas O'Brien, then chief fire officer of County Kilkenny. He had previously been in the army corps of engineering and had held several civilian engineering appointments in various parts of the country. He did not take up his new position until April 1962.

The year 1961 was an unusual one in the absence of a chief officer. For a period following the retirement of Captain Diskin in January, Third Officer Richard Gorman was acting chief officer and Senior District Officer Michael Rogers was acting second officer. Later that year Bernard Larkin (who was chief fire officer of County Louth) returned to the Dublin Fire Brigade as second officer and acting chief officer. During that time the corporation was once more seeking legislation to provide that insurance companies contribute substantially towards the upkeep of fire brigade services. As in the past, not surprisingly, the government refused to act on the matter.

For the first time a bell was provided on one of the new ambulances and the corporation announced to the press that it would "put bells on the others if the experiment is regarded as successful". Before then ambulance drivers depended on the car horn to warn pedestrians or other motorists of the oncoming ambulance.

The year 1961 was also a year when the city experienced some of its biggest fires in years. In February an explosion and fire caused extensive damage to the premises of Irish Pharmaceuticals of Mount Brown. The explosion took place as a lorry carrying methylated spirits was being unloaded, hurtling burning debris through windows and onto part of the roof. Terrified operatives rushed from the factory, leaving most of their personal belongings behind. When the fire engines arrived some firemen concentrated on releasing twenty stall-fed cattle from a nearby yard. Beams, rafters and much of the factory roof crashed down sending flames, smoke and debris out onto the roadway. For hours as the fire crews battled the flames a great pall of pungent smoke covered the entire district.

July saw one of the most potentially dangerous fires when a warehouse containing 4,000 hogsheads of maturing whiskey at the Powers Distillery was destroyed. It was lucky that this fire occurred during the "silent season" when the distillation plant is not fuelled and no liquor is actually being distilled. As the begrimed firemen fought the flames two local factories, John's Lane school and some tenants in Oliver Bond flats were evacuated.

Every now and then the huge crowd of onlookers cheered or scattered as caskets exploded, changing the hue of the flames and heightening the danger. Hundreds of thousands of gallons of water was poured into the fiercely rising flames as the firemen fought desperately to control the spread and prevent involvement of a bonded warehouse beneath the blaze.

The attending city engineer, E. J. Bourke, informed newspaper reporters that a terrible catastrophe had been averted: "a marvellous piece of work .... the best save we have made in this city in thirty years". John Power and Son Ltd took an advertisement in the National Papers conveying its thanks and appreciation "to the members of the Dublin Fire Brigade, the outstanding efficiency of the Fire Service and the City Engineer's Water Works Department". It was reported that the damage "was confined almost exclusively to one of the bonded warehouses containing whiskey (216,000 gallons) in casks and to the portion of the premises which contained empty casks only". The *Evening Herald* editorial went even further in its congratulations, drawing a parallel with the recent fatal whiskey fire in Glasgow where nineteen firemen were killed.

In August six men and one woman were rescued by fire brigade ladder when a huge smokey fire destroyed tons of files, office equipment and a printing plant at the CIE depot in Broadstone. Up to eighty buses had to be removed from the adjacent garage as the fire was prevented from spreading to the nearby petrol and diesel storage tanks. Most of the old departmental accounts buildings was destroyed before the fire was extinguished.

In September a huge tragedy was averted as the brigade responded to a call from Dublin Airport where a DC4 aircraft with sixty-nine passengers and a crew of four overshot the runway to crash onto the main Dublin-Belfast road, narrowly missing passing cars. Twenty passengers received minor injuries while three sustained serious injuries and were removed for treatment by brigade ambulances.

The Belfast blitz was recalled when a dust explosion and massive fire destroyed the Johnson, Mooney and O'Brien bakery in Russell Street. Tons of debris and a fallen wall blocked the North Wall railway line while masonry and twisted girders blocked the adjacent roads and the Royal Canal. Seven workers were removed to hospital suffering from burns, shock and abrasions as the premises was engulfed in flames within minutes of the explosion taking place. Very little could be done by the fire crews to save any part of the building but, watched by the usual mass of onlookers, this destructive fire was prevented from spreading.

Brian Larkin had not long been reappointed second officer when the brigade tackled a huge fire that destroyed the furniture and carpet warehouse of Holroyd and Jones of Princes Street. Such was the dense smoke pouring from the burning premises that the first fire crew, on arrival at the scene, could not make out which building was on fire. Huge volumes of thick black smoke enveloped the area as the linoleum department burned and a new and truly hazardous situation of hugely burning plastic materials was added to the ongoing risks facing firemen. A new deadly era of fire fighting had now arrived.

Thomas O'Brien became chief in April 1962, taking over a service with 146 full-time personnel in four fire stations that was expected to provide all emergency and ambulance cover for a population of 535,488 in Dublin city and 132,865 in Dublin county. Belfast then employed 198 personnel covering less than one-third of this number of people from six fire stations.

Soon after O'Brien took over the Dublin Fire Brigade he submitted a report on revised staffing and the purchase of extra equipment. With approval for expenditure of £34,000 a new turntable ladder, a fire engine and a "Gypsy" Land Rover for chimney fires were ordered. Approval was also given for spending of £15,000 on the demolition of the top storey of the condemned former married quarters in Townsend Street and the carrying out of alterations to convert the rest of the building to a lecture hall, toilet facilities, a projection room and ancillary rooms for storage and repair of breathing apparatus. Although the height of the building was lowered the other essential alterations were not completed. This was a serious omission; the fire brigade, in an age of rapid scientific advance, needed to be backed by heightened technical skills, efficiency, organisation and ongoing training, and desperately needed the lecture hall facilities.

A complete review of the fire service was urgently needed at this crucial time and the city manager was assuring the elected councillors that it would be undertaken by the new chief officer. He agreed that

> the last general re-organisation of the service had been in 1938 and while maintaining that the Brigade could deal with any contingency agreed that in the changed conditions which now exist a complete review of the service was deemed necessary.

In December Dublin Corporation adopted the chief officer's report. New fire stations would be built in better locations to provide fire cover. Unfortunately the issue of a training centre would be ignored for many years. At this December meeting the city manager also accepted that "the 1958 Office Premises Act is not being or capable of being fully complied with in every section of the Corporation". The fire brigade had responsibility under Section 18 of this Act to see that all offices were immediately openable from the inside, that means of escape were kept clear of obstruction, that all emergency exits were clearly marked and that all staff were made familiar with fire drills.

With retirements or transfers to more lucrative positions in the corporation, there had been a growing intake of new fire brigade staff since 1956. When the numbers in the service were increased in 1963 in line with the chief's revised staffing requirements some 70% of the 160 firemen had less than eight years service. Those of the old rigid disciplined regime of Major Comerford, with vast fire fighting experience because of the long hours worked, were quickly becoming a tiny minority. There was an absolute need in these circumstances to initiate a comprehensive programme of ongoing training and education in the brigade with proper quarters and facilities to advance this essential work. Throughout the UK training centres had become an essential feature of fire brigade organisation. Under the British Fire Services Act 1947 a duty was placed on each fire authority to secure the efficient training of fire fighters and the fifteen major brigades in Britain all had training centres by this time. As well as these, Britain also had a national training centre updated in 1963 and renamed the Fire Service

Technical College at Moreton-in-Marsh. There was also a fire service college for officer training in Surrey.

Training in Dublin was still conducted as it had been in Captain Purcell's time, with regular drill on station dependent on the frequency of fire or other emergency calls. With a fifty-six-hour week due to commence in 1964 it was imperative that updated recruit training was introduced along with continuation, specialist, promotion and officer training. But only recruit training was updated.

Since Major Comerford's resignation recruit training had been conducted in the overcrowded conditions of Tara Street station. Drills or lectures were conducted by officers who remained available on duty at the time. No written copies of the substance of these drills or lectures were provided and the class was regularly disrupted when either the officer conducting the class or the fire appliance being used had to respond to an incident. Some officers with a genuine interest in training strove to pass on as much knowledge as possible to the eager recruits and it was due to them alone that those who passed through this inadequate system learned how to operate equipment and picked up the first rudiments of fire fighting. All other training was garnered on the fire ground from more experienced firemen or from the oft-interrupted drills on the fire station. The first batch of recruits to receive a formal six-week training course under officers seconded from operational duty were those trained for the introduction of the fifty-six-hour week, in the newly opened fire station at Dolphins Barn under the recently promoted Third Officer Larry Carroll.

Under Captain O'Brien's reorganisation the district officers were all recalled to headquarters and removed from fire prevention work, and the district stations were manned by station officers. A separate fire prevention section was put in place to be operated by two engineers, a seconded district officer and a clerk typist. Captain John Williams BE and Tom Ruddy BE became the first fire brigade officers recruited specifically for fire prevention work.

This approach of regarding fire extinction and fire prevention as two different problems rather than two aspects of one problem was a new and questionable approach. In 1950 a committee set up by the Minister for Local Government had produced a code of standards as a guide to fire officers in carrying out their duties of fire prevention. The code was finally issued in a revised format in a handbook published in 1967. It set out minimum requirements regarding structure, general fire precautions and internal fire protection, under the headings a) how to eliminate the usual cause of fire, b) how to limit the spread of an outbreak should it occur, c) what facilities are necessary for the prompt and efficient extinguishing of fire, and (d) how buildings can be evacuated promptly and with the minimum danger to life from fire, smoke, panic or structural collapse. This handbook reiterated that all assembly halls should in all respects conform with the byelaws relating to "Places of Public Resort – Protection from Fire" made by Dublin Corporation in 1934. These were the protection standards already referred to which contained the statement that "panic rather than fire is more likely to result in loss of life". Those in charge of these places of public resort had a responsibility "to impress on all occupants the necessity for orderly evacuation and for calmness, even in the dark in the event of an emergency". The handbook became the bible for the new fire prevention officers but was largely ignored by the owners of places of public resort.

The brigade took delivery of three new ambulances in 1963. These vehicles were built on special chassis with twin stretchers one of which was fitted onto Lomas gear which allowed the stretcher to roll out on an extended frame for easy removal or loading. Cream in colour with red flashings the ambulances were provided by McCairns Motors Ltd. With chimney fire fighting equipment removed and extra locker space they carried instead resuscitation equipment and a wider range of blankets and first-aid items.

Administering the Dublin fire service meant a huge increase in responsibility from being chief in Kilkenny and Tom O'Brien must have wondered very early on what other major problem he would be faced with. In the first two years of his stewardship he was faced with serious and growing industrial relations problems, wanton attacks on his fire crews, major fires in Townsend Street, Mary Street Furnishings, Bective Rugby Club, a ship fire at Ocean Pier and a massive blaze at the Fiat Motor Plant on Upper Mayor Street where three firemen had a miraculous escape from death. In this incident an exploding acetylene cylinder in a smoke-filled and flame-filled paint store hurled itself with massive force towards the door where the three firemen were working. Their lives were spared when the cylinder struck a steel girder in front of them but the force of the blast threw the men against a wall on the far side of the road. All three were removed to and detained in hospital with their injuries.

On 3 June 1963 the brigade had to respond to the collapse of two three-storey tenements which killed an elderly couple and injured seven others at Bolton Street. Here and at Fenian Street on 12 June firemen risked serious injury working in clouds of choking dust amid tons of fallen masonry in a vain search for survivors. At Fenian Street, where two young girls died in similar circumstances, firemen searched inside damaged and treacherous old buildings for trapped victims as timbers and slates came down on them. Having successfully dealt with these two tragic incidents the fire brigade was then forced by a panicking corporation to allow tenants into the former married quarters in both Dorset Street and Buckingham Street Fire stations. These tenancies were supposed to be temporary but, years later when the firemen vacated these stations, moving on to new premises, tenants were still ensconced in the old fire stations.

The murder trial of Shan Mohangi, a South African medical student charged with the killing of sixteen year old Hazel Mullen, made huge headlines for weeks. It also created staff shortage problems in the brigade owing to the fact that two fire engine crews were called as witnesses. The firemen had been called to the Green Tureen restaurant in Harcourt Street on 7 August 1963 when passers-by saw smoke issuing from the basement. The crews were turned away from the premises by an unidentified foreigner who showed them some smoking rags which he had been burning. Unknown to the firemen the accused was actually dismembering and burning the body and the clothing when they responded to the fire call.

In 1963 agreement was reached between Dublin Corporation and the two unions representing the firemen for a reduction in the working week from seventy-two hours to fifty-six hours and the abolition of twenty-four hour shifts. The proposed new shorter working week would operate on the basis of nine-hour day shifts and fifteen-hour night shifts commencing from 1 July 1964. However, needing to recruit and train fifty-nine recruits as well as promoting twelve new officers caused a delay and the new system did not commence until the autumn of 1964. The chief reported 2,605 fires attended by the brigade in the year 1963-64 with the cost of providing the service rising to £233,500 by the end of the year.

Some remarkable public servants have served in the Dublin Fire Brigade, most of whose exploits remain unrecorded. One who deserves to be remembered is Richard Gorman who retired in 1964. He joined the Pembroke Fire Brigade in 1918 and served throughout the war of independence and the civil war. Following the merger of Pembroke with Dublin he served as an officer, becoming the first third officer in the Dublin Fire Brigade. In 1941 he travelled to Belfast with the Dublin contingent on the night of the first blitz. Richard Gorman maintained a tremendous rapport with all those who served with him and at a time of crisis in the brigade he took on the role of chief fire officer for some months. In that period he showed outstanding skill as an operational officer at many spectacular fires. He won the respect of all personnel and retired in relatively good health after a lengthy, stress-filled and successful career.

By contrast with such stories of successful longevity which form part of the history of the brigade, there are stories of men killed on duty with many years left to serve. A year after Richard Gorman's retirement, in September 1965, young fireman John Finbar Hearns was struck by a car at Finglas, having responded to what turned out to be a bogus call. Tragically, he died ten days later in St Laurence's Hospital.

The full-year figure for bogus calls in 1965 was 841 recorded. Dick Hearns, while suffering the tragic loss of his young son, issued a statement outlining how ten bogus calls were received by the brigade on the night of his son's fatal accident and appealing to those responsible

> to end their fiendish game .... in God's name, stop making these calls before other tragedies occur. My son has made the supreme sacrifice in the performance of his duties for the people of Dublin. If his premature demise was instrumental in the elimination of bogus calls then his mother and I would find a little consolation in the knowledge that his sacrifice was not in vain.

Progress for his surviving colleagues continued as Dublin Corporation was found to be in breach of the Holidays (Employees) Act at the high court following a test case taken by the author Tom Geraghty in 1965. As a result of this decision firemen were paid an extra day's pay for each Bank Holiday worked. Negotiations for a further shortening of the working week resulted in agreement to bring the brigade into line with the major British fire brigades then on a forty-eight hour week. With the employment of forty-five extra men, the new system came into operation on 14 November 1965. The month before, Dublin Corporation earmarked sites at Swan's Nest in Kilbarrack and Mellowes Road in Finglas for the erection of new fire stations while progress was being made on a Phibsborough replacement site for Dorset Street station.

It seems to be in the nature of the fire service that progress is always matched by mounting problems for the firemen. The year 1965 outstripped even the frenetic year of 1961 with a spate of massive fires in the city.

More than forty people found themselves out of work when an arson attack badly damaged the Rehabilitation Institute's production centre at Basin Lane in January. Some days later, wearing their newly issued short waterproof jackets with luminous bands and overtrousers which had replaced the long Liverpool overcoats the firemen fought a disastrous blaze at Hanna's two-storey factory in Finglas which put 110 employees out of work. Later the same month fire swept through the top floor laboratory at University College Dublin in Earlsfort Terrace.

The destruction continued when in May the five-storey premises of Thomas Mason and Sons, photographic wholesalers and opticians, was consumed in flames. It was a prominent business premises established in 1780 but the firemen could not prevent the destruction once the photographic and laboratory equipment became involved. As thousands of parishioners looked on, the Dublin Fire Brigade assisted their Dun Laoghaire colleagues at the spectacular fire which destroyed St Michael's Church in September. Flames roared along the pitch-pine timbers of the roof and up the 130 ft spire as the landmark for boats approaching Dublin Bay for all of 140 years was consumed by the fire. Water pumped up from the harbour helped to contain the fire to the church building. On the following morning the area was a four-walled shell surrounding smouldering debris. Fortunately the tower and spire remained standing.

But it was in October 1965 that the greatest challenge confronted the Dublin firemen when the CIE goods depot at North Wall went on fire. This one hundred year old building exploded in a mass of filthy acrid smoke and flying sparks as strong winds fanned the roaring flames in the 45,000 square feet enclosure. Train carriages loaded with explosives were shunted from the blazing building as up to thirty of them caught fire. As the explosives train was shunted down the tracks the Westport train was pulled from the sheds with carriages blazing. Thousands watched as the two Port and Docks Board fire fighting tugs joined the brigade in the battle to prevent the fire from spreading. All down the stone-settled streets of the north docks glowing remains of highly inflammable materials drifted to earth like burned-out rockets and the night sky became alive with a fury of intense heat as hardened dockers with pints in their hands stood outside the local pubs calmly watching the ever-developing spectacle. Caretakers and security men were called in to check their premises as exhausted firemen gathered whatever adrenaline they could muster to make a final assault on the blazing enemy. News reporters might wax eloquent with comparisons to Vesuvius or the burning of Atlanta in *Gone with the Wind*, but the working fire crews, singed and blackened, tired and emotional, still pulled heavy water-filled hose into outrageously dangerous positions to get at the heart of the flaming monster. Always in

Fire in CIE depot, 1965 – the biggest fire in the city during that decade. Photograph,
© Independent Newspapers Ltd, Dublin. Collection Dublin Fire Brigade Historical Society

imminent danger, the crews edged forward, finally catching sight of other crew members through the sodden grey smoke and knowing then that the terrible genie was at last safely back in his box. All hands were on duty; there was no relief or respite until the job was done.

Walls collapsed and huge buckled steel girders lined with rows of steel rods were a spectacle of eerie danger above the smoking mass of debris and chaos that a few hours before had housed organisation and mundane routine. After that night the young firemen believed they could handle anything that would confront them in the future.

On 30 November Second Officer Larkin chaired a meeting in Tara Street station to establish an Irish Branch of the Institution of Fire Engineers. The meeting was well attended and was to prove the basis for the present thriving organisation of the IFE in the Irish Republic. Visiting senior representatives from England had an opportunity to see the Dublin firemen in action when the extensive premises of Norton's china stores and warehouse in Moore Street caught fire that night. The blaze spread through all three storeys causing the roof to collapse and hurtle sparks into O'Connell Street. The fire spread rapidly because of the hundreds of crates of delphware wrapped in straw and woodwool. A strange year ended with the destruction on Christmas morning of the premises of James North and Sons in Dartry, the company providing the new fire fighting gear for the firemen. This was another of the growing infernos in the city involving large amounts of plastics. The four-storey premises were almost entirely destroyed.

On 8 March 1966 at 1.30 am an ear-shattering explosion echoed through the city and the responding firemen were met with hundreds of tons of granite in O'Connell Street and half of Nelson's Pillar gone. No members of the public were injured and the admiral's head was rescued. It is now on view in the Civic Museum.

During 1968 the sites were purchased by Dublin Corporation at Mellowes Road in Finglas and Swan's Nest in Kilbarrack for the construction of new fire stations and a replacement site for Buckingham Street was procured at North Strand. Priority was given to the two new stations because Dublin Corporation had entered into an agreement with Dublin County Council pursuant to Section 59 of the Local Government Act 1955 "whereby the Corporation would exercise the powers, functions and duties of Dublin County in relation to fire services". A loan of £140,000 was agreed and allocated to the construction of the two stations.

Since 1963 the majority of firemen were pursuing a claim for a pay scale that would be related to some group of workers outside of the corporation. In 1964 strike notice had been served on Dublin Corporation by the Workers Union of Ireland following the rejection of their claim for parity with the gardaí. Following a labour court hearing a special pay increase was awarded to the firemen but the court failed to accede to the parity claim. Again in 1968 the question of a pay relationship agreement was pursued. This time both the Workers Union of Ireland and the Irish Municipal Workers Union supported the claim. When it was rejected the firemen went on strike.

The strike lasted from 5 August to 22 August when a labour court settlement was agreed. The basis of the agreement was that

> ... the Corporation should concede percentage increases on the pay scales thus adjusted equal to whatever increases are granted to the Garda Síochána arising from their present pay claim.

When all gardaí ranks were granted a 9% increase with effect from 1 June 1968 the maximum of the firemen's pay scale was brought on parity with the new maximum of the gardaí scale thus establishing a fixed pay relationship at the maximum of both scales. It was an agreement that still exists into the new millennium.

During the period of the strike the army provided fire cover in the city under the direction of senior officers of Dublin Fire Brigade. The fire brigade control was operated by these officers who kept in contact with army crews on Green Goddesses (civil defence fire engines) using despatch riders who carried short-wave radios. The army operated from its local barracks. A number of large fires took place during the dispute but potentially the greatest disaster occurred on the last night of the strike when the three-storey Valbern Clothing Company at Capel Street went on fire. Luckily, with a strike settlement imminent, firemen volunteered to immediately return to work in Tara Street in order to man one engine and the turntable ladders. All three appliances were involved in stopping this dangerous fire from engulfing the whole block of premises adjacent. The army and fire crews confined the blaze to the top floor in spite of the total collapse of the roof.

The dispute marked the coming together of the Dublin and Dun Laoghaire firemen to fight for their right to have a direct pay relationship with a body outside of Dublin Corporation. This unity in the strike, supported by the retained firemen in County Dublin, led to a complete victory and changed pay rates in the fire service nationally in the interest of all firemen. The last word on the bitterly fought dispute appeared in a thank you note sent to all the newspapers from Matthew Macken, the city manager.

> *I wish to express sincere appreciation of the wonderful assistance which the Army gave to the Corporation in the maintenance of essential services during this difficult period. I want to say publicly, through you, thank you to all the Officers and men who came to our assistance in the operation of the Sanitary Services [also on strike], Fire Brigade Services and Ambulance Services.*

Arrangements to implement the integration agreement of the fire services of Dublin city and county were agreed in September. Up to then the Dublin Fire Brigade had provided a limited service in portions of the south and west of the county. The new agreement entailed the building of the two additional fire stations, at Finglas and Kilbarrack, and a third at a south county location, on a site to be determined. Within a few years Nutgrove station was built. The total annual cost of the combined service was fixed at £487,000 to be divided on a basis of proportion to gross valuation of all buildings in the two areas. This resulted in Dublin Corporation paying £393,000 and Dublin County Council £94,000 in the first year. The full implementation of the agreement was to be completed over three years. The corporation decided that the Howth retained station would be replaced by the new full-time station at Kilbarrack.

On 5 March three sections of the brigade tackled a major fire at a partially completed block of flats in the new Ballymun housing complex. The outbreak in one of the eight-storey blocks on Shangan Road caused considerable damage to three flats being used by the builders to store pitch, paint, timber and insulation board. It was to be the first of many deadly fires in this newly developed high-rise flats complex on the north side of the city.

In conjunction with the new expanded responsibility of the Dublin fire service the brigade was provided with a new control room. The old watch room, in use since 1907, had long been inadequate for day-to-day operations. Work on converting the old harness room and spare engine house had been ongoing since 1964 but it was not until 1968 that it was fully equipped with three separate consoles for receiving emergency calls and a separate master control console for use by the mobilising officer when required. The new area was more spacious and away from the ever increasing noise of the traffic flow outside the station. It included automatic fire alarm systems from many high risk sites such as the docks, power stations, major city stores, the art galleries and public buildings and had tape recording equipment.

The decade could not pass out quietly for the brigade. Before the ink was dry on the agreement between Dublin Corporation and Dublin County Council a group of county councillors tried to have it nullified. They succeeded in delaying its implementation and, reading the report on their reasons for this change of heart, it is hard to believe they were motivated by concern for the people living in the county. Their opposition was helped by the minister's decision to abolish Dublin Corporation in April 1969 when the councillors failed to strike a rate for the city. This time John Garvin was appointed as commissioner and the brigade was back to the 1920s type administration.

The year 1969 also turned out to be another year of damaging fires, mounting numbers of road traffic deaths and lives lost in domestic dwellings. The firemen provided the guard of honour at the funeral of (young) Jim Larkin on a cold February morning. As General Secretary of the Workers Union of Ireland he had presented the firemen's pay claim at the labour court the previous year. A short time before his untimely death the firemen were involved in the rescue of over a hundred psychiatric patients when a two-storey block in St Brendan's Hospital, Grangegorman, was destroyed by fire. As the fire was tackled the patients were led or carried out to the lawn by vigilant staff from where they were transferred by a fleet of ambulances to other parts of the hospital. During the early stages of the fire fighting, brigade crews were hampered by a shortage of water but the fire was confined to two units and prevented from spreading to other sections of the affected wing.

On 10 February a massive explosion ripped through the corporation sewage pumping station at Ringsend killing one man, injuring another and sending the 120 foot red-brick chimney crashing to the ground. The first explosion blasted a half-ton iron man-hole cover into the air and seconds later a flash explosion ripped the building apart taking down the chimney with it. With the threat of a further explosion still possible firemen entered the smoke filled rubble to rescue the trapped men. In this work Station Officer Smart was injured. One of those taken from the rubble died in hospital. At the inquest the engineer-in-charge of the pump house maintained the explosion was caused by Light Virgin Naptha, commonly called "distillate", which was accidentally discharged earlier into the sewers at the gas company premises and ignited by a spark at the sewer station.

Two hundred boys were evacuated from Artane Industrial School on 15 February when a huge fire swept through the roof of the recreation hall and theatre. Before the fire crews extinguished the blaze hundreds of chairs, band equipment and instruments were consumed, the roof collapsed and the massive two-storey section of the old granite building was destroyed. On 7 March the new £2.5 million factory of Irish Paper Products on the Naas Road disappeared in flames along with 40,000 tons of paper. Twisted girders, a crumpled roof and partially felled walls running up to the office section was all that was left surrounding the devastated 200,000 square feet interior area. It was the biggest and fastest spreading fire as well as the most costly ever seen in Dublin.

With other major fires destroying the Raheny Inn, Insulations Ltd in Ranelagh, McGrath Brothers tea wholesalers of Bachelors Walk, Shine Brothers Bacon Factory on Crumlin Road and Gings Costumiers in Exchange Court, insurers faced massive losses for the year. It was claimed that 1969 was the costliest year for fire claims, loss of production and employment as the crippling bill for damages climbed steadily into millions of pounds. Modern factories tend to be built with light roofs and exposed metal surfaces which buckle like cardboard in a fire – there are generally no firebreaks, so once a large fire occurs there is total loss. Water supplies were inadequate at the time for major fire fighting and few premises installed sprinkler systems or employed a safety officer. It was clear that a new approach to improving fire prevention was absolutely necessary. In Britain a major effort was being made to increase awareness of waste from fire. In Ireland very little was being done in spite of the disruption to business, loss of employment, loss of production and fall in exports that were such a burden on our economy. In 1969 Dublin Fire Brigade got one extra fire prevention officer and an extra third officer who would have some fire prevention duties.

Seven elderly women were rescued from the Protestant Retreat at Drumcondra in November 1969 when a fire broke out in a room on the second floor of the three-storey building and trapped them in the smoke. The hall door was forced by a garda officer and some civilians but because of the dense smoke and flames they were unable to get all the occupants from the building until the fire crews arrived. Most of the people rescued were elderly, feeble and deaf but all survived their ordeal and were discharged from hospital on the same day. The fire was confined to one room.

In December when a stolen car which was parked with its headlights on near the Central Detective Unit in Ship Street blew up, the incident was reported as an explosion. When the fire brigade had dealt with the fire that followed, gardaí discovered parts of a clock under the front seat. Army explosives experts were called. The incident led to the issuing on 23 December of a major accident plan for the Dublin area. This plan set out

> a procedure to be adopted for dealing with an incident in which seriously injured casu-
> alties of the order of twenty-five or more may be expected. The responsibility for
> putting the plan into operation rests with the senior garda officer on duty at Dublin
> Castle.

Fire brigade responsibility in the event of the major accident plan becoming operational was to 1) send out whatever was considered necessary to deal with the incident, 2) to inform brigade senior officers, 3) inform all the major accident hospitals, 4) inform the health board ambulance service, 5) inform the Stillorgan private ambulance services. Brief as the document was it conveyed a sense of problems to come.

### The 1970s

With the new city and county fire fighting service finally in place definite moves were initiated for the construction of the two new fire stations in the northern suburbs. An additional sum of £110,000 had to be borrowed before the work on the stations could commence and, when tenders of £127,820 for Kilbarrack and £97,632 for Finglas were approved, building started in 1971. By the end of 1970 the number of fire calls had increased to 7,257. Ambulance calls were 14,970 and about 2,000 of these calls were false alarms. At the same time the city and county engineer had to admit that water pressure in the county was kept low but was still adequate although water was cut off in certain areas from 9 pm to 5.30 am as a conservation measure. The brigade staff whose duty it was to deal with all of this consisted of 64 officers of various grades and 252 firemen. Only eight of these had fire prevention responsibilities.

In 1969 the Northern Ireland government had brought in a new Fire Services Act to consolidate the 1947 and 1963 Acts. This Act laid a duty on the fire authorities to provide for "the efficient training of firemen" and established a fire precautions section dealing with "potentially dangerous buildings". In Britain, parliament enacted the Fire Precautions Bill 1971 which made it compulsory for a whole range of premises to have a "fire certificate" and provided "power to impose requirements as to the provision of means of escape from buildings in case of fire and means for securing that such means of escape can be safely and effectively used at all material times". The Holroyd report in 1970 contained the following statement.

> Fire prevention implies measures to reduce the risk of fire breaking out while fire protection denotes measures other than fire fighting operations designed to protect life/and or property when a fire occurs.

The new Fire Precautions Act in Britain was an important step towards marrying these two safety responses. A major new emphasis on the growing need for enforced fire precautions in the greatly expanding Dublin area was obviously necessary to save lives but it could not be enforced without new legislation and the use of men with practical fire fighting experience to assess the necessary fire prevention provisions for particular premises. However, even with the looming political chaos in Northern Ireland, which would obviously affect Dublin, no new measures were put in place to help the firemen deal with the situation.

As a harbinger of things to come a gas explosion demolished two houses and damaged two others at Finn Street off Oxmantown Road in January 1970. Many people had lucky escapes. The fire crews searched the wrecked buildings for victims but, despite the severity of the blast, no one was killed or seriously injured. Another gas explosion injured five people when a shoe shop with upstairs accommodation at Glasnevin Hill was wrecked and set on fire in March 1970. A huge fire and explosion on 9 April burnt out the famous landmark three-storey building of McKenzie's Hardware Stores in Pearse Street, directly opposite Tara Street fire station. When the bells went down at 5.15 am for the turn-out to this massive blaze firemen rising from their dormitory beds thought at first that the station was on fire. The fierce heat cracked windows in both the fire station and the Cork Examiner offices on the opposite corner of Tara Street. At the height of the blaze explosions sent red hot rubble hurtling through the air, crashing on parked cars nearby or narrowly missing busy firemen.

All south city traffic was disrupted. The 20,000 square foot premises had stood on the spot since 1860 but, under the pressure of the explosion, the fire and internal collapses, the extensive front wall suddenly moved out from the abutting walls sending firemen fleeing for their lives before it crashed down in a crumbled mass exploding into chunks of flying masonry as it smashed into the roadway below. Tons of concrete, bricks and slates hit the spot which the second brigade turn-out had sped through, just seconds before, on its way to an extension of the blaze into the University Press at Trinity College. All sections of the brigade were involved in fighting this massive fire and retained sections from Howth and Skerries were called into city stations to provide cover in the event of another fire call. Much more damage might have been caused had the Queen's Theatre been still standing adjacent. However that famous home of tradition and thespian fervor had been demolished some months earlier. As pedestrians gingerly made their way to work through the masonry-strewn streets, weary firemen stood by new lines of hose laid out after the collapse of the building, viewing the scene of battle and marvelling at how lucky they were to be alive to hear the emerging sounds of the city once again coming to life.

The month of April was not yet over before a huge explosion and fire took place at the Alexandra Road oil jetty on board a barge loaded with oil and containers of gas. The area was transformed into a mass of flames as burning oil floated on the water and an expanding pall of deadly black smoke engulfed the oil depot. Fearing that the roaring flames would spread to adjacent storage facilities and shipping, tug boats towed away two oil tankers and the Dublin firemen disappeared into the thick fog of belching smoke intent on surrounding and extinguishing the treacherous blaze. Luckily the crew of the barge had all jumped into a nearby launch and headed down river before the flames had time to cause the explosion. The blackened oil-storage tanks showed just how close the port area was to a major disaster but the combination of the city and port firemen assisted by Irish Shell workers had again triumphed over the deadly enemy.

Following three huge fires in June – at McMahon's timber yard, Holroyd and Jones wholesalers and the diesel stores of Casin Air Transport – newspapers speculated on possible involvement by the Ulster Volunteer Force. A hue-and-cry went up and there was widespread activity after a brown Hillman car, displaying Northern Ireland plates and containing three or four men, was seen driving away from one of the fire areas. However, no such car was apprehended by the gardaí either in the north city suburbs or in the border region. As more premises were burned down gardaí and army experts probed the debris for clues but rejected the notion of UVF involvement. Later on, as some youths were indeed charged in connection with one or two of the outbreaks, Second Officer Larkin issued a press statement that in his opinion the recent large fires were coincidental: "just a run of bad luck". Whatever the causes, the firemen had again demonstrated their abundant courage and versatility, and thir capacity to successfully tackle any major blaze in the city and extinguish it in the shortest possible time.

When British paratroopers shot and killed thirteen unarmed civilians during a civil rights march in Derry on 30 January 1972 a new reality was shaped and it would take many years and thousands of lives before sanity would again determine political policy on the island. Fire brigades would not be exempt from the overall confrontation. On Wednesday 2 February 1972 the first major effect of

Derry occurred when the Dublin Fire Brigade tried unsuccessfully to tackle a fire in the British Embassy in Merrion Square. The fire crews attempted to deal with the growing blaze while crowds of rioters broke through police cordons and flung petrol bombs at the building. When gardaí could no longer guarantee the safety of the firemen, they were forced to withdraw and look on helplessly while the building was consumed in flames.

Nine days later the new Kilbarrack fire station opened and a decision was made to close Howth retained station. On 12 April Finglas station also opened and the first element of the new expanded brigade was in place. Just before 3 pm on 27 March the brigade responded to a fire call at Noyeks of Parnell Street. Within minutes a "district call" was sent back to headquarters and a further pump, two ambulances, the emergency tender and a senior officer joined the two pumps and turntable ladder already at the awful scene. The entire three-storey building was engulfed in flames, the corrugated timber-sheeted roof had collapsed and gusts of flames swept right across the street stopping all traffic flow and adding to the chaotic spectacle. Soon the top two floors would crash to ground floor level and parts of the supporting walls would spread their debris around the stricken site.

Between forty and fifty people were in the building when a quart-size tin of Bostic adhesive fell from a shelf in the ground floor shop and burst open on the floor. The central heating system in the premises was out-of-order at the time and gas fires had been substituted. Fumes from the highly inflammable petroleum mixture burst into flames. The quantity stored in the premises was well in excess of the minimum quantity of three gallons for which a licence under the Petroleum Acts of 1871 and 1879 was required. Such a licence had not been applied for however. The flames spread rapidly and, as those in the shop ran for safety outside, one brave female member of the staff rushed upstairs to alert her work mates. She was to survive the holocaust that followed but sustained a serious injury to her back when she had to jump from an upstairs window.

Some staff were rescued by climbing along a ladder extended from a shop opposite, others got out onto the roof of a van parked alongside the building while others jumped from upstairs windows. Seven women and one man perished in the flames. Even though the fire crews had the fire under control within forty minutes from receipt of the initial call, the extensive destruction was testimony to how ferocious the conflagration had been. With the assistance of staff from the Dangerous Buildings Section of Dublin Corporation, the fire crews set about the unenviable task of searching the smouldering debris for bodies. By 5.30 pm eight bodies had been recovered and twelve other staff members had been removed to hospitals; five were detained suffering from burns, injuries or shock.

Over the following weeks the newspapers supported calls for an inquiry into all aspects of the tragic fire. The jury at the inquest into the deaths heard twenty-three witnesses including the state pathologist who stated that all victims had died from carbon monoxide poisoning. The jury called for a complete investigation of the fire and revision of the 1940 Fire Brigades Act. They were supported

Fire at Noyek's, Parnell Street, 1972 in which eight people lost their lives.
Photograph, © Independent Newspapers Ltd, Dublin. Collection Trevor Whitehead

by the Chief Fire Officers Association and by the Workers Union of Ireland. Following months of questioning in the Dáil, the Minister for Local Government finally set up a working party on the fire service in November. In the meantime Chief Officer Tom O'Brien furnished a detailed report on the fire to the city manager. With the abolition of the city council however, citizens had no means of pressing for assurances regarding inspections of potentially dangerous buildings. Within days of the fire the Workers Union of Ireland had produced a report criticising the lack of equipment and the inadequacy of the Dublin Fire Brigade compared to brigades in UK cities of similar size and population. The report urged the setting up of a regional fire service, a national training centre, a national fire advisory board, standardisation of drills and enactment of a fire precautions act. When the minister appointed his working party there were no Dublin fire officers or members of the Workers Union of Ireland on board.

On 20 April Dublin experienced the first widespread placing of incendiary devices in major department stores in the city. The brigade dealt with a blaze in Dunnes Stores in North Earl Street and saved most of the building – in the process firemen discovered acid-balloon cigarette packet devices. Following immediate warnings to keyholders more of these devices were found and minor damage was caused in Roches Stores and Woolworths. In September when the chief's report on the Noyeks fire was finally released to the press the minister was outlining powers he would have under the Dangerous Substances Act 1972 which was to become law in 1973.

On 9 September Noyeks new timber storage depot at Santry went up in flames. Some 100,000 tons of timber in high stacks was involved as firemen fought to extinguish the fire. Much local criticism had been directed at Noyeks for locating this huge amount of timber so close to a large residential area.

A few weeks later Dublin Fire Brigade took delivery of a new £20,000 Merryweather turntable ladder. Built on an AEC chassis with built-in pump and one hundred foot extension capability it became the brigade's third turntable ladder and eventually replaced the Metz.

The troubles in Northern Ireland again involved the brigade when on 29 October a gelignite bomb was defused as crews stood by in Connolly railway station and three hotels were evacuated after incendiary devices caused small fires in upper-floor bedrooms. The first Dublin casualties of the troubles involving the fire brigade happened on 26 November when a small bomb at the side door of the Film Centre near O'Connell bridge exploded, injuring innocent picture goers. It was to be followed on 1 December by two car bombs which killed two CIE workers in Sackville Place and injured over a hundred others. Fire crews dealt with car fires started by the explosions and removed many of the wounded to hospital.

Dublin firemen had received no briefing on how to deal with this barbaric development so they treated the incidents as routine. Liberty Hall, badly damaged by the first Sackville Place explosion, was searched for victims as were buildings on the far side of the quays. Ambulances and fire engines drove right to the scene of the explosions not knowing whether other bombs would go off. When the bombs exploded many of the business fire alarms in the control room in Tara Street were activated and control staff had a lot of checking to do as crews dealt with the real life and death incidents outside.

Firemen everywhere deal with horrific incidents. It was a time when brigade

The aftermath of car-bomb explosions at Sackville Place, 1972. Photograph, © *The Irish Times*

colleagues in Northern Ireland were confronted with this type of slaughter on a regular basis as the cynical scoundrels who perpetrated these horrors continued on with their thwarted ambitions. The quaysides littered with glass blown from Liberty Hall, wrecked cars, debris strewn across the streets, fires burning unchecked – this was not what mainly bothered the fire crews at the time they responded to the call. It was the indelible sight of gloomy dust-filled streets, the sickly smell and, above all, the human chaos. The cries of the wounded, the helplessness of the shocked, the panic of those lucky to have escaped injury. The men arriving in their fire appliances and ambulances were fully aware that, somehow, they had to restore some semblance of normality as rapidly as possible.

It is said that the difference between bravery and courage is that bravery is having no fear, while courage is having fear and still doing the work to be done. The courage of the firefighters, together with the amazing resilience of survivors and bystanders, meant that the well-nigh impossible was achieved. The dead and wounded were successfully removed to hospital in the shortest possible time. For the next two and a half hours incident followed incident throughout the city centre as hotels were evacuated, suspect cars detonated, arrests made and false bomb scares investigated.

The city's major accident plan worked well in almost all circumstances during these incidents but a major review was now essential to further co-ordinate all of the front-line emergency services. The plan was updated the following year but the Dublin Fire Brigade role was not changed.

The old wheel escapes then in operation in Dolphin's Barn, Dorset Street and Buckingham Street stations were coming to the end of their use in the brigade. Four new forty-five foot ladders were purchased in 1972 and in a short time they were to become the main working ladders in the brigade. Meanwhile the Dublin firemen submitted a demand for representation on the minister's review body but this demand was not acceded to until April 1973 following the defeat of the Fianna Fáil government in a general election. In March the working party had issued an interim report taking into consideration the recommendations made by the McKinsey consultancy in relation to organisation of the fire service in its report *Strengthening the Local Government Service*.

The interim recommendations of the working party concentrated on fire prevention and the need for additional staff

> *to carry out fire prevention duties in a reasonable manner ... not only to undertake those inspections and controls for which they have specific statutory cover but to do everything they can, by providing information, and advice, and exhortation, to increase awareness of fire risks and the value of fire preventative measures .... to improve the capacities of local authorities to engage in enforcement activities in relation to fire prevention measures in existing legislation.*

Action in this area was "directed mainly at management of premises in the public resort category". Accepting that the report *Strengthening the Local Government Service* was principally focused on the issue of a national fire advisory body and on the chief fire officer's reporting relationship to the county manager, the working party supported the McKinsey recommendations on both of these.

The *Report of the Working Party* was published in December 1974 and consisted of thirty-nine primary recommendations. The main thrust of these recommendations was for "an investment programme" for the improvement of fire services, the Minister for Local Government to take overall responsibility for the fire services provided by the local authorities, a review of the Fire Brigades Act 1940, promotion of standardisation in fire fighting equipment, all fire appliances to carry breathing apparatus and road rescue equipment, a fire prevention council, places of public resort to be brought under a comprehensive system of inspection and control, the Minister for Local Government to take responsibility for assessing the efficiency of fire brigades, the provision of a new technological qualification for senior grades, entry to the fire service to be on the basis of a single-tier entry, the chief officer to report directly to the manager, the formation of a fire service training council, the establishment of a national training centre, chief fire officers to be consulted on water improvement schemes, fire service control centres with modern mobilisation systems to be established and county councils or county boroughs only to be fire authorities. At the time of writing many of these essential recommendations for a proper professional fire service have still not been implemented.

The biggest fire in the city in 1973 was on 3 May when many people had to be evacuated from a hotel, guesthouse and flats as Dockrells three-storey shop and warehouse was destroyed by fire. This landmark hardware and builders providers premises in South Great Georges Street was discovered to be on fire just before 1 am on what was a very windy night and needed up to seventy firemen using breathing apparatus, all fire engines and two turntable ladders before it was contained. City traffic came to a halt as fire engines, thousands of feet of fire hose, burning debris and sweeping flames closed one of the main thoroughfares to the south. Much more damage would have resulted had the firemen not succeeded in isolating almost 500 gallons of "white spirits" and removing them from the blazing building. As the exhausted men continued the fight into the brightening morning, staff from the corporation's dangerous buildings section braved the hot smoke-filled area to carry out an inspection of the premises and finally decide that the entire front of the damaged building would have to be demolished immediately. Winston's drapery next door suffered severe damage in the fire but there were no casualties other than the loss of employment.

The following month Macey's store in George's Street was extensively damaged by fire, and on the same night a turntable ladder was used to rescue five men from their third-floor flats above McGovern's pub on the corner of Kevin Street and Wexford Street.

Some time previous to this the *Irish Times* published a series of six articles on the fire fighting service in Ireland and the scope of its task. This was a remarkably wide ranging, in-depth study of the unsatisfactory situation surrounding the alarming increase in fire damage losses, particularly in Dublin. The comprehensive and costly measures needed to be taken by the government were clearly stated. In 1973 a catastrophic fire in the Isle of Man cost fifty lives, the repercussions of which affected fire prevention law throughout the UK. Further Irish newspaper articles questioned the adequacy of the Dublin Fire Brigade. It was lucky that by then an Irish government working party was examining the fire services.

In the single most tragic fire in a domestic dwelling to date in this country both parents and ten of their children died in a fire at their home on Carysfort Road in Dalkey on the night of 11 March 1974. The other three members of the Howard family all received serious injuries on the occasion. The fire in the family two-storey semi-detached house erupted so suddenly that neighbours were not even aware of it until they heard the terrified screams of the dying children. Locals confirmed that the house was an inferno within minutes of them arriving at the scene and they could do nothing because the front door was securely locked and barred. An hour after the fire had been extinguished by the Dun Laoghaire Fire Brigade, bodies were still being removed from the gutted building. The inquest that followed this terrible tragedy concluded that the family members had died from carbon monoxide poisoning or shock following extensive burns. Mrs Howard had very recently bought a new three-piece suite which was lined with polyurethane foam. This highly inflammable padding was to cause the deaths of many civilians and firemen in the succeeding years.

On 20 January 1973 a no-warning car bomb explosion at Sackville Place killed a bus driver and injured seventeen people. It was a repeat of the earlier Sackville Place bombings. On a day when the city was full of rugby supporters for the

Ireland versus the All Blacks fixture the firemen were again in the midst of the trauma caused by this senseless attack on ordinary people going about their normal Saturday chores. Much worse was to follow on Friday 17 May 1974 when the greatest mass murder of innocent people took place on the streets of Dublin. With the buses on strike, parking restrictions had been lifted in the city. After 5 pm many streets were filled with office workers finished for the weekend or people completing their city shopping and preparing to make their way home. In Parnell Street a green Avenger and in Talbot Street a blue Ford Escort almost simultaneously at 5.27 pm exploded, transforming utterly two busy shopping streets into devastated killing grounds. The explosions were heard in both Tara Street and Buckingham Street fire stations so the men on duty hurried to the engine rooms. Soon the control room in headquarters was alive with calls and fire engines and ambulances were being dispatched to the scenes of the carnage. Some two minutes after the initial explosions as the fire and rescue crews sped through the north city, another murderous blast took place in South Leinster Street causing more death and destruction.

When the immediate shockwave of the ferocious blasts began to subside, the rescue crews arriving in the smokey darkness heard the first sounds and movements of the stunned victims. The scenes of carnage unfolded; there were wrecked cars, windowless buildings, debris-strewn streets, massive amounts of dust and floating paper. In the midst of all this mayhem were the cries of the maimed, the injured, the shocked and the traumatised. Due to a change of shift at 6 pm crews were coming on duty to empty stations and those in Tara Street were dispatched in any vehicles available to assist at the dreadful sites. The emergency plan was immediately activated. All ambulances were routed to the terror-filled streets to be loaded with victims and rushed to the designated hospitals.

Dublin Fire Brigade vans and transporters were assisted by private cars, health board ambulances, taxis and even a bus as fire crews and civilian helpers tried to get as many of the injured away from the bombed locations in fear that other explosions would occur because this horrible crime had been carried out without warning. The no-warning bombs left twenty-eight dead in the shattered streets of Dublin – the largest death toll in a single atrocity in the city since the North Strand wartime bombings – and over two hundred injured or maimed by the vicious psychopaths who planted them. The working firemen had no time to wonder who was responsible as they searched the damaged buildings nearby for further casualties. Within four hours all those dead or requiring medical assistance had been removed to hospitals and the taped-off streets were deserted. Staff going off duty in the Dublin Fire Brigade were finally heading for home and trying to forget as much as possible the terrible trauma they had faced only hours before. Those on duty had at least the camaraderie and its black humour to help quell the anger and revulsion at what they were forced to witness in a normal days work. The sad history of this slaughter and maiming of the innocent on Dublin streets is that it is in many respects the forgotten atrocity of the infamous "northern troubles".

In spite of the city coroner's reference to "this obscene act" and his hope that the perpetrators would be apprehended, the victims have received no justice. All investigations by the Irish and British governments failed to find those who carried out the mass murder. Further acts of terrorism were committed in the city over the years following, mainly in the form of incendiary devices in major stores, but these were generally successfully dealt with by the firemen without major damage.

Successful negotiations between the firemen's trade unions and Dublin Corporation management eventually led to a reduction of the working week to forty-five hours from 1 June 1974, and to forty-two hours from 1 April 1975. This was in line with the working hours of gardaí and other emergency service workers in the country.

The brigade was faced with its own internal tragedy on 11 September 1974 when a serious accident involving one of its ambulances took place in slippery conditions on Clontarf Road. A car and the ambulance collided, causing the death of Fireman Michael Mulligan and serious injuries to his driver. An eye witness reported that on impact the two ambulancemen were pitched through the windscreen. Mick Mulligan was dead on admission to hospital and Kevin Kavanagh received serious injuries from which in time he recovered fully.

With the ongoing problem of incendiary devices being planted in city stores and the destruction of two hotels in large-scale fires, public representatives again began to question the city manager regarding the adequacy of the fire prevention section of the Dublin brigade. At that time there were only six full-time fire prevention officers in the brigade and two others with fire prevention responsibilities. It was agreed to increase the establishment in the area to thirteen personnel. Failure to reorganise and failure to recruit however meant that staffing in this critical area was only increased by two.

The returns of the brigade for the year 1973-74 showed service costs of £827,465 to Dublin Corporation and £298,938 to Dublin County Council. A total of 28,872 emergency calls were attended to – of which 1,639 were bogus. A total of 1,709 inspections were carried out by fire prevention staff and a replacement station for Buckingham Street was agreed at an estimated cost of £250,000 at North Strand. The June 1974 local elections had finally restored an elected council to the city and it sanctioned the purchase of two fire appliances from Riversdale Ltd in October at a cost of £33,994 each. Two further replacement appliances were to be delivered the following year from W. P. Ryan at the same price per vehicle.

The tragic year of 1974 ended with a fire on 6 November at the Calor Kosangas premises on Jetty Road, North Wall. This particularly dangerous and spectacular blaze led to much criticism initially of the fire brigade but eventually, when the facts became known, of the gardaí. The fire took place at the thirty-three acre bulk storage and filling plant for Calor Kosangas liquid petroleum gas. The actual area of the fire was about one acre which contained mainly domestic cylinders. Many of these cylinders exploded during the fire, and were often sent hurtling outside the boundary area. Three pumps, a turntable ladder, an extension ladder and an ambulance were in attendance. Cooling jets at the storage tanks assisted the brigade jets in cooling and eventually extinguishing the spectacularly explosive blaze. The fire was caused by a fracture and major leak at a tanker connection. Calor Kosangas staff succeeded in closing all circuit breakers while turning on the valves to activate the drenching system. A ship discharging at one of the oil berths stopped pumping and was stood off into the bay and a second tanker was moved away down river from the incident. It took almost eight hours before the fire was completely extinguished and, by then, five bulk carriers, a laboratory, paint stores and a large quantity of domestic cylinders had been destroyed.

An interesting situation arose during this fire when gardaí proceeded to evacuate about 200 people from their homes in the East Wall area. It appeared that the gardaí were unaware that, under Section 9 of the Fire Brigades Act 1940, the only person with statutory authority in circumstances such as this "to cause any land or building to be vacated by the occupants" was the fire brigade officer-in-charge!

In April 1975 the city manager compiled a report on developments in the Dublin Fire Brigade for the city council. The report listed seventeen fire engines (four of them retained) and seven ambulances in operation but he intended to upgrade to twenty-three fire engines and ten ambulances by building a two-pump station at Donnybrook to replace Rathmines, a two-pump station at North Strand to replace Buckingham Street and three new pump and ambulance stations at Tallaght, Ballinteer and Blanchardstown – all to be provided over the next five years under the Dublin development plan, at a cost of £2,907,000. Adopting the plan the council agreed to seek authorisation for borrowing the required sum of money from local loan funds.

The fire service had at last accepted the need to upgrade its breathing apparatus and move towards training crews to British Fire Service standards on this now essential breathing equipment. Fourteen Airmaster sets with whistles were purchased for £1,935 and once training was implemented this life-saving apparatus would go on the fire appliances. Previously the new breathing equipment was often kept in storage for a year or more before it was issued for operational work, and very little real breathing apparatus training was taking place. Some modifications were carried out on Kilbarrack fire station so that all recruit training would take place there, a situation that lasted until the training centre was opened in Marino some ten years later. Radio paging units for all senior officers were purchased early in the year in anticipation of the publication of the local government working party report.

Dublin Corporation was issued with copies of the report in July and held a special meeting to discuss its contents and recommendation on 10 September. Many questions were asked by the elected representatives before the council adopted the report in full at a meeting in February 1976. Before then the manager was quizzed on the organisation of fire prevention in the brigade and statistics for the last nine months of 1974 were revealed, showing that there were only five full-time fire prevention staff in the service and that although staff vacancies had been approved it had not been possible to fill these vacancies. The 1974 statistics showed 941 planning applications referred to the section, 424 inspections under the Fire Brigade Act 1940, 714 visits to places of public resort, 184 factories inspected and 39 petroleum storage visits undertaken. There was a desperate need within the brigade to involve more trained personnel in fire prevention activities if the volume of work required was to be properly under-taken and it was obvious that non-engineers on the operational side of the service needed to be trained to do this work and seconded full-time to fire prevention.

The chief fire officer, calling for support for the working party report, contended that the recommendations would go a long way towards solving the staffing problems and creating a reservoir of suitably trained and qualified personnel so long as the proposed fire service training council, a national fire service training centre and a fire prevention council were set up.

The chief fire officer accepted that the fire service must keep pace with the expanding population, increased industrialisation and the increasing complexity of fire and fire prevention. In addition the brigade must recruit and train staff for all positions within the service and maintain an up-to-date emergency fleet and control-room organisation.

In May 1976 a large new fire station was opened at North Strand, intended as a replacement for Buckingham Street. In fact the latter station was not closed until 1984. The new station's appliance bays opened onto Leinster Avenue, and the operational vehicles at that time were two water tenders and one ambulance.

Much prevarication and delay took place before the Fire Prevention Council and the Fire Service Training Council were put in place and, at the time of writ-ing this history, no national training centre has been provided for the fire brigades in this country.

In 1975 the brigade attended a total of 11,057 emergency calls of which 22% or 2,464 were bogus. At year end there was an operational staffing of 20 district officers, 37 station officers, 32 sub-officers and 402 firemen. Much progress on conditions and pay had taken place in the brigade, with working hours reduced to forty-two per week on 1 June 1974 and then to forty per week on 1 April 1975; the first fire service in Europe to achieve the forty-hour week. Chief Fire Officer Tom O'Brien reversed his decision of 1962 and returned to the concept of district officers appointed to district stations with the re-formation of a four-district brigade organisation.

Early in 1976 a new major accident plan was introduced in Dublin Fire Brigade and this time the plan covered Dublin city, Dublin county and Dun Laoghaire borough. The procedure in the plan was in two phases with phase 1 to cover a situation where rescue and removal of persons was well within the combined resources of the fire brigades in the three areas, plus the Eastern Health Board ambulance service and private ambulance services. Phase 2 was to cover a situa-tion where the number of casualties was likely to be beyond these resources or where rescue work was likely to be a long and difficult task. The plan was to be implemented by the senior garda officer at Dublin Castle and involved mobilisation of units of the army, civil defence, voluntary first aid organisations, Dublin Airport emergency services etc. With the introduction of this new plan the brigade order on the 1973 plan was rescinded.

Following protracted tests in Britain in 1975 on plastic-coated uniforms worn by firemen, the British Home Office decided that such uniforms affected a man's safety and should not be issued to fire crews. The relevant report into the matter for the Home Office stated that "in reaching this decision the committee was influenced by the fact that an alternative outer material has come to our notice". This was the first major step in these islands towards providing proper clothing to protect firefighters when working in heat and flames. Accepting the report's findings, Chief Fire Officer O'Brien purchased 800 of the recommended "water-proof serge" coats at a cost of £13,000 for the Dublin firemen. In the years following, much improvement continued in this crucial area of protecting firemen during firefighting operations.

Progress in the brigade was at last taking place after much delay but it was very slow. The North Strand station had opened on 3 May 1976, but the old stations were overcrowded because of the shorter working week. The working conditions at Tara Street headquarters and at Dorset Street and Rathmines fire stations were truly Dickensian and far from conducive to the provision of a quality fire service. However, in spite of the government ignoring the main recommendations of the working party report a move to proper breathing apparatus training was taken in the brigade. Selected officers were sent to Washington Hall, the fire brigade training centre for Lancashire. When they had qualified as instructors, the chief set about renting an excellent breathing apparatus training facility which the Wexford fire brigade had purchased and developed at Castlebridge. More instructors graduated and breathing apparatus training at last took on a whole new meaning.

The training courses in Wexford were continued for up to five years until a sizeable cadre of properly-trained breathing apparatus wearers was available in all stations and on every watch. This came at a time when it was also becoming obvious that many early deaths from cancer were taking place amongst firefighters. The old "smoke eaters" of the brigade who had spent the early part of their firefighting careers dealing mainly with fires in carbonaceous materials without breathing apparatus had become victims of polymers, plastics, adhesives and all the new man-made products, which emitted much more lethal, poisonous cocktails than carbon monoxide.

The 70s was a decade when a number of old hotels were destroyed by fire and others sustained varying degrees of damage. The most spectacular of these blazes was at the famous Spa Hotel in Lucan on 21 March 1977. This recently renovated building was extensively damaged by a fire which began in staff quarters above the kitchens. However all the staff and guests had vacated the building before the arrival of the brigade. In high winds and with a poor water supply the fire crews struggled to contain the blaze which spread rapidly through this old, three-storey, rambling building. Flames and thick black smoke filled the surrounding air as the roof collapsed, venting the dangerous premises. Some six hours after the firemen arrived the fire was a smouldering mass of masonry, slates, floor materials, furniture and bedding.

During that year two new front-line appliances were commissioned. These were Dennis R130 water tenders and were purchased at a cost of £43,624.

Early in 1977 Fireman Willie Bermingham was sent by his officer to climb through a window which had been forced in an old people's chalet in Charlemont Street because the resident had not been seen for a number of days. Firemen, because of the nature of their job, have a special insight into life in the areas in which they work. Breaking down doors or windows in order to enter burning buildings often reveals terrible social problems in a community. At incidents like that at Charlemont Street the plight of poor, elderly people living alone in their seriously deprived conditions often defies belief. Willie Bermingham described graphically the awful scene he encountered in that chalet on the bitter cold February day. A frail old man, blind in one eye, lying dead on a wet bed with an old blanket gripped in his left hand, two cold rooms in that timber chalet, no fuel for the fire and no food in the press and he half naked, stiff in his death sentence, alone and in misery.

Such was the impact of this terrible scene that Willie and three other firemen formed an organisation, ALONE, to highlight the tragic story of Dublin's forgotten old people. Within a year ALONE had become a controversial organisation, outspoken on the plight of old people living alone in the city of Dublin. The slim volume published by Willie and journalist Liam O'Cuanaigh for the ALONE organisation in 1978 was a wholly shocking pictorial indictment of the dire neglect of many old people abandoned in filthy rat-infested conditions in various parts of the city. It was variously described as "a book to stir the conscience" or "a simple cold narrative of the plight of formerly neglected and forgotten people" or "a slap in the face to every Irishman, particularly to everyone living in Dublin". Such was the impact of the ALONE group that government, corporation and health board were embarrassed into taking action. Willie went on to receive an honorary doctorate in Trinity College, gain a man of the year award and die some years later in a painful struggle with cancer. His fire brigade funeral from St. Patrick's Cathedral was attended by the president and many other dignitaries from various walks of life, while the poor of the city stood in their thousands as his funeral cortege passed through much of the poorest areas on its way to his final resting place in Lucan. The organisation he founded still plays a leading role in highlighting poverty and alienation in Dublin.

Just after 1 pm on 19 April 1978 fire engines responded to a fire at Jonathan Richards clothing factory on Merchant's Quay. Luckily thirty of the sixty staff of mainly young female workers were outside of the building at lunch when the disaster struck. As the smokey fire raged through the two-storey building the first three fire crews to arrive on the scene had to engage in life-saving procedures in order to get the trapped girls out of the burning building. Two front windows of the premises had bars, and wire grills were in place on rear windows. Male workers and firemen had to take risks in the smoke-engulfed area to prise the bars

277

apart and pull the wire grills down to rescue those trapped inside. Some fifteen female members of staff had to be removed to hospital by brigade ambulances, suffering from shock, smoke inhalation and abrasions. Luckily there were no really serious casualties.

With eight pumps in place the fire was soon surrounded and extinguished, but even then some firemen had to be treated for smoke inhalation and many locals had to leave their homes for a period of time. Flames were recorded at up to fifty feet high at one point. This and other fires of the period led to questions being asked in both Dublin Corporation and the Dáil on the adequacy of fire prevention legislation and whether sufficient fire inspections were being carried out by fire authorities.

In the Dáil the Minister of State for the Environment said he would like to see fire prevention work stepped up. He maintained that local authorities had been urged to intensify their fire prevention work and added that the enforcement of conditions on premises subject to licensing by the courts, such as dance halls and premises used for singing, music or other entertainment, was a matter for the gardaí.

At the May meeting of the general purposes committee of Dublin Corporation, the principal officer of the engineering department agreed "that there were wide gaps in the city's fire inspection service". For years there had been only four inspectors at work; three other posts had remained unfilled for a length of time because no qualified engineer or architect would accept the job at the salary offered. It was stated at the meeting that the Merchant's Quay premises involved in the recent fire was last inspected by the corporation in 1958 and since then the factory premises had been totally refurbished and altered several times.

Other factories in the city which could be potentially lethal in fire situations had not been inspected under the 1972 Factories Act and nobody knew what type of fire precautions had been requested by the corporation. The general purposes committee went on to call on the minister responsible to tighten up fire-safety legislation. The committee was informed by the Department of Labour that the Merchant's Quay premises were inspected by it in February and the means of escape and other precautions conformed to the fire certificate. Under the Factories Act Dublin Corporation was responsible for the issue or refusal of certificates of escape but enforcement was a matter for the Minister for Labour. This type of divided responsibility had left many gaps in the enforcement of safety legislation throughout the country and the call from the Chief Fire Officers' Association to bring all fire safety into one comprehensive Act with one enforcing authority was ignored.

In May 1977 three notices were served on premises in the city under the Fire Brigade Act 1940. In June 1978 Dublin Corporation finally agreed on the need for a replacement of the 1940 Act and for the implementation of the working party report of December 1974. The elected councillors were not happy with the slow progress of the government in implementing many of the major recommendations and they were in agreement in calling for the introduction of minimum standards of fire cover similar to those existing in Great Britain. The city manager reported in August that

> *pressing problems affecting the fire service which must receive attention immediately is the back-log in driver training … in maintenance and training of men in breathing apparatus and adequate procedures to be devised and implemented in connection with the great expansion recently in transport of dangerous substances.*

He also laid great emphasis on the deficiencies in the fire prevention department.

Total expenditure on the brigade at that time was £3,026,235 – funding an operational firefighting, fire prevention, ambulance and emergency staff consisting of a chief fire officer, second officer and 550 men, one class A officer and two clerk typists. The manager proposed creating two assistant chief officer posts, operational and fire prevention, and improving the number of clerical staff to one senior executive officer, one minor staff officer, one class A officer, one clerical officer and two clerk typists.

On the ground in the brigade major problems had arisen over conditions in the older fire stations and on 12 October 1977 an industrial dispute broke out when the fire crews refused to work in Dorset Street station because of the condition of the building. In a compromise agreement the men operated from there until August 1978 when they would be moved to Buckingham Street and would operate from there until the opening of Phibsborough station in 1984. In 1977 Dublin Fire Brigade received its first foam tender. This was supplied by Carmichael on a Dodge chassis and was stationed at North Strand close to the dock area. This appliance is capable of pumping large quantities of foam to combat fires in oil tanks and in dangerous storage premises. Also supplied by Carmichael, in 1978, was a new Dodge/Magirus turntable ladder. Both of these were major additions to the brigade's fleet. The number of fire and emergency calls had risen to 11,854, of which 3,289 were false alarms.

The year 1978 ended with the city manager informing the corporation that

> *all Places of Public Resort which include night clubs are inspected annually either by a surveyor of Places of Public Resort or by a Fire Prevention Officer for licensing purposes.*

# 17 The Stardust tragedy and the new Fire Services Act

During the 1970s there was a growing realisation of fire hazards related to plastics. In 1971 the British Fire Research Station had investigated the behaviour of polystyrene tiles in a fire and concluded that "expanded polystyrene had a tendency to produce flaming droplets on ignition and burning strips which promoted fire spread if gloss paint had been used on it".

The issue of ignition risk and the contribution generally that plastic products made to rapid fire spread, smoke and toxic hazards in a conflagration was ignored in Ireland and there were no regulations concerning its use. Legislation on liquid petroleum gases (LPG) was put in place in Britain only after a series of major fires involving these products.

It was following the death of ten people and the hospitalisation of forty-one others, together with injuries to six firemen, at the Manchester Woolworths fire in May 1979 that regulations were introduced to ensure that all soft furniture was resistant to being ignited by cigarettes or matches. There had been a massive growth in the use of flexible polyurethane foam in furniture and with it a marked change in the pattern of fire deaths, particularly in the home, due to asphyxiation from the toxic elements now produced in fires. Investigations in other countries had shown that polyurethane foam furniture was easy to ignite, burned rapidly at very high temperatures, produced very dense smoke and toxic gases, giving less time for occupants to escape, and it created a lethal atmosphere for firefighters without breathing apparatus.

This was the background to the new emphasis that was finally put on breathing apparatus training in the Dublin Fire Brigade. Instructors in safety procedures now required qualifications that matched British fire brigade standards. Dublin however did not have a facility suitable for this essential training so in 1979, as we have seen, arrangements were made with Wexford fire brigade for the use of its Castlebridge premises. Because of the intermittent availability of this excellent training facility, progress was slow and it was to take a number of years before all firemen could be trained to the required safety standards.

The details issued by British Health and Safety Commission on the effects of emissions from burning plastics and foams showed that traditional furniture burned more slowly and produced volumes of warning smoke consisting mainly of carbon monoxide. Once ignited, modern foam-filled furniture was burning out of control within minutes while emitting large quantities of carbon monoxide and other toxic gases capable of causing death very quickly. Lethal conditions could develop in areas where very little smoke was in evidence. The rise in the use of polymers, plastics, foams and adhesives represented a new hazard for all firemen and underlined the necessity of wearing breathing apparatus and the rapid evacuation of people in premises where fire broke out. In order to be effective in dealing with these new lethal materials, Dublin needed to improve its response time by having properly trained breathing apparatus-wearers in new strategically-located fire stations. The proposed changes in training and the provision of new fire stations however were not proceeding quickly enough to deal with the growing potential for a massive disaster.

The Fire Research Station tests on the the the fire at Woolworths in Manchester showed that upholstered seating, ignited with a match, had flames impinging on the ceiling within forty seconds but with little smoke in evidence. Within one minute, however, temperatures of 1000°C were reached and a black layer of smoke was obscuring lights on the ceiling. At two minutes the fire had reached maximum severity with sufficient heat to cause ignition to wooden furniture and the emmission of hot toxic gases probably fatal in a brief exposure. The most notable features of this fire investigation were the speed of development and the rapid production of dense smoke and hot gases. Attempts at first-aid fire fighting were not appropriate to this situation and a distraction to those evacuating the building.

In June 1979 Dublin Corporation accepted a tender of £484,043 from Maher and Murphy Ltd and construction of the new fire station at Belgard Road, Tallaght, was agreed. This was to be seen as the first part of a major ten-year development programme for the Dublin brigade in both the city and county to cater for requirements up to 1990. This programme, drawn up by Chief Fire Officer O'Brien, envisaged a new station at Belgard Road, Tallaght, new stations at Donnybrook (to replace Rathmines) in 1982, Nutgrove (1982), Blanchardstown (1983) and Phibsboro (1983), a new headquarters at Tara Street in 1986, a full-time station at Swords and new buildings for the retained crews in Skerries and Balbriggan. The programme was adopted by the city council in July 1980, a time when cut-backs were taking place in most other corporation services. The city manager assured public representatives that no cutbacks would take place in the fire service.

It was accepted that the brigade had thirteen vacancies at fireman level and five in the fire prevention department and that the new Tallaght and Donnybrook stations would require the employment of fifty-six additional staff. The City and County Development Plan at that time provided for major housing development and adjoining industrial estates up to fifteen miles outside the city. The Dublin Fire Brigade expansion programme to bring the brigade in line with this development was estimated to cost £4.65 million.

In the meantime thirty-two Dublin firemen led by Station Officer Jim Sergeant were collecting thousands of pounds for UNICEF by pushing a half-ton trailer pump from Dublin to Cork which gained entry for them in the *Guinness Book of Records*. A year previously, Belfast firemen had failed to beat the record held by the American Fleet Air Arm and Somerset Fire Brigade.

On Friday 16 November 1979 at 8.40 am a city-bound train from Bray crashed into a stationary train just outside Dalkey tunnel injuring over forty passengers and the train driver. The emergency disaster plan was speedily put into operation as the brigade fire engines and ambulances rushed to a scene of derailed carriages where the mangled wreckage of the engine blocked the railway line on a steep banked section. The injured and shocked passengers were removed to hospital by the fire and ambulance crews, assisted by local residents.

It took over three hours and required the assistance of a helicopter before the seriously-injured driver was rescued from his crushed engine cab. Many of the injured were children on their way to school, but fortunately none sustained serious injury. Eight fire engines and twelve ambulances took part in this rescue at what was known locally as the "Khyber Pass". Some criticism on the working of the emergency plan later emerged in the national newspapers but the ability of the firemen to respond to an emergency situation was not in doubt.

The 1980s began with the year of the arsonists. Almost every second fire was alleged to be the result of arson as schools, factories, government property, pubs, buses, shops and hotels were affected. Subversives campaigning for closure of the Curragh prison camp were blamed among others, as were the usual mindless vandals. The people of Dublin had to pay the ensuing bills under the Malicious Injuries Act in spite of the fact that there was no more than a rudimentary and haphazard system of fire investigation in place. This financial burden on Dublin Corporation could have provided for badly needed improvements in the Dublin Fire Brigade.

In the course of awarding a decree against Dublin Corporation for damages caused by one such fire, Judge Noel Ryan felt the need to make the following statement: "It is a shocking situation that ratepayers end up bearing the burden for damages when people were paying insurance companies". The total cost of the brigade for 1980 was £7,827,800 with £4,778,800 coming from Dublin Corporation, £865,000 from the Eastern Health Board (for the ambulance service) and £2,184,000 from Dublin County Council. During the year three Dennis water tenders were purchased for £155,499 from Riversdale Ltd and agreement was entered into with J. Du Moulin Ltd for the construction of the Donnybrook fire station at a cost of £784,102.

In September Chief Fire Officer O'Brien informed a meeting of Dublin Corporation that "all Night Clubs and discotheques in the City are inspected at least annually for licensing purposes". A few weeks later the Fire Prevention Council of Ireland was holding its first National Fire Safety Week, involving the Dublin Fire Brigade and targeting its activities at school children. The most spectacular fires that year included the Phoenix Park Racecourse Stand, the Crumlin Social Services Centre, Botany Weaving Ltd in Cork Street (suppliers of uniforms to the defence forces and gardaí), Chalfont Transport on East Wall Road, the CIE Bus Garage in Summerhill and the Clare Manor Hotel in Coolock.

Regulations were introduced in October for the labelling of all furniture for sale that did not resist a smouldering cigarette or a match and, from December 1982, the sale of all such furniture was to be banned. Ireland now had legislation similar to that introduced in Britain, enacted mainly in order to promote Irish manufactured furniture in overseas markets and to prevent the dumping of outlawed furniture in the Irish market. Despite the fact that trade rather than safety considerations had focused political minds on the need for the safety regulations, the measures were warmly welcomed by Dublin Fire Brigade.

## The Stardust

The terrible Stardust fire in Artane was to make 1981 a defining year for Dublin Fire Brigade. A portion of what had formerly been a food factory in North Dublin, built in 1948, was converted in 1977-78 into an "amenity centre" consisting of a public bar known as The Silver Swan, a restaurant and function room called The Lantern Room and an area, The Stardust, used for cabarets and concerts. In the early hours of the morning of Saturday 14 February, while a St Valentine's disco was in progress, a disastrous fire swept through the building killing forty-eight young people and seriously injuring 128 others.

The brigade received the first call to this tragic event at 1.43 am and despatched two pumps to the scene. Within a few minutes, after several more calls, three pumps, ambulances, an emergency tender and a turntable ladder were sent to the scene. The Kilbarrack crew of the first fire engine to attend observed smoke rising in the area of the fire as they proceeded down the Malahide Road, denoting that at that early stage part of the roof had spalled-off. Turning into Kilmore Road they were confronted by crowds of injured and terrified youths seeking assistance or fleeing from the holocaust.

The total converted portion of the building was 2,944 square metres, while The Stardust, including toilets and kitchen, occupied about 1,853 square metres. The building consisted of rendered concrete blockwork with some brickwork at the front. Inside there was a suspended ceiling of mineral fibre tiles. The walls generally were covered with polyester carpet tiles on a PVC backing attached by a synthetic adhesive. Seats and tables were fixed to the floor on the tiered areas (the north and west alcoves) but were not fixed on the level area. These seats were constructed from steel angle frames with a chipboard base and the back padded by polyurethane foam and covered in PVC-coated fabric. There were also roller blinds made of PVC-coated polyester fabric suspended from the ceiling – these could be pulled down to isolate the two alcoves from the rest of the ballroom. On the night of the fire the west alcove blind was "down" to the level of the tables. The Stardust had eight exits. Six of these were intended to be means of escape in an emergency and, of these, one was the main entrance. The two extra exits were via the kitchen and The Lantern Room. There were several windows in the toilets which could have been used in an emergency, but steel plates had been fixed to them by the management on the inside of the vertical steel bars six weeks prior to the fire; this was to prevent persons handing drink or other articles into the premises.

Some six months before the fire a practice of dropping the chain and lock across the horizontal panic bars was adopted to give the impression that the doors were locked. However, shortly before Christmas 1980 a decision was taken that the emergency exits were to remain locked until approximately midnight on disco nights. The premises had been granted an extension on the night of the fire because of that night's disco. Therefore, on the night of the tragedy, the doors were still locked when the first sign of smoke was noticed at 1.40 am and when flames became visible at 1.41am. It was not until 1.43 am that the first three doors were opened and it took a further two minutes, the lights having failed and with flames all over the west alcove and black smoke filling the ballroom, that a further two exits were opened. Fire was venting through the roof a minute later when the final door was opened. When the first fire engine reached The Stardust at 1.51am, all of the ballroom was on fire and flames were sweeping through three of the exits doors.

The fire crews at the scene were hampered by lack of knowledge of the premises, and by the water supply and the numbers involved. Over 800 people were in The Stardust when the fire broke out. Most who got out were in pain and shock, or anxious for missing friends or relations. The first ambulance to arrive picked up ten to fifteen casualties from the roadway outside the complex. Taxis, private cars and ambulances ferried the casualties to various hospitals throughout the next hour and gardaí arrived on the scene to control the crowds.

Firemen with breathing apparatus entered the premises as soon as fires were driven back from the exits, rescuing seriously injured and trapped victims. Investigation of the fire showed that the conflagration had reached its maximum severity before the fire engines arrived and diminished until completely extinguished not later than 2.54 am. In just over an hour the interior of the building had been destroyed, forty-four young people had died and 214 others were removed to hospitals for treatment. Four of the injured victims were to die in the coming days, 128 required in-patient treatment, 86 suffered minor injuries and were treated as out-patients, while others not hospitalised were to suffer feelings of guilt and psychiatric problems for years afterwards. Amongst the most devastated by this terrible tragedy was the McDermott family whose three children, William, George and Marcella, died in the fire and whose father Jim was a Dublin fireman. He was retired on medical grounds a few years later and died not long after, never having recovered from the dreadful pain of the loss of his children.

On the day following the disaster the government announced the establishment of a tribunal of inquiry, to be conducted by a High Court judge and a number of assessors. The tribunal was appointed on the 20 February, under Mr Justice Ronan Keane (later chief justice), with the following terms of references.

*To examine 1) the immediate and other causes of the circumstances leading to the fire, 2) the circumstances of and leading to the loss of life, 3) the measures and their adequacy on and before 14 February to prevent, detect and to minimise and otherwise deal with the fire, 4) the means and systems of emergency escape from The Stardust, 5) the measures (including Draft Building Regulations 1976) taken on and before 14 February to prevent and to minimise and deal with any circumstances that may have contributed to the loss of life and personal injury, 6) the adequacy of the legislation, statutory regulations and bye-laws relevant to fire prevention and safety in the granting of planning and bye-law permission for the conduct, running, supervision, official inspection and control, and the observance and enforcement of such legislation etc.*

Controversy arose as soon as the tribunal was announced. Noel Browne TD asked whether the taoiseach Mr Haughey insisted on the appointment of Mr Justice Keane instead of Mr Justice Costello who had recently conducted the inquiry into the Whiddy Island disaster. The Chief Fire Officers Association questioned why the Department of the Environment was excluded from the inquiry and why the minister's fire adviser was not one of the assessors. The role of the gardaí on the night was also excluded from the inquiry. In spite of many applications for representation at the inquiry, only the Attorney General, Dublin Corporation, next-of-kin of the deceased, persons who suffered injury, and the owners and occupiers of the building were allowed representation.

Evidence was given by 363 witnesses and the gardaí took over 1,600 statements. The tribunal sat for 122 days, including two preliminary sittings, between 2 March and 26 November and Justice Keane presented his report to the Minister for the Environment on 30 June 1982. By that time some legislation had been put in place, the Fire Services Act 1981 and sections of the building regulations which had remained in draft form for twenty years, ignored by various governments during that period.

Justice Keane's conclusions were

> *that approximately four minutes after the fire was first seen, virtually the entire alcove and all its contents were on fire ... [when the lights failed] this also increased the panic .... within five minutes after the failure of the lights, virtually all the contents of the ballroom went on fire. By the time the first fire brigade appliance arrived the fire had begun to diminish. [The fire was probably caused deliberately] the most likely mechanism being the slashing of some of the seats with a knife and the application of a lighted match or cigarette lighter to the exposed foam. [The rapid fire spread was caused] by tiers of combustible seats at least one of which was ignited against a wall completely lined with carpet tiles having a relatively high surface spread of flame rating and heat evolution. The presence of a low suspended ceiling in the alcove increased the radiation downwards ... [the ceiling collapse and venting of the fire through the roof] probably resulted in a substantial mitigation in the number of casualties .... there were no hose-reels in the building and the use of portable extinguishers was entirely ineffective. [Many of the deaths were caused by carbon monoxide but] none of the victims suffered significant crush injuries resulting from trampling "panic" or exit accidents.*

The report went on to state that planning permission for the conversion of the building was complied with. However

> *the use of carpet tiles on the walls of the ballroom and the main entrance foyer was a breach of the C.F.O.'s requirements .... There were a number of serious instances of failure by the owners to comply with the Public Resort Bye-laws: a) the owners did not take due precautions for the safety of the public, the performers or the employees, b) timber partitions were used to enclose portions of the building which should not have been so enclosed, c) exit doors were persistently locked and chained in breach of the Bye-Laws while the public were on the premises, d) the means of escape were not kept in an unobstructed condition, e) employees were not allotted specific duties to be performed in the event of fire, fire drills were never held on the premises ...*

It was also found that electrical installation was defective and that no building regulations had been made by the Minister for the Environment under the

Planning Acts. There were a number of serious failures to comply with the requirements of the Draft Building Regulations, an example being that "materials not of a suitable nature and quality" were used. Furthermore, "structural precautions were inadequate in respect to the main entrance, travel distance in relation to some escape routes exceeded those permitted under the regulations". The following findings also emerged in relation to the Dublin Fire Brigade and to the enforcement of legislation.

> The unsuitable nature of the main entrance and its lack of compliance with the Draft Building Regulations was not appreciated by the Fire Prevention department of the Dublin Fire Brigade [which department] was grossly understaffed and such staff as there were had not the appropriate specialised training or qualifications .... The condition subject to which the Planning Permission was granted was of dubious legal validity and was not enforced. The Inspection of the building carried out on behalf of Dublin Corporation was gravely inadequate. No inspections of the building were carried out by any member of the Fire Prevention or fire fighting Department of the Fire Brigade from the time it opened to the public until the fire. No proceedings were taken by the Corporation in respect of breaches of the Bye-Laws relating to the locking and obstructions of exit doors revealed by inspections as were carried out.

With regard to rescue operations the report maintained that

> the members of the Fire Brigade carried out their duties on the night of the fire in a dedicated manner and, in the case of those Firemen and Fire Officers who took part in the actual evacuation of the building, at considerable risk to themselves. The evidence clearly established however, that the management (including the general organisation), the training and the equipment of the DFB has not been modernised so as to enable it to cope with a disaster of these proportions in a satisfactory manner .... serious shortcomings in the training of officers and men, no proper command structure, lack of knowledge of the buildings and location of hydrants or static water tanks in the grounds of the building, lack of incident control point, the out-of-date nature of the Central Control Room, only two B.A. [breathing apparatus] sets on the fire appliances.

The report asserted that injury and discomfort to a number of survivors would have been avoided by a more efficient rescue operation and that the fire brigade was seriously hampered by the hysterical behaviour of large crowds gathered outside the building. The provision of ambulances was adequate but there were serious deficiencies in the operation of the major accident plan.

In relation to the garda investigations, the taking of statements "was carried out in an exhaustive manner and with meticulous attention to detail", but "there were serious shortcomings in the forensic investigation by the gardaí and the Department of Justice, in some instances samples of material which could have been critical were not taken and important tests were not carried out". However, responsibility for the disaster was firmly placed on the owners and their advisers, the corporation and the Department of the Environment. The report drew attention to the low level of professional expertise availed of that led to carpet tiles being used on the walls and the unsuitability of the main foyer entrance.

Attention was also drawn to the owners' attitude that their responsibilities in relation to ensuring the fire safety of the premises were discharged once they satisfied Dublin Corporation requirements.

> *The inadequate consideration given to the drawings by Dublin Corporation, the inadequate inspection of the building and the serious deficiencies in the fire prevention and firefighting services contributed significantly to the disaster. The responsibility for these matters ultimately lay with the Chief Fire Officer and the City Manager and Assistant Managers with responsibility in these areas in the immediately preceding years. The Department of the Environment bore a share of the responsibility for the scale of the disaster because of failure to make building Regulations under the Planning Acts, notwithstanding the lapse of a period of nearly twenty years since the enactment of the legislation, a failure which greatly increased the workload on an already overburdened Fire Prevention department in Dublin Corporation, failure to respond to a crisis of morale and efficiency in the Dublin Fire Brigade although warned of its seriousness by the then City Manager in 1978, and the failure to ensure that adequate training facilities existed for firemen and officers in Ireland, although they had been advised as far back as 1975 of the need for establishing of a National Training Centre for such personnel.*

### The recommendations

The report's recommendations dealt with "the shortcomings in the approach to fire safety in Ireland" and the urgent need "for a modern fire safety code embodied in appropriate legislation and regulations implemented by effective fire safety organisations". The report maintained that the degree of knowledge expected in the case of a fireman is manifestly different from that required of senior officers and, in particular, the chief fire officer. Similarly what is required of an inspector concerned to ensure that emergency exits in any crowded disco are unlocked is different from what is required of a senior fire prevention officer assessing all fire aspects of a large complex design. Since heightened awareness of fire safety is crucial to fire prevention a continuous programme in this area should be maintained particularly in schools and the media.

> *In the case of all fire fighting personnel, knowledge can only be obtained by a combination of training and experience. An essential feature of any training programme is the existence of a National Training Centre which is recommended and is in the opinion of the Tribunal, the most important and seriously overdue reform which is required in order to improve morale and efficiency. The curriculum, should necessarily include instruction in all aspects of fire safety engineering of particular relevance to the work of firemen and fire officers. In the case of senior officers the knowledge imported at the Training Centre should be supplemented by attendance at more specialised courses in the universities or other third-level institutions. Attendance at courses of this nature is essential for fire prevention officers ...*
>
> *Since there is no legally enforceable system in place in the Republic of Ireland which requires the fire safety of all new buildings to be considered before permission*

> for their erection is granted, the Minister of the Environment under Section 86 of the Planning Act should make building regulations including relating to fire precautions, such regulations should incorporate the recommendations of the Tribunal. These Building Regulations when introduced will of course need to be regularly reviewed. With the growing number of disco and similar premises in the Dublin area the Tribunal recommends that those sections of the Building regulations which deal with means of escape should be applicable to existing places of assembly in buildings such as theatres, cinemas, restaurants, discos, public houses and dance halls which present a high level fire risk.
>
> The Tribunal further recommends that the Fire Services Act 1981 should be amended so as to include the case of any places of assembly with an occupant capacity in excess of one hundred; it shall be unlawful for the premises to be open to the public until such time as a Fire Certificate has been issued signed by the Chief Fire Officer ... . the District Court should have power, in the case of any breach of the Building or Management Regulations, to order the immediate closing of a place of assembly on an application by the fire authority .... where management fail to take all reasonable steps to ensure that exit doors required by law to be unlocked and unobstructed do not remain locked or obstructed, this offence should be punishable by a fine not exceeding £10,000 or imprisonment for a term not exceeding six months.

The fire prevention department was grossly understaffed, having decreased in numbers since 1973 when there were four technical officers (graduates) and four operational officers plus a typist to just the four graduates and a typist at the time of the fire. Chief Fire Officer O'Brien had told the tribunal that "not alone was the staff inadequate but the permitted establishment was also inadequate". In his view there should have been at least twelve officers in the fire prevention department. The tribunal recommended that the vacancies in the department should be filled immediately and members of the fire fighting branch should also be employed, particularly for carrying out during-performance inspections. As early as 1971 O'Brien had replied as follows to a ministerial circular of that year.

> Plans submitted for approval under the Local Government (Planning and Development) Act 1963 are referred to me when necessary. The number has grown from 458 in 1963 to 1,937 in 1973 and involved 93 on-site inspections. The figure quoted for 1973 refers only to the actual number of plans received at FB Headquarters and gives little indication of the work involved in this section .... . The work occupies over 75% of the time of the Fire Prevention Section and in my opinion places an unfair burden on it. It would be a considerable relief if a Code of Buildings Regulations were available and the bulk of this work transferred to Building Surveyors in the Planning Department.

At that time the Department of the Environment had exclusive statutory function, in relation to planning appeals, of dealing with fire precaution matters on such appeals.

The working party report on the fire service which was laid before the Oireachtas by the government in July 1975 included a recommendation that the sections of the building regulations dealing with fire standards and safety should be brought into operation "at the earliest date possible", but nothing had happened. Again in October 1976 when the minister invited local authorities to submit reviews of the strength and adequacy of the fire service in their respective areas, Dublin Corporation had written on the fire prevention department as follows: "apart from difficulties experienced in filling positions in this section the activities in this field are seriously hampered by … the absence of building standards which include structural fire precautions, means of escape in case of fire and means of assistance for the Fire Brigade". In April 1978 the city and county manager sent a letter to the Department, accompanying the review of the Dublin Fire Brigade, which contained the following.

> *The repercussions of the non-filling of non-operational posts and the restrictions on the payment for unsociable hours to those normally receiving it, is felt throughout the service particularly in relation to training, maintenance of equipment, and fire prevention. I am informed by the Chief Fire Officer that there is a great deal of dissatisfaction among senior officers about their present pay structure and that morale is seriously affected. The review proposes a new organisation structure for the fire service which will require an adequate pay differential between the different levels responsibility. This matter needs urgent attention.*

Unfortunately the department failed to address this serious problem up to the time of the tragedy of The Stardust.

The final conclusion of The Stardust tribunal was that the Department of the Environment was obliged to assess with greatest care the introduction of building regulations having the force of law. However, a delay of almost twenty years in introducing such regulations was wholly unacceptable. The attention of the department had been drawn on more than one occasion during that time to the serious additional workload being imposed on the fire prevention department of Dublin Corporation because of the absence of such regulations.

> *The Department failed to treat the introduction of the regulations as an urgent matter … . During a period when there was the most rapid expansion in the history of the area, the staff of the Fire Prevention Department actually declined to what can only be regarded, in terms of workload, as a derisory complement of four Officers and a Typist. In the view of the Tribunal, the crisis in Dublin Fire Service ceased at that stage to be a local problem and became a national scandal …. The problems of the Dublin Fire Brigade were not merely a reflection of management/union difficulties common to other branches of the public service, they were inseparably bound up with the absence of training in the Brigade, with all its attendant implications for morale and efficiency.*

The Working Party in 1975 had expressly drawn the attention of the minister to this problem and had recommended the establishment of a national fire service training centre. At the time of writing no such training centre has been provided in this country. The tribunal report noted the limited role assigned to the minister under the 1940 Fire Brigade Act and concluded that such a limited role " … was not adequate in modern conditions and amending legislation was urgently required to ensure a more effective supervision by Central Government of the discharge of their fire safety duties by local authorities". In addition it presupposed the introduction of a fire inspectorate. Neither of the above conclusions have been implemented by government over twenty years on from The Stardust horror.

The tribunal report was received with mixed feelings in the Dublin Fire Brigade because it contained little that was not already known in the brigade. However, it did highlight the following major issues which had held back the necessary development of the fire service: the failure of politicians to update safety legislation which had been completely inadequate for so many years; the failure to pass a building control Act governing the construction of buildings and changes in use of existing properties so that the fire service and other regulatory bodies could protect the lives of the citizens. Such legislation was not enacted until 1990, under which a fire safety certificate is now required for all new commercial buildings, extensions or alterations, and change of use.

At the March 1982 meeting of Dublin Corporation, Councillor Paddy Dunne a long-time advocate of improvements in the Dublin Fire Brigade, proposed that "a steering committee representative of all groups on the Council be set up to examine our own Fire Brigade service with particular emphasis on equipment, communications and other facilities necessary to meet current needs etc". This resolution was carried but when he moved a motion to suspend standing orders "to enable the Council to consider as a matter of urgency a motion calling on the Government to implement the recommendations of the Fire Service Report 1975" it was voted down, in spite of the fact that the same council had in 1976 endorsed the very same report in its entirety.

In April the proposed construction of the fire station in Donnybrook ran into problems when the application to the Department of the Environment for £1,350,000 from the Local Loan Funds for this building project was not granted by the minister. The minister however declared that he was prepared to sanction a loan obtained from another source. Ulster Investment Bank agreed to the loan, repayable over seven years at 16.0625% interest.

In June 1982 a new Irish-built pumping appliance was purchased. Up to this time all the brigade's fire fighting vehicles were bought from one of the well-known British fire engine builders such as Leyland, Merryweather, Dennis, HCB-Angus and Carmichael. But now, for the first time, the Dublin Fire Brigade bought a water tender with chassis and bodywork entirely built in Ireland by the Timony company. Timony had been originally established to research and design diesel engines and had progressed to the manufacture of armoured personnel carriers with a unique suspension system, and airport crash tenders. This work was followed by the production of a water tender designed for municipal fire brigades. Dublin initially ordered two at a total cost of £147,377. They were built at the company's plant in County Meath and the first went into service in 1982, with the second one arriving the following year. A third was bought in 1985. Somewhat surprisingly, the bodywork design resembled the company's airport crash tenders and some problems were apparently encountered with the steering and brake systems. These three machines did not prove popular and Timony began to build its own bodies on Ford, Bedford or Dodge standard commercial chassis. At the same time as the first Timoney appliance arrived the brigade obtained 432 smoke canisters for use in breathing apparatus training. At this time all fire appliances and brigade ambulances requiring overhaul or major repair were being serviced by Smithfield Motors or CRV Engineering.

A new fire station was opened at Tallaght in April 1982, followed by another at Donnybrook – replacing Rathmines – in December of the same year. An agreement was signed with Brock and Sons Ltd for the construction of yet another new station, at Nutgrove Avenue in Rathfarnham, for the sum of £707,872. This one, of a similar design to that at Tallaght, was opened in July 1984 and it was followed by a large new station at Phibsboro in September. That year saw the closure of the Buckingham Street station, built in 1900.

Following The Stardust fire, both Dublin Corporation and Dublin County Council inspectors carried out inspections on a considerable number of so-called "places of public resort" (mainly discos and dance halls). Within a month over 150 premises in Dublin city had been visited and no major faults were discovered. However, in the Dublin County Council area up to fifty-one premises used for dances were discovered not to be licensed for that purpose. During 1982 there were very destructive fires at the Ashtown Tin Box Company, the Four Courts, O'Connell Schools in North Richmond Street, T.C. Carpets in Walkinstown, and Jordan's Warehouse and Showrooms in Marks Lane.

While awaiting the publication of The Stardust report the brigade suffered two major tragedies. On the morning of 25 January 1982, Station Officer Jim Brady was killed when the fire engine he was travelling in to a fire near Ashbourne crashed into a ditch injuring two other members of the crew. Damage estimated at £200,000 was caused by this fire which destroyed a barn, 1,300 bales of hay, three tractors and other machinery of a Dutch national. Station Officer Brady was crushed beneath the fire engine when it skidded into a ditch on the narrow country road. He was buried with full fire brigade honours on 28 January at a funeral attended by Chief Fire Officer O'Brien, other uniform officers and men of the Dublin Fire Brigade.

Three months later the brigade lost its chief officer, Thomas O'Brien, when he died suddenly from a heart attack on 6 April at sixty-two years of age. He had been chief for twenty years. The Stardust fire and the tribunal had aged him prematurely. A considerate man with a dry sense of humour, he had borne the burden of the many problems of the Dublin Fire Brigade with outstanding humanity. The tragedy of The Stardust weighed heavily upon him, surrounded as he was in that last year by those who isolated him as a scapegoat for all the fire brigade's ills. Firemen turned out in their hundreds to march behind his coffin to Sutton graveyard. Now that political minds were focused on the Dublin Fire Brigade, advances would be made that he had sought but could never achieve. He died a victim of The Stardust fire.

Michael Walsh BE, then assistant chief fire officer and the former chief officer of Limerick Fire Brigade, was appointed acting chief fire officer from 6 April 1982 and held the position until early in 1987.

At the time of The Stardust tragedy, Dublin Corporation operated under government restrictions on expenditure and, in framing the annual estimates, were limited on rate increases. In spite of these restrictions the city manager had provided for special consideration to the needs of the fire service: "The level of expenditure provided in the draft estimates for 1981 in respect of the Fire Brigade Service was as recommended by the Chief Fire Officer and represents an increase in excess of that for other services". These estimates included provision for equipping the new Tallaght station and for the recruitment and training of the additional personnel for the station.

Both the 1909 Cinematograph Act and the 1940 Fire Brigades Act were repealed when the Fire Services Act 1981 became law. This Act was based on the 1940 Act but, for breaches, liability was now "to a fine not exceeding £500 or, at the discretion of the Court, to imprisonment for a term not exceeding 6 months or to both the fine and the imprisonment". But for breaches of Section 18 (2), which covered places of entertainment or "use for any purpose involving access to the premises by members of the public, whether on payment or otherwise", a person guilty of an offence "shall be liable to a fine not exceeding £10,000 or, at the discretion of the Court, to imprisonment for a term not exceeding 2 years or to both fine and imprisonment".

The new Act did not fix standards of fire cover for fire authorities nor did it increase the role of the minister in the running of the fire service, provide for a national training centre or training standards for fire brigades, and it failed to introduce a fire inspectorate. In the light of The Stardust disaster and the short-comings in the Dublin Fire Brigade exposed in the tribunal report, the new Act fell short of what was required to bring the fire service into line with that provided in Northern Ireland or Britain. Talks were entered into between Dublin Corporation and the trade unions representing the firemen to allow operational fire officers to carry out fire prevention duties. Following prolonged negotiations, provision was made for eight fire prevention district officers. The men selected were sent to the UK for a six-month fire prevention course and on 26 May 1983 they were appointed to their positions. Their main task was to visit buildings in a fire prevention capacity and to carry out regular checks on late-night entertainment venues up to 2.00 am. These would be the first regular checks on night entertainment centres by the brigade since 1973. In 1983 the fire prevention section had only five officers (engineers) because it was unable to recruit qualified staff due to the salary scales on offer.

In April 1982 a £3 million claim for malicious damage by WRE was served on Dublin Corporation following a massive blaze which caused extensive damage to seven major business premises on Henry Street and Liffey Street. About this time moves to have a gas storage tank situated in Dublin Bay and growing concerns in the city regarding the petrochemical area in Dublin Port and the transportation of dangerous substances through the city, prompted the corporation to support the firemen's call to make the internationally recognised Hazchem Code compulsory on all containers of dangerous substances as it was in Britain.

Aerial photographs of the Stardust complex, following the fire.
© *The Irish Press*

# 18 The modern fire brigade

Dublin Corporation purchased the former O'Brien Institute and twenty-two acres of land in 1981 for housing and for use as a recreational/cultural centre. In the autumn of the following year that decision was changed and part of the site, adjoining the national monument known as the Casino, was handed over to the Commissioners of Public Works. Another portion was set aside for community homes for senior citizens, while the balance including the institute premises was to be developed as a training centre for the Dublin Fire Brigade. This was a major step forward in raising the standards of professionalism in the brigade: it was probably the single most significant development since the opening of Tara Street in 1907.

Over the years this excellent facility was refurbished to provide for the full range of breathing apparatus training, including a purpose-built "ship". Classrooms were upgraded, a canteen was provided and eventually living-in accommodation for students was made available. There is a large drill yard with drill towers, an oil-fire training facility and a crash-rescue area. All of this has created an establishment where the integrated emergency service which is the Dublin Fire Brigade can cater for all aspects of its training needs.

With a developed cadre of qualified instructors and access to graduate engineers or architects all practical and technical training programmes in fire prevention, fire fighting, ambulance and emergency procedures and techniques can be provided on site. The campus could have been developed as the essential national training centre for the fire service if it had been fully supported by both the Department of the Environment and the other fire authorities. However its reputation has grown through the provision of courses for overseas fire officers, industry, the port authority, management safety officers – and through its training in relation to hazardous materials, manual handling and lifting techniques, hydraulic platform training and instructional techniques, etc. It has been a success story for the brigade, with selected courses accredited by the National Ambulance Advisory Council, the Health and Safety Authority, the National Safety Council, the Department of the Marine, the National Registry of Emergency Medical Technicians, the American Heart Foundation, etc.

In 1989 the Dublin Fire Brigade became one of the three centres selected by the Department of the Environment and Local Government when it was decided to initiate what was known as the Computer Aided Mobilisation Project (CAMP) The object of this project was to rationalise emergency 999 call-taking for the fire service nationally in three control centres. A new building, designed for the purpose, was constructed as part of the Dublin brigade headquarters complex and was called East Control Centre. CAMP East comprises the fire authorities of the twelve counties of Leinster plus Cavan and Monaghan. The Dublin centre also provides A&E ambulance service for the city and county of

Loreto Convent School, St Stephen's Green, destroyed by fire 1986.
Photograph, © Collection Dublin Fire Brigade Historical Society

Dublin while the Eastern Region Ambulance Service area – comprising of Dublin, Kildare and Wicklow – although operationally and administratively independent, is serviced from the same control room.

The interim system, installed in the Dublin Fire Brigade in 1993, involved a change from the much criticised manual, paper-based call-taking to the use of a computer for address location and mobilisation. Almost 600 fire fighters were trained to use the computer system before it became fully operational. The original system has been modified quite substantially since 1993 and is due to be replaced by a new command and control system which interfaces to a fire service standard GD92 communications system. Both of these systems process a 999 call from time received from Eircom to bells or pagers sounding in the stations. At present the fire services of Dublin, Meath, Wexford, Longford and Laois are catered for fully by the control centre. A further nine counties await their integration into the system.

As part of this system an accurate and comprehensive database of addresses and other information required for efficient mobilisation has been built up. The address database uses streets, townlands, towns and villages as the main searchable fields. Local knowledge is tagged to each address held for street and townland by the fire crews who operate in the local station. The new command and control system will also present geographical information to the call taker to help match address and plan and track resource deployment. The system is designed specifically for emergency service requirements and various Ordnance Survey map scales are used. Functions include address location/verification with caller, data capture and updating, pre-fire planning, mobilising, management information systems and radio propagation analysis. All this control room operation is carried out by operational trained firefighters.

For the year 2000, CAMP East fire control room dealt with 140,875 call-outs for the Dublin brigade, 497 for Longford, 746 for Laois, 1,433 for Wexford plus calls for Meath, totalling somewhat in the region of 150,000. The calls to the Dublin brigade break down into 5,084 domestic fires, 25,033 other fires, 11,561 road traffic accidents, 183 other rescues, 106 hazardous substances, 94,303 ambulance calls and 1,605 miscellaneous, of which 600 were false alarms with good intent and 2,180 were malicious.

The total cost of the brigade for 1999 was £49,894,981 covering 11 full time stations and 4 retained stations with an operational staffing level of 1 chief, 2 assistant chiefs, 6 senior executive fire prevention officers (FPOs), 8 executive FPOs, 3 assistant FPOs, 6 third officers, 27 district officers, 73 station officers, 70 s/officers and 615 firefighters full time; plus four retained s/os, 2 retained s/officers and 50 retained firefighters. With the employment of female firefighters in 1994 the name "fireman" was dropped in favour of the non-gender name of "firefighter" and in 1999 there were 6 female firefighters in the brigade. The brigade has 32 standard fire engines, a foam tender, 4 turntable ladders, 2 hydraulic platforms, 3 emergency tenders, 4 water tankers, 2 prime movers, 1 coach, 14 ambulances, 7 vans, 11 cars, 4 mini-buses, a decontamination unit and 252 breathing apparatus sets, as well as a whole range of lesser items of equipment. The two prime movers are used to transport de-mountable pods which contain special equipment for specific situations: control, foam, major incident, hose, breathing apparatus.

In 1985 a group of Dublin firemen held a meeting to form a pipe band. It was a momentous decision. Within three years the band was participating in the St

Patrick's Day Parade and two years later it was performing in Scotland. Since then it has become the great ambassador of the Dublin Fire Brigade, travelling to perform in many of the major cities of the USA and in Moscow in 1993 as the guest of the Russian fire service on the occasion of its 75 anniversary. The band performs regularly at venues in County Dublin, at "pass-out" functions for recruits, on civic occasions, at the Rose of Tralee, and at charity and sporting events. Continually seeking new blood it has maintained its deserved image as a unifier and a true brigade ambassador. In 1999 it was honoured by the Northern Ireland fire service by being invited to attend the passing-out ceremony for recruits in the Belfast Training Centre and again to play at the memorial service in Lisburn in June 2000 at the unveiling of the fire service memorial statue to those firefighters who lost their lives in the fire service in Northern Ireland.

In 1994 the Dun Laoghaire Fire Brigade was merged with the Dublin Fire Brigade, which now provides fire, rescue and emergency ambulance cover for practically the whole of the former county of Dublin. This area is now divided into four independent local authorities, namely Dublin City Council, Fingal, South Dublin and Dun Laoghaire-Rathdown.

Tragedy again struck the brigade on 26 August 1994 when Stephen (Timmy) Horgan, a fine sports man and very popular firefighter with a young family, was killed when the ambulance he was driving to an emergency call overturned when it struck a loose manhole cover on the Dublin-Belfast road. Timmy repre-sentented the best traditions of the fire brigade, a great family man who loved working in the service. His ambulance assistant, young firefighter Gary Burke, sustained serious injury in the same accident. Tragedies like this are a constant reminder of the day-to-day dangers faced by firefighters every time they respond to an emergency call.

With the latest census figures for Dublin city and county showing a rise to 1,056,666 the workload on the brigade is ever-increasing. In 1998 the fire preven-tion section dealt with 974 new applications for fire safety certificates, inspected 305 premises under the Fire Services Act 1981, 395 under Section 24 (Licensing) and 76 retail and private petroleum stores. This growing volume of work completed, along with all the other fire prevention work undertaken, is a major advance on the pre-Stardust era. At the end of the twentieth century much improvement has taken place in all aspects of the Dublin fire service and this progress needs to be maintained in order that an emergency service capable of protecting the people of Dublin in the future continues to be provided.

## The Dublin Corporation Fire Brigade Act, 1862

# 25 *Victoria, cap. xxxviii*

*An Act to extend and define the Powers of the Right Honourable the Lord Mayor, Aldermen, and Burgesses of Dublin in respect to the extinguishing of Fires, and the Protection of Life and Property against Fire; and for other Purposes. 3 June 1862.*

Whereas by "The Dublin Corporation Waterworks Act, 1861" the Right Honourable the Lord Mayor, Aldermen and Burgesses of *Dublin* (in this Act called "the Corporation") are authorized and required for the Purpose of providing a Supply of Water for better Security against Fire, and other the Purposes therein mentioned, in the Month of *December* in every Year, by Precept under their Common Seal, to order and direct the Collector General to applot, collect and levy upon and from the Owners of all rateable Property within the City or Borough of Dublin a Rate called "the Public Water Rate:" And whereas certain other Powers and Provisions in reference to an efficient Water Supply, and to the Maintenance and Repair of Fireplugs are granted to and imposed on the Corporation by the said Act: And whereas the Corporation have for some Time past, at considerable Expense, established and maintained a Fire Brigade, Fire Engines and other Appliances in the said City, but on too limited a Scale to afford sufficient Protection to the Lives and Property of the Citizens; and Doubts are entertained whether the Funds under the Control of the Corporation are legally applicable to such Purposes: And whereas it is expedient that such Doubts should be removed, and that, in addition to the Powers already conferred on the Corporation, Provision should be made for the Formation and Maintenance by them of an efficient Fire Brigade, and the Supply of all proper Steam and other Engines, Machinery, and Appliances for the Protection of Life and Property from Fire: And whereas the Objects aforesaid cannot be effected without the Authority of Parliament: May it therefore please Your Majesty, by and with the Advice and Consent of the Lords Spiritual and Temporal, and Commons, in this present Parliament assembled, and by the Authority of the same, as follows:

1. This Act may be cited for all Purposes as "The *Dublin* Corporation Fire Brigade Act, 1862."

2. The recited Act and this Act shall be read and construed together as One Act, and, except as is by this Act otherwise provided, the several Words and Expressions to which in the recited Act Meanings are assigned shall have in this Act the same respective Meanings, unless excluded by the Subject or Context.

3. It shall be lawful for the Corporation to erect and provide, in some convenient Situations in the Borough, such Houses, Lands, and Buildings for the Accommodation of Firemen, and for the Reception and Custody of Fire

Engines, Fire Escapes, Water Carts, and other Apparatus and Appliances, as the Corporation shall think necessary, and from Time to Time to alter the Situations of the present and all future Fire Engine Stations; and for those Purposes the Provisions of "The Lands Clauses Consolidation Act, 1845" with respect to the purchase of Lands by Agreement shall be incorporated with them and form Part of this Act.

4. It shall be lawful for the Corporation to purchase or provide Steam and other Engines and Machines, Water Carts, Water Buckets, Pipes, Hose, Fire Escapes, and other Implements and Apparatus for extinguishing Fire, and to purchase, keep or hire Horses, and generally to employ a Superintendent and Officers to instruct, train, organize, and control a Fire Brigade, and regulate their Operations, and also to employ a proper Number of Persons to act as Firemen, for directing and working the said Engines and Fire Escapes, and as Drivers of the said Engines and Water Carts, and out of the Income arising or to arise from "the Public Water Rate" authorized to be levied by "The *Dublin Corporation Waterworks Act, 1861*" to take on Lease or purchase all such Houses, Lands and Buildings, and purchase and provide such Fire Engines, Machines, Water Carts, Water Buckets, Pipes, Hose, Fire Escapes, Horses, and other necessary Apparatus and Appliances, and allow and pay all such Superintendents, Officers, Firemen, and other Persons who may be so employed such Salaries or Wages as the Corporation may think proper, with full Power and Authority to displace and remove such Superintendents, Officers, Firemen, and other Servants from Time to Time, and to appoint others in their Stead, and also to regulate the Expenditure of the said Establishment, and from Time to Time to frame such Rules and Regulations for the Government of the Fire Brigade, and for the Government and Control of such Members of other separate or associated Company or Companies, Parish Officers, or other Party or Parties who may be dispose to assist in the Extinction of Fires, and the Preservation of Life and Property therefrom, and attend for that Purpose; and also to frame Rules and Regulations for the more effectual Prevention of Disorder, Neglect, or Abuse, and the rendering such Superintendents, Officers, Firemen, and other Persons employed in the said Fire Brigade efficient in the Discharge of their Duties, and to impose and recover Fines or Penalties for any Breach or Nonobservance of Such Rules and Regulations, and generally to do all other Matters and Things which they may deem expedient and necessary to do, with a view to the good Government and Utility of the Fire Brigade: Provided always, that any such Rules and Regulations as aforesaid shall be subject, as to Publication, Confirmation, and Evidence, to the Provisions with respect to Byelaws contained in the Eighteenth, Nineteenth, Twentieth, Twenty-first, and Twenty-second Sections of "The Dublin Improvement Act, 1849."

5. In all Cases where any Superintendent, Officer, Fireman, or other Person who shall have been employed by the Corporation in any Capacity in the Fire Brigade Establishment, and shall have been discharged therefrom, continues to occupy any of the Houses or Buildings so to be provided as aforesaid, or

any Part thereof, after One Week's Notice in Writing from the Corporation, signed by the Town Clerk, to quit and deliver up the Possession thereof, it shall be lawful for any Divisional Justice, on the Oath of One Witness of such Notice having been so given, by Warrant under his Hand to order and direct any Constable or other authorized Bailiff or Officer usually employed to execute the Orders of such Justice to enter into and upon the House or Building occupied by such discharged Superintendent, Officer, Fireman, or other Person as aforesaid, and to remove him and his Family and Servants therefrom, and afterwards to deliver the Possession thereof to the Corporation as effectually to all Intents and Purposes as the Sheriff of the County of the City of *Dublin* could or might lawfully do under and by virtue or a Writ of Possession or a Judgment at Law.

6. If any Superintendent, Officer, Fireman, or other Person who may from Time to Time be employed by the Corporation, either temporarily or permanently, in extinguishing any Fire, shall happen to suffer or sustain any material Damage or Injury in the Discharge of such Duty, it shall be lawful for the Corporation, if they think fit, to grant him or his Representatives reasonable Compensation, either by a Sum in gross or by Annuity, for such Damage or Injury, out of the said "Public Water Rate;" but no such Superintendent, Officer, Fireman, or other Person, or his Representative, shall be entitled to claim Compensation as a Matter of Right, in case the Corporation shall not deem it expedient to grant the same.

7. The Fire Brigade constituted under this Act may, under the special Order and Direction of the Superintendent or other superior Officer present in charge thereof, on the Occasion of a Fire take any Measures that appear expedient for the Protection of Life and Property, and may on any such Occasion shut the Water off from the Mains and Pipes of any District in order to give a greater Supply and Pressure of Water in the District in which the Fire has occurred.

8. It shall be lawful for the Corporation, if they think fit, by or out of the said Public Water Rate, to make and establish and from Time to Time extend, by means of Telegraphic Wires or other approved Means, such rapid Communication between the Houses and Buildings in which their Fire Engines, Fire Escapes, and Fire Brigade may be stationed and maintained, and the more distant Parts or Quarters of the City or Borough, and do all other necessary Acts and Things for the furtherance and facilitating of such Communication as to them may seem expedient.

9. It shall be lawful for the Corporation, as Occasion may require, to permit such Engines, Escapes, Implements and Apparatus, and every or any Part of the Fire Brigade Establishment, to proceed beyond the Limits of the City or Borough for the Purpose of extinguishing Fire happening to Property in the Neighbourhood thereof; and the Owners of such Property shall in such Case defray the actual expense that may be thereby incurred, and shall also pay to the Corporation a reasonable Charge for the Use of such Escape or Engines, Implements, and Apparatus, and for the Attendance of such Fire Brigade; and in case of Difference between the Corporation and the Owners of such

Property, the Amount of the Expenses and Charges, as well as the Propriety of the said Engines, Escapes, Implements, Apparatus and Fire Brigade having proceeded as aforesaid for the Purpose of extinguishing such Fire (if the Propriety thereof be disputed), shall be summarily determined by any Divisional Justice, whose Decision shall be final and conclusive on all Parties, and the Amount of the said Expenses and Charges in dispute shall be fixed by such Justice, and may be recovered in like Manner as any Penalty imposed by the recited Act is or may be recoverable, and the Amount of all such Expenses and Charges paid or recovered shall be applied in Payment or in Aid or the Costs of maintaining such Fire Brigade Establishment.

10. The Owner and Occupier of any House, Building, or Premises, Ship, Goods, or other Property within the Limits of the City or Borough in which a Fire breaks out, shall be jointly and severally liable to pay and shall pay to the Corporation, as a Contribution towards the Expenses incurred or to be incurred in extinguishing such Fire, the Sum of Fifteen Pounds Sterling, or whatever lesser Sum is equal to One Half of the said Expenses; and the Amount of such Contribution or Proportion of Expenses shall in the case of Difference be determined, and be recoverable and applied in manner provided by the immediately preceding Section of this Act with respect to the Expense of extinguishing Fires beyond the Limits of the City or Borough.

11. The Corporation may receive any Sum or Sums of Money either by way of Donation or annual Subscription, which Insurance or other Companies, Societies, or Individuals may agree to allocate or contribute towards the Expense of the Establishment and Maintenance of the Fire Brigade, Engines, Implements, and Appliances requisite for the Prevention and Suppression of Fire and the Protection of Life and Property; and the Corporation may, if they think fit, purchase and acquire from any such Company or Society now existing in *Dublin* any Fire Escapes or other Implements or Machinery they may be possessed of for the Extinction of Fire or the Protection of Life and Property from Fire for such Consideration as may be mutually agreed on.

12. Provided always, That the Proportion of "the Public Water Rate" which the Corporation may appropriate and apply for the Purposes of this Act, shall not exceed in Amount Three Halfpence in the Pound of the annual Value of the Property in respect of which "the Public Water Rate, " shall from Time to Time be applotted and levied for each and every of the First Ten Years, and One Penny in the Pound for each and every succeeding Year.

13. The Costs, Charges, and Expenses attending on or incident to the applying for and obtaining this Act shall be paid by the Corporation out of the said Public Water Rate.

# Pembroke Fire Brigade

The Pembroke township was created in 1863 and under the 1898 Local Government Act it became the Pembroke Urban District Council. It covered the area of Donnybrook, Ballsbridge, Sandymount and Ringsend – formerly part of the estate of the Earl of Pembroke. Like Rathmines it became part of Dublin from October 1930. From 1863 to 1899 the township was run by fifteen commissioners, five of whom would retire each year and an election would take place. One of these commissioners was Edward Henry Carson who served from 1863 to 1881 and was the father of "Coercian Carson" who led the Unionists in their efforts to successfully partition the country. To qualify as a commissioner one had to live in the township and have a poor law rating from £30 upwards or, if non-resident, have land or other holding worth £200 or upwards. In 1892 the commissioners opposed the Municipal Franchise Bill which would establish virtually full manhood suffrage, claiming that under certain conditions if this Act became law "we might have the curious spectacle of a Board of Commissioners composed of the inhabitants of cabins". They preferred to maintain a system where less than 3,000 of the approximately 25,000 population were allowed to vote. Their housing policy was regularly criticised because of the huge contrast between the great houses of the wealthy in Donnybrook/Ballsbridge and the cottages and slums of Ringsend and Irishtown. The Council was dominated by pro-Unionists until the 1919 Local Government elections. The commissioners formed a fire service of sorts in 1865 when they directed the

> surveyor to keep tank and water carts full of water, same in readiness for fire, to employ Commissioners' men and horses for the purpose of extinguishing fires should such occur and that half the expenses so incurred shall be charged to the owner or occupier of any house, building or other property in which fire breaks out.

In spite of Captain Ingram's correspondence in 1869 setting out the position of the Dublin Fire Brigade in responding to fire calls from the township, the commissioners refused to purchase a fire engine. In 1870 Dublin Corporation rejected their application for telegraphic communication with the Dublin Fire Brigade. When they proposed in 1877 to pay Dublin Corporation £50 for use of the brigade they were told "that 1½ pence in the £ on their rates would provide a fire station in the Township and the necessary appliances". It was not until 1881 that they finally accepted the need to establish a fire service of their own, ordering "caps and red shirts for six men to constitute a fire brigade, 600 feet of canvas hose, a set of harness, four bells, two horses, lamps and stand pipes". The following year they purchased two telescopic ladders from Brown's Dublin Wheel Company for £4–16 each. In May 1884 they bought "two bay geldings for £110. The following month P. J. Graham, who had carried out the duties of superintendent on a part-time basis, retired.

Fire engines belonging to Pembroke Fire Brigade. Photograph, © Las Fallon

By 1892 the commissioners had attached to the town hall (built in 1880) a fire brigade supervisor and four men with a hand pump (Merryweather) plus the horses and one of the telescopic ladders etc. In the early years of the twentieth century the numbers increased to two officers and seven men with motor driven vehicles. A Merryweather pump purchased in 1910 was joined by two BSA/Knight ambulances in 1913, by a Daimler Hose Tender the following year, and later by an Armstrong Siddeley Fire Tender, a Ford motor and a Sulzer trailer pump. The latter saw service at the Belfast Blitz in 1941. The Pembroke brigade was so proud of its early abolition of horses that each vehicle from 1910 was lettered "Petrol Motor No –".

The rivalry with Dublin ended in 1930. The Pembroke fire station was closed in March 1931 following Captain Connolly's recommendation that "having considered the number and nature of fires attended by the Pembroke section during the past five years, the total of which is 189, and the proximity of both Rathmines and Tara Street Stations, I am confident that equal protection may be assured to the Pembroke Area by these two sections". All plant and men were distributed to Rathmines or Tara Street and their engines and ambulance were painted over with the Dublin Fire Brigade colours and logos.

The final chapter on Pembroke ended in the autumn of 1931 when the town-hall and fire station were leased to the Vocational Education Committee and the Earl of Pembroke was paid £2,200, which the original lease required once the premises passed out of the hands of Pembroke council.

The hand-drawn fire escape was removed from the corner of Lansdowne Road and Dublin Corporation sanctioned the erection of five new street alarms, all of which were to be connected to Tara Street station, including those formerly installed by both Pembroke and Rathmines.

# Rathmines and Rathgar Fire Service

The township of Rathmines was created by Act of Parliament in 1847. Under the Improvement Act 1862 the areas of Rathgar and Ranelagh were added. A further extension took place in 1866, with Milltown being added in 1880. This then was the area that came under Dublin Corporation in 1930. Originally the township was governed by commissioners but under the Local Government Act 1898 the Rathmines and Rathgar Council was established as the governing body.

In 1861 the Royal Irish Society for the Protection of Life from Fire provided two of its fire escapes to the township, one being placed in a commissioner's yard. These escapes were different to those allocated in the city in that they were narrower so that they could be trundled "through the small gates leading to the houses in the township". Many of the stately houses in the township were built back from the roads with long front gardens. Captain Ingram on his appointment as chief officer in Dublin sent an order to Rathmines, with the agreement from the Royal Irish Society for the Protection of Life from Fire, that the escapes be handed over to him. The township, along with the Kingstown Improvement Commissioners, had earlier resolved "to support the Corporation (Fire Brigade) Bill and oppose the other".

*As Parliament has entrusted the Corporation with exclusive powers for the supply of water to the city of Dublin, your petitioners consider that it would be to the advantage of the city, that the entire control over all officers and men and over all engines and other instruments used for extinguishing fires within the city would be entrusted to that body in preference to the police, whose duties in preserving order are sufficient to occupy their lives and attention ... as much as from motives of humanity, the Corporation will be required to send their engines and staff to places outside their jurisdiction. Your petitioners are favourable to the clause providing that such services are paid for by the parties benefited.*

It is clear from this petition that the township had motive in supporting the Dublin Bill and believed they would have a right to avail of Dublin Fire Brigade services.

The Rathmines and Rathgar Act of 1880 included sixteen sections enabling the township commissioners to establish a fire brigade and provide a station and fire fighting equipment. Following additional powers conferred in 1883 respecting the borrowing of money, further progress was made in forming an adequate fire brigade for the township. A new superintendent of the fire brigade, Charles Smith, was appointed in November 1886 with a salary of forty shillings per week plus uniform. In 1893 the fire station was connected with the central telephone office in the townhall, and in 1899 an offer by the National Telephone Company for the erection and maintenance of fire alarms was accepted – "cost not to exceed £30 per annum". Chimney fires were obviously a problem because in 1900 the superintendent was reporting on the prosecution of "persons leaving chimneys on fire".

The brigade already had a Shand Mason steamer and a manual engine, and in 1901 a fire escape was purchased. But a more modern fire appliance was now required. In the early years of the twentieth century, while the petrol driven combustion engine was becoming ever more widespread, motor propulsion by electricity was being introduced. Henry Simonis and Co of London was producing fire brigade vehicles propelled by motors mounted on the two front wheels and supplied with current from batteries stored in the bonnet. The chassis was built by the Austrian Daimler Motor Co. At a meeting of Rathmines and Rathgar Urban District Council on 7 June 1911 the following recommendation of the Public Health Committee was accepted: "That Messrs. Henry Simonis' tender of £900 for Electric Motor be accepted as being the most suitable". The appliance purchased was a Hose Reel Tender (Reg IK 874) equipped with two twenty-five gallon soda acid/water tanks at the rear and carrying a large ladder. With the fairly rapid increase in the reliability of the petrol motor, electric commercial vehicles with their heavy weight of batteries requiring frequent charging did not remain popular for many years.

The next major appliance bought by Rathmines Fire Brigade was destined to be its last. It was a Leyland 500gpm motor pump (Reg IK 4246) and cost £2,000. It was officially tested on 29 September 1921 and remained with the brigade until the city boundary extension of 1930 when it became part of the Dublin Fire Brigade fleet. In 1960, by a stroke of good fortune, it was obtained on permanent loan by the Transport Museum Society – the first Dublin fire engine to be preserved as part of the city's history. It now awaits total refurbishment, time consuming and costly!

In November 1914 the Commissioners had contracted with Joseph Pemberton Builders for the conversion of 67 Rathmines Road into a fire brigade station and it was this premises which was to serve that district as a fire station for the next seventy years. The superintendent lived in Homeville.

The Rathmines firemen were represented by the Irish Stationary Engine Drivers' and Firemens' Union but had major problems gaining recognition from the town council. In October 1913 the council finally agreed to recognise the "employees' Trade Union provided it is confined to workmen in the employment of the Council. The Council reserve the right of employing any person they may think fit", and in March 1915 the council agreed "a 1/- a week war bonus to all employees earning less than 25/- per week, on the distinct understanding that the allowance cease on the termination of the War". This was increased to two shillings in October. In December 1917 the council "resolved that an increase in wages of 5/s per week be granted to all able bodied men, save Fire Brigade men, in the employment of the Council, at present in receipt of weekly wages not exceeding 40/s per week".

In November 1916, following negotiations with the Rathmines and District Workmens Union, "all employees with the exception of Electricity, Fire Brigade and Refuse Destructors" were to have their working hours reduced to forty-five

Rathmines Fire Brigade, with chief officer in civilian clothes.
Photograph, © Collection Dublin Fire Brigade Historical Society

hours per week in winter. This led the firemen in 1918 to make a claim for £1 on pre-war wages and 12½% not paid on bonus. When this was rejected the firemen threatened strike action, so the claim was referred to an arbitrator. The firemen then rejected the arbitrator's award and served strike notice to expire on 31 May 1918.

At the special meeting of the council held on 28 May the members present unanimously decided that "in as much as the members of the Irish Stationary Engine Driver's and Firemen's Union refuse to accept the award of the Arbitrator, after signing an agreement to accept same, this Council accepts the notice of above Union to withdraw their membership from our employment on 31st May 1918". The council then "referred the Lighting Committee to take all necessary steps to deal with the strike if it comes". The Union withdrew the strike notice. However all Rathmines employees including the firemen went on strike in August. This strike led by the United Corporation of Dublin Workers Trade Union lasted for one week before a settlement was agreed. The fire engine was manned by Lieutenant Whyte but did not have a call-out, while the ambulance work was performed by a military detachment. Such was life in Rathmines fire service; the senior officers were given a £5 bonus for staying at work during the strike and were further rewarded by an increase of £26 per annum from 1 April 1920.

# Chief Officers of Dublin Fire Brigade from 1862

| Name | Dates in Office |
|---|---|
| Robert Ingram | 1862-1882 |
| John Boyle | 1882-1892 |
| Thomas Purcell | 1892-1917 |
| John Myers | 1918-1927 |
| John Power | 1927-1929 |
| Joseph Connolly | 1929-1938 |
| James Comerford | 1938-1948 |
| Valentine F. Walsh | 1949-1951 |
| Patrick Diskin | 1952-1961 |
| Thomas O'Brien | 1962-1982 |
| Cathal Garvey | 1987-1990 |
| P.A. Gillick | 1990-2001 |

Michael J. Walsh was Acting Chief Fire Officer from 1982-1987 and from 2001-2004

# Principal Mobile Firefighting Appliances in use in Dublin Fire Brigade, 1863–2000

| Type* | Year | Reg. number | Make |
|---|---|---|---|
| M | 1863 | | Shand Mason |
| | | | Four "London Brigade" types purchased |
| S | 1864 | | Shand Mason |
| | | | 300gpm Single Vertical |
| S | 1867 | | Shand Mason |
| | | | 300gpm Single Vertical |
| S | 1893 | | Shand Mason |
| | | | 750gpm Equilibrium |
| AL | 1899 | | William Rose |
| | | | Designed by T. Purcell |
| AL | 1904 | | William Rose |
| | | | Designed by T. Purcell |
| AL | 1908 | | William Rose |
| | | | Designed by T. Purcell |
| P | 1909 | RI 1090 | Leyland |
| | | | The first fire engine built by Leyland |
| | | | Designed by T. Purcell |
| | | | The first motor fire engine in Dublin |
| | | | The first appliance lettered "Dublin Fire Department" |
| P | 1912 | RI 2080 | Leyland |
| TL | 1924 | YI 5204 | Morris/Magirus |
| | | | 85 ft wooden ladder |
| P | 1924 | YI 5637 | Leyland |
| AL | 1925 | YI 7587 | Peugeot/Rose |
| | | | One of the original horse-drawn Aerial Ladders |
| | | | Re-built on motor chassis |
| AL | 1925 | Not known | Peugeot/Rose |
| | | | One of the original horse-drawn Aerial Ladders |
| | | | Re-built on motor chassis |
| P | 1930 | ZI 4234 | Albion/Merryweather |
| P | 1931 | IK 686 | Merryweather |
| | | | Bought new in 1910 by Pembroke |
| P | 1931 | IK 4246 | Leyland |
| | | | Bought new in 1921 by Rathmines |
| TL | 1931 | ZI 7528 | Albion/Merryweather |
| | | | 85 ft wooden ladder |

* for abbreviations see p. 312

| Type | Year | Reg. number | Make |
|------|------|-------------|------|
| TL | 1936 | ZA 7706 | Merryweather |
| | | | 100 ft steel ladder |
| P | 1937 | ZC 2386 | Morris/Merryweather |
| PE | 1939 | ZC 9696 | Leyland Cub |
| | | | The first modern wheeled escape |
| PE | 1940 | ZD 2147 | Leyland Cub |
| P | 1941 | ZD 2902 | Leyland Cub |
| | | | The last open-bodied appliance |
| | | | The last appliance lettered "Dublin Fire Department" |
| ET | 1940 | ZD 2275 | Dodge |
| | | | Colour blue |
| PE | 1953 | ZO 8056 | Dennis F12 |
| | | | The first motor appliance to be lettered "Dublin Fire Brigade" |
| P | 1954 | KRI 670 | Dennis F8 |
| TL | 1955 | YRI 584 | Dennis/Metz |
| | | | 100 ft ladder |
| PE | 1956 | MIK 678 | Dennis F12 |
| PE | 1956 | MIK 809 | Dennis F12 |
| P | 1956 | NIK 749 | Dennis F8 |
| P | 1956 | NIK 888 | Dennis F8 |
| ET | 1963 | WZD 265 | Bedford/HCB |
| | | | Colour silver (unpainted) |
| | | | Lettered "Dublin City Fire Brigade" |
| L4P | 1963 | JZE 864 | Austin Gypsy |
| PE | 1964 | RZE 116 | AEC/Merryweather |
| | | | The first diesel appliance |
| TL | 1964 | RZE 117 | AEC/Merryweather |
| | | | 100 ft ladder |
| P | 1965 | MZH 885 | AEC/Merryweather |
| P | 1967 | GZL 809 | AEC/Merryweather |
| | | | The first appliance to have two-tone horns |
| L4P | 1969 | LYI 850 | Land Rover |
| | | | Bought new in 1958 by Dublin County Council |
| L4P | 1969 | GZA 685 | Land Rover |
| | | | Bought new in 1960 by Dublin County Council |
| L4P | 1969 | GZA 686 | Land Rover |
| | | | Bought new in 1960 by Dublin County Council |
| L4P | 1969 | GZA 687 | Land Rover |
| | | | Bought new in 1960 by Dublin County Council |
| WrT | 1969 | IZU 967 | AEC/HCB-Angus |
| | | | The first appliance to have a siren |
| WrT | 1970 | 6166 RI | AEC/HCB-Angus |
| WrT | 1971 | 3285 Z | AEC/HCB-Angus |
| WrT | 1972 | 5129 ZC | AEC/HCB-Angus |

| Type | Year | Reg. number | Make |
|------|------|-------------|------|
| WrT | 1972 | 5220 ZC | AEC/HCB-Angus |
| | | | First appliance with 45ft alloy ladder |
| TL | 1972 | 6524 ZE | Merryweather |
| | | | 100 ft ladder |
| WrL | 1973 | 6295 ZI | Dodge/HCB-Angus |
| WrL | 1973 | 6296 ZI | Dodge/HCB-Angus |
| WrL | 1973 | 6297 ZI | Dodge/HCB-Angus |
| | | | The last appliance to be equipped with a bell |
| WrL | 1975 | 42 NIK | ERF |
| WrL | 1975 | 43 NIK | ERF |
| WrT | 1976 | 977 GYI | Dodge/Carmichael |
| WrT | 1976 | 978 GYI | Dodge/Carmichael |
| WrL | 1977 | 637 DZA | Dennis R130 |
| WrT | 1977 | 251 BZA | Dennis R130 |
| FoT | 1977 | 441 EZC | Dodge/Carmichael |
| TL | 1978 | 453 PZC | Dodge/Magirus |
| | | | 100ft ladder |
| WrT | 1978 | 290 XZC | Dodge/Carmichael |
| WrL | 1978 | 815 CZD | Dodge/Carmichael |
| ET | 1979 | 9 HZH | Dodge/Carmichael |
| WrT | 1979 | 916 KZH | Dodge/Carmichael |
| WrL | 1979 | 261 MZH | Dodge/Carmichael |
| WrL | 1980 | 302 BZH | Dennis RS |
| WrT | 1980 | 303 BZH | Dennis RS |
| WrT | 1981 | 206 GZL | Dennis RS |
| WrT | 1981 | 207 GZL | Dennis RS |
| WrL | 1981 | 208 GZL | Dennis RS |
| WrL | 1982 | 40 NZU | ERF (Ex-UK) |
| WrT | 1982 | 41 NZU | Dodge/HCB (Ex-UK) |
| WrT | 1982 | 518 YZU | Timoney |
| | | | The first Irish-built appliance |
| WrL | 1983 | XS 157 | Timoney |
| WrL | 1983 | AZG 272 | Bedford/Carmichael (Ex-UK) |
| WrL | 1983 | AZG 274 | Bedford/Carmichael (Ex-UK) |
| WrT | 1984 | UZG 232 | Dodge/Hughes |
| WrL | 1984 | GZS 86 | Dennis RS |
| WrT | 1984 | GZS 87 | Dennis RS |
| DU | 1984 | IZS 917 | Dodge/Hughes |
| | | | Colour white |
| WrT | 1985 | PZS 88 | Magirus/Hughes |
| WrL | 1985 | YZS 433 | Timoney |
| WrL | 1986 | ZS 8254 | Dodge/Hughes |
| WrL | 1986 | ZS 8255 | Dodge/Hughes |
| WrL | 1987 | 87D 3823 | Dodge/Timoney |

| Type | Year | Reg. number | Make |
|---|---|---|---|
| WrL | 1987 | 87D 3947 | Dodge/Timoney |
| WrL | 1987 | 87D 5573 | Dodge/Champion |
| TL | 1987 | 87D 19999 | Dennis/Magirus |
| | | | 100 ft ladder |
| HP | 1987 | 87D 26262 | Dodge/Hughes |
| | | | The first Hydraulic Platform |
| WrL | 1988 | 88D 4368 | Mercedes/Hughes |
| WrL | 1988 | 88D 18860 | Volvo/Timoney |
| WrL | 1988 | 88D 18861 | Volvo/Timoney |
| WrL | 1988 | 88D 18862 | Volvo/Timoney |
| WrL | 1988 | 88D 18864 | Volvo/Timoney |
| PM | 1989 | 89D 19102 | Dennis/Hughes Multilift |
| | | | Pod Prime Mover 1 |
| | | | For Major Incident Unit |
| WrL | 1989 | 89D 999 | Volvo/Timoney |
| WT | 1990 | 90D 41377 | Mercedes |
| HP | 1991 | 82D 1620 | Dennis (Ex-UK) |
| WrL | 1991 | 91D 21865 | Mercedes/Hughes |
| WrL | 1991 | 91D 30561 | Mercedes/Hughes |
| PM | 1991 | 91D 32476 | Mercedes/Hughes Multilift |
| | | | Pod Prime Mover 2. Foam Support Unit. |
| WrL | 1992 | 92D 920 | Mercedes/Hughes |
| WrL | 1992 | 92D 9347 | Mercedes/Hughes |
| WrL | 1992 | 92D 25161 | Mercedes/Hughes |
| WrL | 1992 | 92D 17771 | Mercedes/Hughes |
| WrL | 1994 | 1580 ZJ | Dennis RS |
| | | | Ex-Dun Laoghaire |
| WrL | 1994 | TZS 944 | Bedford/Alexander |
| | | | Ex-Dun Laoghaire |
| ET | 1994 | 87D 8129 | Iveco/Hughes |
| | | | Ex-Dun Laoghaire |
| WrL | 1994 | 87D 8130 | Dennis/Hughes |
| | | | Ex-Dun Laoghaire |
| RT | 1994 | 94D 34868 | Mercedes/IDT |
| RT | 1994 | 94D 35131 | Mercedes/IDT |
| WrL | 1995 | 84D 6032 | Dennis SS (Ex-UK) |
| WrL | 1995 | 84D 6799 | Dennis SS (Ex-UK) |
| WT | 1996 | 80D 1210 | Bedford (Ex-UK) |
| TL | 1997 | 81D 1862 | Dennis/Magirus |
| | | | 100 ft ladder (ex-UK) |
| WrL | 1997 | 97D 29531 | Dennis Sabre/JDC |
| TL | 1997 | 97D 57009 | Dennis/Magirus |
| HP | 1998 | 86D 8002 | Dennis/Saxon (Ex-UK) |
| WrL | 1998 | 86D 8154 | Dennis/Carmichael (Ex-UK) |

| Type | Year | Reg. number | Make |
|---|---|---|---|
| WrL | 1998 | 86D 8155 | Dennis/Carmichael (Ex-UK) |
| WrL | 1998 | 85D 7913 | Dennis DS (Ex-UK) |
| WrL | 1998 | 86D 8156 | Dennis RS (Ex-UK) |
| WrL | 1998 | 86D 8157 | Dennis RS (Ex-UK) |
| WrL | 1998 | 87D 38423 | Dennis RS (Ex-UK) |
| RT | 1998 | 85D 8093 | Dennis/Saxon (Ex-UK) |
| WrL | 1999 | 99D 57140 | Volvo/Browns |
| WrT | 1999 | 86D 8216 | Dennis SS (Ex-UK) |
| WrL | 1999 | 99D 78151 | Dennis Sabre/Browns |
| WrL | 2000 | 00D 10148 | Dennis Sabre/Browns |
| WrL | 2000 | 00D 101876 | Dennis Sabre/Browns |
| WrL | 2000 | 00D 101879 | Dennis Sabre/Browns |
| TL | 2000 | 83D 4280 | Dennis/Magirus 100 ft ladder |
| WT | 2000 | 89D 51977 | Volvo |
| WrL | 2000 | 87D 38621 | Dennis (Ex-UK) |
| WrL | 2000 | 87D 38622 | Dennis (Ex-UK) |
| WrL | 2000 | 87D 38623 | Dennis (Ex-UK) |

**Abbreviations**

AL    Aerial Ladder: 66 ft: horse-drawn
DU    Decontamination Unit
ET    Emergency Tender
FoT   Foam Tender
HP    Hydraulic Platform
L4P   Light 4-wheel drive pump
M     Manual Pump: horse-drawn
P     Pump
PE    Pump Escape (carries wheeled escape)
PM    Prime Mover
RT    Rescue Tender
S     Steamer: horse-drawn
TL    Turntable Ladder
WrL   Water Tender Ladder (carries 45 ft ladder)
WrT   Water Tender
WT    Water Tanker
YEAR  Indicates the year in which the appliance entered service
Note: This list excludes ambulances, cars, A.F.S. pumps, personnel carriers, etc.

# Bibliography

*Documentary Sources*

**Dublin City Archives**
The Chain Book of Dublin
Correspondence Files from Dublin City Manager's Department, 1940-1951
Records relating to bombings in Donore and North Strand, Dublin 1941
Dublin Corporation Reports, 1854-1985
Dublin Corporation Water Supply Contracts, 1910-1960
Dublin Corporation Waterworks Committee Register of Supplies to the Dublin
    Townships, 1913-1938.
Dublin Fire Brigade Chief Fire Officers Annual Reports, 1863-1960
Minutes of the Municipal Council of the City of Dublin, 1854-1985
Minutes of the Waterworks Committee, Dublin City Council, 1861-1985
Rathmines & Rathgar Council Minute Books, 1917-1927.

**Dublin Fire Brigade Museum**
Correspondence of Major J.J. Comerford in connection with The Cavan Fire
    Tribunal, 1943.
Fire Brigade Order Books, Records of Personnel, Disciplinary Code, Drill Book
    1940, and various Booklets on Auxiliary Fire Service and Wartime Regulations
    stored in Fire Brigade Museum.
Log Book of Captain Purcell, Chief Fire Officer, 1892-1917
Personal Papers of Firefighters in Fire Brigade Museum.
Scrap Books of Chief Fire Officers Ingram 1862-1882
Boyle 1882-1892, Purcell 1892-1917.
Various newspaper cuttings kept by Chief Fire Officers or by Ms. Vera Hutchin,
    Secretary to the Chief Fire Officers for the 1940s, 1950s, 1960s and 1970s.

**Engineering Department, Dublin City Council**
Scrap Book of Spencer Harty Dublin City Engineer from 1860s to 1890s.

*Legislation*

The Licensing Acts. 1833 and 1988.
The Towns Improvement (Ireland) Act. 1854
The Dublin Corporation Waterworks Act. 1861
The Dublin Corporation Fire Brigade Act. 1862
The Petroleum Acts. 1871 and 1879
The Factory & Workshops Acts. 1878 and 1901
The Dublin Corporation Act. 1890
The Dublin Corporation Metropolis Water Supply Bill. 1891
The Local Government (Ireland) Act. 1898

313

The Cinematograph Acts. 1901 and 1909
The Government of Ireland Act. 1920
The Local Government Acts. 1925/1946/1960
The Public Dance Halls Acts. 1935 and 1977
The Fire Brigades Act. (Westminster) 1938
The Fire Brigades Act (Eire) 1940
The Fire Services Act (Westminster) 1947
The Fire Services Acts. (Northern Ireland) 1947/1969
The Factories Act. 1955
The Local Government (Superannuation) Act. 1956
The Office Premises Act. 1958
The Local Government Planning & Development Acts. 1963/2000
The Fire Precautions Act. (Westminster). 1971
The Dangerous Substances Act. 1972
The Local Government Superannuation & Pensions Act. 1976
The Fire Services Act. 1981.
The Safety, Health & Welfare at Work Act. 1989
The Building Control Act. 1990

## Newspapers and periodicals

Belfast Telegraph
Brigade Call
Commercial Motor
Cork Examiner
Dublin Evening Post
Dublin Historical Record
F.B.S. Newsletter (Ireland)
Fire
Firecall
Fire Cover
Fire Journal
Fire Protection Review
Fire and Water
The Fireman
Freeman's Journal
Irish Builder and Engineer
Irish Firefighter
Irish Independent
Irish News
Irish Press
Irish Times
Journal of the Insurance Institute of Ireland
The Kerryman
Sunday Independent
Sunday Press
Weekly Examiner

## Printed Sources

Bailey, Victor (editor). *Forged in fire. The history of the Fire Brigades Union.* (London, 1992).
Bermingham, Willie and O'Cuanaigh, Liam. *ALONE.* (Dublin, 1978).
Barton, Brian. *The Blitz. Belfast in the war years.* (Belfast, 1989).
Bell, J. Bowyer. *In dubious battle. The Dublin bombings 1972-1974.* (Dublin, 1996).
Blackstone, Geoffrey Vaughan. *A History of the British Fire Service.* (London, 1957).
Broadhurst, William and Welsh, Henry. *The flaming truth. A history of the Belfast Fire Brigade.* (Belfast, 2001).

Chartered Insurance Institute. *British Fire Marks*. (London, 1971).

City of Dublin Vocational Education Committee. *The Old Township of Pembroke, 1863-1930*. (Dublin, 1993).

Conroy, J.C. *Report on Remuneration and Conditions of Service in An Garda Siochana*. (Dublin, 1970).

Devine, Francis. "Trade Union Records in the Registry of Friendly Societies, Dublin" in *Saothar* (1986), no. 11.

Doherty, James. *Post 381. The memoirs of a Belfast air raid warden*. (Belfast, 1989).

Dublin Corporation. *Opening of the new Central Fire Station*. (Dublin, 1907)

Dublin Corporation. *Historical Souvenir. The Dublin Fire Brigade 1862-1937*. (Dublin, 1937).

Ellis, Peter Berresford. *A history of the Irish working class*. (London, 1985).

Farrell Grant Sparks Consulting. *Review of Fire Safety and Fire Services in Ireland*. (Dublin, 2002).

Fetherstonhaugh, Neil and McCullagh, Tony. *They never came home. The Stardust story*. (Dublin, 2001).

Fire Station Centennial Committee. *Newry Aflame. 1877-1977*. (Newry, 1977)

Fisk, Robert. *In time of war. Ireland, Ulster and the price of neutrality 1939-45*. (London, 1983).

Gilbert, Sir John T. and Lady. *Calendar of Ancient Records of Dublin*. 19 vols. (Dublin, 1889-1944).

Gilligan, H.A. *A History of the Port of Dublin*. (Dublin, 1988).

Harkness, David and O'Dowd, Mary. *The Town in Ireland*. (CHECK, 1981).

Holloway, Sally. *Courage high! A history of firefighting in London*. (London, H.M.S.O., 1992).

Holroyd, Sir Ronald. *Report of the Departmental Committee on the Fire Service*. (London, 1970).

*Journal of the Royal Society of Antiquaries of Ireland*. (from 1896 onwards).

Keane, Ronan. Report of the Tribunal of Inquiry on the Fire at the Stardust, Artane, Dublin on the 14th February, 1981. (Dublin, 1982).

Kravis, Judy and Morgan, Peter. *When the bells go down. A portrait of Cork City Fire Brigade*. (Cork, 2001).

Klopper, Harry. *The Fight against Fire. The history of the Birmingham Fire and Ambulance Service*. (Birmingham, 1954).

Macardle, Dorothy. *The Irish Republic. A documented chronicle of the Anglo-Irish conflict and the partitioning of Ireland, with a detailed account of the period 1916-1923*. (London, 1937).

McKinsey & Company, Inc. *Strengthening the local government service. A report prepared for the Minister for Local Government*. (Dublin, 1972).

Merryweather, James Compton. *The Fire Brigade Handbook*. (London, 1888).

Mullan, Don. *The Dublin and Monaghan Bombings*. (Dublin, 2000).

O'Farrell, Padraic. *Down Ratra Road. Fifty years of civil defence in Ireland*. (Dublin, 2000).

Poland, Patrick. *Fire Call*. (London, 1977)

Quaney, Joseph. *A Penny to Nelson's Pillar*. (Portlaw, 1971).

Redmond, Sean. *The Irish Municipal Employees Trade Union 1883-1983*. (Dublin, 1983).

Roetter, Charles. *Fire is their Enemy*. [An account of the Fire Service]. (London, 1962).

Shaw, Sir Eyre Massey. *The Business of a Fireman*. (London, 1945).

Somerville-Large, Peter. *Dublin – The Fair City*. (London, 1979).

Smith, Mike. *Irish Fire Engines*. (Dublin, 1985).

Stationery Office, Dublin. *Fire Protection Standards for Public Buildings and Institutions*. (Dublin, 1950 and 1967).

Stationery Office, Dublin. *Report on the Fire Service*. (Dublin, 1975).

Stationery Office, Dublin. *Report of the Tribunal of Inquiry into the Fire at Pearse Street*. (Dublin, 1937).

Whitehead, Trevor. *Dublin fire fighters. A history of fire fighting, rescue and ambulance work in the city of Dublin*. (Dublin, 1970).

Yeates, Padraig. *Lockout. Dublin 1913*. (Dublin, 2000).

Young, Charles Frederick T. *Fires, Fire Engines, and Fire Brigades*. (London, 1866).

# Index

ALONE , 277

Air Raid Precautions Act 1939, 220

Assembly Rolls, 1,2

Association of Professional Fire Brigade Officers, 137

Auxiliary Fire Service, *see* Dublin Fire Brigade

Barbon, Nicholas (first fire insurance office 1680), 2

Barrington, Thomas (engine keeper c. 1751), 6

Belfast Blitz 1941, 225–32

Betagh, Robert (fire master c. 1740), 5

Blackrock Township, 40, 132

Boundaries Extension Act 1900, 111

Bourke, E.J., 253

Boyle, John, 23, 50–90 *passim*

British Insulated Wire Co, 111

Brooke, Basil, 226

Brown, Stephen (engine keeper c. 1838), 10

Building Labourers Union, 163

Byrne, Alfie, 195

Byrne, James, 71–98 *passim*

Calendar of Ancient Records of Dublin, 4

Cameron, Charles, 96, 115

car bombs
Parnell Street, 272
Sackville Place, 269–70, 271–2
South Leinster Street, 272–3
Talbot Street, 272

Carson, Edward, 111

celluloid, 188

chief fire officers, 307

Chief Fire Officers Association, 276, 278, 284

Cinematograph Act 1909, 135

Cinematograph Association, 147–8, 196

city water supply, *see* Dublin Corporation

Civil War, 171–8

Clontarf Township, 56, 111

Collins, Michael, 174, 175

Comerford, Major J.J., 215–39 *passim*

*Commercial Motor*, 131

Computer Aided Mobilisation Project, 294–6

Connolly, Joseph, 157, 180, 186, 189–215 *passim*

conscription, 159

Cosgrave, William, 132, 137, 162, 176, 181

Crofton, Mervyn, 18, 20, 23, 25, 26, 28, 32, 38

Daly, P. T., 88

de Valera, Eamon, 159, 171, 226, 227, 229

Devlin, Joe, 159

Dillon, John, 159

Diskin, Captain P.J., 237, 240, 250

*Dublin Builder*, 25

Dublin Civic Museum, 9, 34, 44, 84, 260

Dublin Corporation
commissioners 1924–30, 182–94
city water supply, 6–7, 13, 14–17, 20–1, 23, 24, 39–40, 42, 49–50, 56–8, 93, 98, 121, 144, 183, 202, 213
dissolution of, 181, 262
earliest fire service, 2–7
establishment of Dublin Fire Brigade, 25–41
waterworks committee chairmen
Briscoe, J. M. C., 122–3
Dwyer, John, 29, 32
Gray, John, 30, 40, 51, 56, 57
O'Meara, John, 80, 82, 83, 84, 90, 96

Dublin Corporation Act 1890, 92

Dublin Corporation Act, 1897, 102, 108, 109, 110

Dublin Corporation Fire Brigade Act 1862, 31, 33, 93, 109, 298–301

Dublin Corporation Markets Bill 1901, 113, 117

Dublin Corporation Superannuation Bill 1905, 118

Dublin Council of Trade Unions, 87, 103, 212

*Dublin Evening Freeman*, 11

*Dublin Evening Post*, 73

Dublin Fire Brigade
accommodation, *see* fire stations in this entry
ambulance service, 102–3, 111, 112, 116–17, 120, 135, 138–9,143

145, 147–53, 165, 176, 180, 184, 192, 256

annual costs, 39, 65, 69, 77, 83, 180, 199, 261, 279, 282, 296

annual reports, 35, 38, 41, 46, 49, 52, 55–6, 64, 69, 74, 84, 92, 98, 118, 124–5, 135, 141, 148, 158, 160, 161, 180, 219, 224, 239, 250

Auxiliary Fire Service, 219, 222–3, 224, 230, 235, 249–50
abolition of, 236, 237

award system and awards, 38, 79, 80, 87, 92, 105, 109, 116, 121, 155, 159, 207, 218

breathing apparatus, 112, 120, 138, 193, 216, 274, 294

equipment, 27, 29, 41, 48–9, 50, 53–4, 58, 64, 93, 94, 112, 125, 179, 183, 196, 206, 214, 243, 250, 274, 282, 296

establishment of, 25–41

Fire Brigade Act 1862, 31, 33, 93, 109, 298–301

Fire Brigade Act 1940, 221–3, 237, 249, 270, 274, 278, 289, 292

fire prevention, 251, 255–6, 263, 264, 270, 275, 278, 279, 281, 292

fire service personnel
Barry, Patrick, 77, 78
Birmingham, Willie, 277
Blake, E., 121
Brady, Station Officer Jim, 291
Bruton, Patrick, 161
Burke, Gary, 279
Burke, Peter, 77, 78
Byrne, Henry, 105
Byrne, William, 141
Carroll, Third Officer Larry, 255
Clancy, Turncock, 117
Coleman, Tom, 232, 247–8
Conway, Station Officer James, 198
Crowe, Leslie, 207, 208, 210
Cullen, Richard, 74, 113
Cummins, Fireman, 84
Curran, Fireman, 190
Darmon, John, 217

Doherty, Inspector
  Christopher, 76, 77, 78
Dowling, Station Officer
  Martin, 176, 177, 216
Doyle, Edward, 159
Dunphy, Tom, 103, 112, 113,
  116–17
Gibney, John, 204
Gorman, Third Officer R.,
  202–3, 216, 219, 230, 250,
  252, 257
Guildea, Inspector James,
  74, 109, 112
Hearns, John Finbar, 257
Hines, John, 74
Hogan, Fireman, 84
Horgan, Stephen (Timmy),
  297
Jennings, Second Officer
  Martin, 121, 159, 162
Kavanagh, Inspector, 74
Kavanagh, Kevin, 273
Kelly, C., 121
Kelly, M., 121
Kelly, Station Officer, 173,
  176, 216
Kennedy, James, 184
Kiernan, Station Officer
  Joe, 103, 109–10
King, Bernard, 172
Kite, John, 73, 74, 75
Lambert, Fireman, 120, 121
Lancaster, R., 155
Leech, District Officer, 219
Lynch, Joseph, 159
McArdle, Peter, 206, 209,
  210, 213, 214, 216
McDonald, District Officer
  A., 248–9
McEvoy, James, 74
Malone, Robert, 209, 210,
  213, 214, 216
Markey, Station Officer
  James, 108–9, 113
Mansfield, Fireman, 117
Markey, John, 197–8
Matthews, Bernard, 172,
  190, 204
Mulligan, Michael, 273
Murphy, John, 161
Myers, William, 135
Nugent, Tom, 206, 209, 210,
  213
O'Brien, William, 103
O'Farrell, 121
O'Hara, R., 121
Potts, Tom, 209
Redmond, Francis, 156–7

Reilly, Patrick, 157
Rogers, District Officer,
  204, 206, 252
Ruddy, Tom, 255
Ryan, James, 74
Sarsfield, Joe, 117–18
Sergeant, Jim, 281
Smart, Station Officer, 262
Walsh, James, 157
Walsh, Station Officer, 217
Williams, Captain John, 255
fire stations
  Belgard Road, 281, 291
  Blanchardstown, 281
  Buckingham Street, 108,
    112, 113, 124, 125, 142, 180,
    201, 218, 260, 274, 275, 279,
    291
  Clarendon Row (later
    Chatham Row), 71–3, 76,
    80–1, 83, 97, 108, 119, 122
  Cook Street, 17, 35, 81, 129,
    138
  Dolphin's Barn, 217, 219,
    235, 236
  Donnybrook, 274, 281, 282,
    290, 291
  Dorset Street, 109, 113, 118,
    124, 125, 142, 180, 218, 251,
    258, 279
  Finglas, 263
  Kilbarack, 261, 263, 266
  North Strand, 274, 285
  Nutgrove, 281, 291
    125, 143, 179, 197, 219, 236,
    254, 255, 281
  Phibsboro, 279, 281, 291
  Rathmines, 196, 218, 274,
    281, 291
  South William Street, 33, 35,
    41, 46, 53, 64–5, 71–2, 73
  Thomas Street, 108, 124,
    129–30, 135, 137–8, 150, 151,
    168, 180, 217, 237
  Townsend Street (Great
    Brunswick Street / Pearse
    Street / Tara Street), 108,
    109, 121–4,
  Winetavern Street, 17, 18,
    20, 34, 35, 37, 41, 50–1, 53,
    58, 64, 70, 73, 81, 83, 96,
    108, 110, 116, 124, 125, 129,
    138
life insurance, 83, 84
major emergency plan, 263,
  269, 275–6, 281–2
motorisation, 127–8, 130–4, 136,
  138

out-of-area attendance and
  charges, 51, 64, 85, 97, 142–3,
  162, 195–6, 221, 222
pay and conditions, 35, 36, 37,
  65–113 *passim*, 125, 127, 128,
  136–7, 145, 154–5,
  160–71 *passim*, 176, 178, 183, 205,
  220, 221, 223
pensions, 83, 89, 97, 110, 113,
  114, 117–18, 183, 184, 198, 212,
  219, 221
pipe band, 296–7
senior officers
  Boyle, John, 23, 50–90 *passim*
  Byrne, James, 71–98 *passim*
  Commerford, Major J.J,
    215–39 *passim*
  Connolly, Joseph, 157, 180,
    186, 189–215 *passim*
  Diskin, Captain P.J., 237, 240
  Howard, James, 188, 189,
    209
  Ingram, James Robert,
    32–70 *passim*, 118
  Larkin, Brian, 240, 250, 251,
    252, 253, 259, 265
  Myers, John, 98, 125–87
    *passim*, 188
  O'Brien, Thomas, 252–91
    *passim*
  Power, John, 173, 176, 177,
    178, 183, 187, 188, 189
  Purcell, Thomas, 86–159
    *passim*, 188
  Walsh, Michael, 291
  Walsh, Valentine, 218, 229,
    237, 240
sprinkler systems, 160
staff complement, 28, 29, 35,
  36–7, 53, 48–9, 60, 74, 76, 80,
  93, 101,125, 162, 179, 192, 218,
  223, 243, 250, 254, 279, 296
street fire alarms, 69, 83, 96,
  118, 119, 129, 135, 157, 219, 236
supernumeries, 35, 37
telegraph system, 29,34, 35, 64,
  70, 73, 76
training, 82, 254–5, 274, 275,
  280–1, 287, 289, 292, 294
Dublin Fire Brigade Museum, 15
Dublin Firemen's Trade Union,
  87, 110, 163, 167, 176, 182, 184,
  219, 221
Dublin 'lock-out' 1913, 139
Dublin Metropolitan Police,
  17, 19, 30, 39, 46, 50, 65, 70, 73,
  79, 133
  abolition of, 186

Atwell-Lake, Mr, 25-6
Sheahan, Constable, 120-1
Dublin Southern Districts
Tramway Company
Murphy, Martin, 104
Robinson, Clifton, 104
Dublin Steam Packet Company, 12
Dublin Trade Union Committee,
111
Dublin Trades Council, 88, 89,
118, 155, 191
Dublin United Tramways
Company, 103
Easter Rising 1916, 148–53
General Post Office, 151, 152
'Emergency', the, 219–39
*Evening Herald*, 253
*Evening Mail*, 121, 243
Factories Act 1875, 73
Factories Act 1955, 249
Factories Act 1972, 278
Factory and Workshop Act 1897,
102
Fianna Fáil, 197
*Fire*, 150
*Fire and Water*, 86
Fire Brigade Act 1862, 31, 33,
298–302
Fire Brigade Act 1940, 221–3, 237,
249, 270, 274
fire extinguisher, first portable
type, 160
fire fighting appliances, 308–12
makers
Argyle Motors, 130
Bedford, 250
Bolton, John (c. 1750), 6
Carmichael, 279, 290
Callow and Sons, 184
Clayton, William, 54, 79
Dennis, 130, 243, 244, 245, 290
Dodge, Magirus, 279
HCB-Augus, 290
Henry Simonis and Co,
127–8
Jessop Brown, 76, 99, 100,
102, 105
Jones, Roger (patentee,
1625), 1
Keeling, John (c. 1711), 3
Leyland, 130, 131, 135, 136,
138, 139, 144, 160, 290
Mallet, 17
Mather and Platt, 131
Merryweather, 76, 105, 127,
134, 189, 290
Newsham, Richard (c.
1750), 5

Oates, John (c. 1711), 2, 3, 6
Shand Mason, 35, 41, 48, 93,
94, 105, 127
Smith, John (c. 1750), 6
Timony, 290
William Rose, 105
operated by insurance
companies, 8, 9, 10, 20, 23,
25, 27, 37, 38, 41, 42, 43, 46, 133
operated by parishes, 4, 5, 9, 10,
18, 22, 23, 27, 30, 37, 38, 41, 42
fire insurance, 2, 8, 9, 11, 30, 39
companies
Atlas, 20
Commercial, 27
Globe, 9, 12, 43
Hibernian, 7, 9, 12
Imperial, 12
London Union, 12
National, 9, 10, 12, 15, 20, 38,
41, 43, 46
Patriotic, 28, 69
Phoenix, 9, 11
Royal Exchange, 8, 9, 10, 12,
20, 27, 37, 38, 41
St Patrick, 27
Scottish Union, 12
Shamrock, 27
Sun, 20, 43
West of England, 9, 10
fire marks, 9
first fire office (1680), 2
proposed contributions to
support fire brigade, 47,
60–61, 69, 115, 135, 170–1
firemen, *see* Dublin Fire Brigade
Fire Prevention Council, 275, 282
fire protection byelaws, 200–1
Fire Service Training Council, 275
Fire Services Act 1981, 133, 288, 292
fires
Abbey Theatre, 240–1
All Hallows College, 97
Alliance Gas Company, 193
Archer's timber stores, 134
Armstrong's paper factory,
145–6
Arnott and Co, 94–5
Artane Industrial School
Belfast Blitz 1941 (during),
225–32
Bohemian Cinema, 187–8
British Embassy, 266
Calor, Kosangas, 273–4
Cavan orphanage, 236–7
Central Hotel, 245
CIE depot, 258–9
Civil War (during), 171–8

Clonsast bog, 248
Crampton's Works, 133
Cuffe Street, 247–8
Custom House, 168–70
Custom House Docks, 10–12
Dalkey, 271
Dockrell, 270–1
Drumcollogher, 188
Dublin and Drogheda Railway
Workshop, 37
Dublin Granaries, 127
Dunsinea, 97
East Indiaman cutter, *Nagpore*,
55
Easter Rising (during), 148–53
Edward Lee drapery, 132–3
ESB stores, 201–2
Four Courts, 171–2
Gibbstown House, Navan, 137
Grafton Street, 126
Great Britain Street, 48
Great Brunswick Street gas
works, 126
Great Fire of London, 2, 8, 14
Green Street, 112–13
Grennell drapery, 119
Guinness Brewery, 240
Hammond Lane Foundry,
61–4
Hibernian Gas Company, 22
Irish Bus Company, 190–1
Irish House of Commons, 7–8
Irish Paper Products, 262
Irish Pharmaceuticals, 252
Irish Times building, Fleet
Street, 161, 242–3
Jacob's biscuit factory, 65–6
Johnson, Mooney and
O'Brien, 253
Jonathan Richards, 277–8
Kennedy's bakery, 235, 239
Kildare Street Club, 22–4, 25,
27, 42
Killeen Paper Mills, 193
Killua Castle, Clonmellon, 191
Kingstown Pavilion, 146
'La Mancha', Malahide, 185–6
Lincoln Place, 20
Loreto Convent, 295
Lower Sackville Street, 37–8
McKenzie's hardware, Pearse
Street, 264–5
Malone bonded warehouses,
58–9
Maple's Hotel, Kildare Street,
161
Marine Station Hotel, Bray,
156

Moore Abbey, Monasterevan, 238
National Service Garage, Hawkins Street, 197
Navan Flour Mills, 161
North Strand bombing, 232–6
Noyek and Sons, 238, 266–8
Pearse Street, 208–13
Pile's timber yard, 98–100
'Plaza', Middle Abbey Street, 202
Portmarnock House, 244
Powers Distillery, 252–3
Reilly's, Henry Street, 24
Royal Arcade, 12–13
Rowntree's factory, 207
Royal Hibernian Military School, 185
St Mary's Abbey, 1
St Michael's Church, Dun Laoghaire, 258
St Patrick's College, Maynooth, 65
South City Market, 90–2
Spa Hotel, 276–7
Stardust, 283–93
State Apartments, Dublin Castle, 224
tanker *Columbo*, 244–5
Terenure plane crash, 203
Theatre Royal, 66–9
Todd Burns, 114–15
Trinity Street, 73–5
War of Independence (during), 163–71
Wellington Quay / Eustace Street, 38
Westmoreland Street, 42–8, 77–9
Whitly Company, 93–4
fire stations, *see* Dublin Fire Brigade
*Freeman's Journal*, 6, 7, 8, 12–13, 44, 63, 64, 94, 121, 132, 153
Garda Síochána, 187, 189
Garvin, John, 262
Gavin, Charles (fire plug designer c. 1850), 15
Grattan, Henry, 7
Greater Dublin Bill, 192
Gregg, John, 60
Griffith, Arthur, 159
Hammond Lane Foundry, 15
Hardy, Keir, 140
Harty, Spencer, 77, 80, 82, 83
Hawksley, Mr, 21
Hazchem Code, 292
Healy, T. M., 88, 113, 197

*Herald*, 103
Hernon, P.J., 207, 226, 230
Hibernian Gas Company, 22
Hobson, Bulmer, 189
housing conditions of the poor, 16, 42, 65, 69, 111, 115, 139, 155
Ingram, James Robert, 32–70 *passim*, 118
Institution of Fire Engineers, 184, 193, 199, 206, 214, 215, 259
*Irish Independent*, 196, 198
Irish Municipal Employees Trade Union, 163
Irish Municipal Workers Union, 260
Irish Society for the Protection of Life from Fire, 27, 30–1, 35
*Irish Times*, 37, 44, 45, 77, 114, 161, 163, 210, 224, 228, 243, 271
Irish Trade Union Congress, 87, 159, 178
Irish Transport and General Workers Union. 163
Keane, J.P., 207, 211
Keane, Justice Ronan, 284, 285
Kingstown Township fire service, 56, 98, 132, 146, 162
Carroll, Captain, 146
Labour Party, 87, 168, 197
Larkin, Brian, 240, 250, 251, 252, 253, 259, 265
Larkin, Jim, 220
Larkin, Jim (young), 262
Local Authorities (Officers and Employees) Act 1926, 207
Local Government Act 1955, 260
Local Government Appointments Commission, 215
Local Government Officers (Ireland) Trade Union, 163, 167
Local Officers Superannuation Act 1869, 110
McDermott, John, 226, 229
Mackey, Matthew, 261
Magnetic Telegraph Company, 29, 34
Myers, John, 98, 125–87 *passim*
Nedley, Thomas, 65
North Strand bombing, 232–6
*Northern Whig*, 229
O'Brien, Thomas, 252–91 *passim*
O'Connell, Daniel, 11
O'Connor, Rory, 172
O'Duffy, Eoin, 187
Parke Neville, Mr, 21, 32, 62, 68
Pembroke Township fire service, 40, 51, 56, 111, 112, 118, 119, 128,

132, 133, 146, 153, 155, 162, 257, 302–3
Hutson, Captain J. C., 128, 133, 134, 183, 189, 196
Petroleum Act 1871, 65, 73
plastics, 259, 280–1
polymers, 280–1
Power, John, 173, 176, 177, 178, 183, 188, 189
Powers Distillers, 90
Purcell, Thomas, 86–159 *passim*, 188
Rathmines Township fire service, 51, 75, 111, 112, 118, 153, 155, 160, 162, 189, 204–6
Whyte, Captain, 183, 193, 196
Redmond, John, 145
Richardson, E. L., 88
Robinson, George (water engineer c. 1750), 6
Robinson, Leo (fire extinguisher c. 1750), 6
Royal Irish Constabulary, 19
Royal Society for the Protection of Life from Fire, 87
St John Ambulance Association, 176, 183
Shaw, Captain Eyre Massey, 32, 33, 34, 48, 118
Sherlock, Gerald J., 195
Simmons, John, 88
Sinn Féin, 162, 163, 164, 168
Smith, C. G., 76, 87
Stardust tragedy, 280–93
statistics, 27, 35, 51, 61, 65, 69, 80, 92, 101, 115, 124–5, 148, 155, 201, 273, 275, 297
street fire escapes, 19, 21–2, 23, 27–8, 31, 35, 36, 96, 105–7, 114, 115,118, 129, 138
Symonds, Lieutenant-Colonal, 216
Townsend, John, 7
Trinity College Dublin, 12
United Corporation Labourers Union, 110
Vagen-Bader head protector, 112, 120
Walsh, Michael, 291
Walsh, Valentine, 218, 229, 237, 240
War of Independence, 163–71
Waterworks Act 1861, 24, 27, 40
Wivell, Abraham, 28
Workers Union of Ireland, 260, 267, 268